The Ecosystem Concept in Natural Resource Management

Contributors

EGOLFS V. BAKUZIS

F. H. BORMANN

CHARLES F. COOPER

R. T. COUPLAND

JAMES K. LEWIS

G. E. LIKENS

JACK MAJOR

E. A. PAUL

ARNOLD M. SCHULTZ

STEPHEN H. SPURR

GEORGE M. VAN DYNE

FREDERIC H. WAGNER

R. Y. ZACHARUK

THE ECOSYSTEM CONCEPT IN NATURAL RESOURCE MANAGEMENT

Edited by George M. Van Dyne

COLLEGE OF FORESTRY AND NATURAL RESOURCES
COLORADO STATE UNIVERSITY
FORT COLLINS, COLORADO

 1969

ACADEMIC PRESS   NEW YORK and LONDON

ACADEMIC PRESS, INC.
111 Fifth Avenue, New York, New York 10003

United Kingdom Edition published by
ACADEMIC PRESS, INC. (LONDON) LTD.
Berkeley Square House, London W1X 6BA

LIBRARY OF CONGRESS CATALOG CARD NUMBER: 72-86367

Second Printing, 1971

PRINTED IN THE UNITED STATES OF AMERICA

List of Contributors

Numbers in parentheses indicate the pages on which the authors' contributions begin.

EGOLFS V. BAKUZIS (189), School of Forestry, University of Minnesota, St. Paul, Minnesota

F. H. BORMANN (49), School of Forestry, Yale University, New Haven, Connecticut

CHARLES F. COOPER (309), Resource Planning and Conservation Department, University of Michigan, Ann Arbor, Michigan

R. T. COUPLAND (25), Department of Plant Ecology, University of Saskatchewan, Saskatoon, Saskatchewan

JAMES K. LEWIS (97), Animal Science Department, South Dakota State University, Brookings, South Dakota

G. E. LIKENS* (49), Department of Biological Sciences, Dartmouth College, Hanover, New Hampshire

JACK MAJOR (9), Botany Department, University of California, Davis, California

E. A. PAUL (25), Soils Department, University of Saskatchewan, Saskatoon, Saskatchewan

ARNOLD M. SCHULTZ (77), School of Forestry, University of California, Berkeley, California

STEPHEN H. SPURR (3), School of Graduate Studies, University of Michigan, Ann Arbor, Michigan

* Present address: Biology Division, Cornell University, Ithaca, New York.

v

GEORGE M. VAN DYNE (327), College of Forestry and Natural Resources, Colorado State University, Fort Collins, Colorado

FREDERIC H. WAGNER (259), Department of Wildlife Resources, Utah State University, Logan, Utah

R. Y. ZACHARUK (25), Biology Department, University of Saskatchewan, Regina, Saskatchewan

Preface

Man's rapidly developing technology provides him with increasing ability to manipulate the environment, e.g., traces of pesticides are now found in organisms throughout the world. But man has not had sufficient understanding of the many long-term consequences of environmental manipulation. In part, this has been because he often has not taken an ecological viewpoint; but the ecological basis of natural resource management is becoming more clearly understood, especially in multiple-use management of natural resources to increase productivity. Understanding the ecological basis of productivity in nature means understanding ecosystems. An ecosystem results from the integration of all of the living and nonliving factors of the environment for a defined segment of space and time. It is a complex of organisms and environment forming a functional whole. The study or management of such complexes requires more than one individual; no one man can encompass all of the required specialties and knowledge. This has led to the concept of interdisciplinary teams for research on and management of natural resource ecosystems. This diversity of skills and specialties is reflected in the variety and kinds of training received by the authors of the chapters presented in this volume (see section openings).

Natural resource ecosystems have been observed for some time in research and management studies, but only a few components have been measured or considered in most instances. Several of the chapters in this volume, particularly in Section I, include brief reviews of the development of the ecosystem concept and its adaptation in natural resource management fields. Examples of the concept in research on natural resource phenomena are considered in Section II. Evaluation of ecosystem applications and implications in several natural resource management fields is considered in Section III. The implications of even more intensive resource use in the future and of the implementation of ecosystem concepts on training tomorrow's resource managers and scientists are considered in Section IV.

This volume is based on a symposium held at the annual meeting of the American Society of Range Management in Albuquerque, New Mexico on February 12–15, 1968. The papers, of course, are not restricted to one natural resource field. They cover range, forest, watershed, fishery, and wildlife resource science and management. Collectively, a large number of scientists, educators, and technicians are employed in these professions. For example, approximate membership in three important resource-oriented societies in North America is 4300 in the American Society of Range Management, 16,000 in the Society of American Foresters, and 6100 in The Wildlife Society. These scientists and managers have an increasingly important role in developing and utilizing the biosphere for human welfare.

This volume will be timely because of the widespread current interest in ecosystem approaches. This interest is present in research, management, and academic areas. For example, in the evaluation of topics for the annual meeting at which this symposium was presented, the "ecosystem approach" was second in priority among some fifteen topics evaluated by society membership for inclusion in the program. When I was asked to organize the symposium I found the authors shared this interest, especially because of the opportunity to evaluate the ecosystem concept's applicability to both research and management in several natural resource management fields. Perhaps this reflects the fact that in many resource management fields there is a dearth of good textbooks on the conceptual basis of the field. This volume is not written as a textbook, but it should be a useful reference in resource management courses, especially providing a comparative evaluation of the ecosystem concept in several resource management fields. This volume has particular pertinence to research in the International Biological Program whose theme includes understanding biological productivity to enable adequate estimates of the potential yield of new, as well as existing, natural resources. The ecosystem concept is central to the planning, conduct, and analysis of many of these studies.

I am indebted to the authors and the publisher for their assistance in the preparation of this work.

GEORGE M. VAN DYNE

Fort Collins, Colorado
September, 1969

Contents

SECTION I
The Meaning, Origin, and Importance of Ecosystem Concepts   1

Chapter I. The Natural Resource Ecosystem

Stephen H. Spurr

Chapter II. Historical Development of the Ecosystem Concept

Jack Major

SECTION II
Examples of Research Development and Research Results Applying Ecosystem Concepts

Chapter III.   **Procedures for Study of Grassland Ecosystems**

R. T. Coupland, R. Y. Zacharuk,
and E. A. Paul

Chapter IV.   **The Watershed-Ecosystem Concept and Studies of Nutrient Cycles**

F. H. Bormann and G. E. Likens

Chapter V.   **A Study of an Ecosystem: The Arctic Tundra**

Arnold M. Schultz

SECTION III
Ecosystem Concepts in Natural Resource Management Fields

SECTION IV
Instilling the Ecosystem Concept in Training

Chapter X. **Implementing the Ecosystem Concept in Training in the Natural Resource Sciences**

George M. Van Dyne

The Ecosystem Concept
in Natural Resource
Management

SECTION I THE MEANING, ORIGIN, AND IMPORTANCE OF ECOSYSTEM CONCEPTS

The first two chapters define the evolution, development, and application of ecosystem concepts. In Chapter I, S. H. Spurr briefly introduces the concept of natural resource ecosystems. This chapter provides a framework for those that follow. In this introductory chapter, Spurr brings to bear a long background in natural resource fields. Forester, ecologist, resource analyst, and educator, Spurr has had an interesting career. In 1938 he received his bachelor's degree in botany from the University of Florida, and in 1940 he received his master's degree in forestry from Yale. He then worked at the Harvard Forest before continuing his formal education and was awarded a Ph.D. by Yale in 1950. His next position was with the University of Minnesota School of Forestry. He transferred to the University of Michigan in 1952. There he was first professor, then Dean of the School of Natural Resources, and in 1965 he was appointed Dean of the Rackham School of Graduate Studies. Spurr is the author of four books on forestry and was the founder and first editor of the journal, *Forest Science.*

The second chapter, by Jack Major, reviews the historical development of the ecosystem concept. Major's chapter reflects his broad knowledge of the literature and especially his command of some European languages and Russian. Major reads avidly a broad range of literature in several languages on ecological and conservation topics. Major obtained his undergraduate degree in range management at Utah State University in 1942 and then worked for the Intermountain Forest and Range Experiment Station of the Forest Service, USDA, before a period in the Armed Services in World War II. Major completed his Ph.D. in soil science at the University of California at Berkeley in 1953. Subsequently he was employed by the University of California, first as a weed control assistant specialist and later as a lecturer, assistant, and associate professor of botany at the Davis campus of the University of California. His major activities include teaching plant ecology with special interests in synecology, alpine vegetation, and the flora of California.

Chapter I The Natural Resource Ecosystem

STEPHEN H. SPURR

I. MAN'S ROLE

A natural resource ecosystem is an integrated ecological system, one element of which is a product of direct or indirect use to man. The product may be biological as in the case of forests, ranges, agricultural products, fish, and wildlife; physical as in the case of water, air, and soil; or both. In all cases, the distinguishing facet of a natural resource ecosystem is that man has a direct involvement in the complex set of ecological interactions.

Management is defined as the manipulation of the ecosystem by man. Beneficial management involves manipulation to maximize the returns to man, while exploitation is management that results in the reduction of the productivity of the ecosystem to mankind over a period of time. The ecological principles of natural resource ecosystems are generally applicable regardless of the particular natural commodity. So, too, are the tools of management and the basic rules governing their application. The principles of ecosystem management apply equally to wilderness and to the urban environment, but they are most clearly understood today with regard to the wildland resources of forest, range, wildlife, and the like.

The purpose of this chapter is to introduce and provide a framework for the nine chapters that follow in this volume. Applied ecosystem ecology has a general philosophical validity of its own. Whether general prin-

3

ciples are deduced or whether they are induced from specific studies in specific subfields, they have a general applicability to all natural resource management fields. Thus, the history of the ecosystem concept has a general relevance and forms the appropriate basis for the following paper. The other chapters in this volume are concerned with the applications and implications of ecosystem concepts in natural resource management. The contributors have chosen to look at their aspects of the problem from various levels of integration. This is as it should be. We need to understand the problem at the level of the entire natural resource field, whether forestry, range management, wildlife management, or watershed management. Within this broad conceptual framework, we need detailed consideration of specific ecosystems, whether bounded by the limits of a small watershed or by other natural delineations. At a still more specific level, we need to concentrate our measurements on finite experimental plots or areas. Within these areas, individual plants and animals become experimental units, and even these will be treated in parts as they contribute to particular cycles within the broader ecosystem.

In the following chapters, emphasis is given to terrestrial natural resources: ranges, forests, wildlife, and watersheds. Examples are given to show how the understanding of ecosystem concepts can aid in developing plans for wiser and even more intensive use of our natural resources as our world population continues to increase at near exponential rates. Not only do we need to derive general concepts applicable to the study and management of our natural resources, but also we must introduce them into our undergraduate and graduate training programs in the natural resource sciences.

II. DEVELOPMENTS IN NATURAL RESOURCE SCIENCES

I shall say no more about the chapters that follow, because to do so would be anticipatory. Rather, I should like to take advantage of my role as first author by developing the reasons which make me think that this volume is particularly topical and that convince me that we are at the threshold of major developments in the field of ecology, especially as applied to natural resource management. My argument will be illustrated by examples from my own field of forest ecology, but I believe that it will hold for ecosystems other than the forest alone.

Classic ecology has long been divided into autecology and synecology, and the difference between the two major approaches to the understanding of biological systems has been marked. Look at any textbook dealing with the subject and see how distinct are the treatments of autecology and synecology, and how information developed from one approach has had little effect on the development of principles from the other. Historically, autecology has been experimental and inductive, while synecology has been philosophical and deductive. Only toward the middle of the twentieth century have we developed concepts, experimental techniques, and analytical procedures permitting the inductive study of anything approaching an actual ecological system. Only in the electronic and atomic age are we able to deal with the ecosystem as a whole on a firm, scientific basis. Let me elaborate.

Autecology has long been on a sound experimental basis. The effects of two or three environmental variables on the growth, development, survival, and reproduction of individual biological organisms have been studied scientifically for two or more centuries. Forest trees, for example, may be subjected to a variety of environmental conditions in the nursery, greenhouse, or forest environment itself. Our knowledge of their reaction to these variables is well documented in eighteenth-century horticulture, nineteenth-century plant physiology, and twentieth-century forest autecology. The classic Ph.D. thesis in silviculture during the first half of the twentieth century could be identified by the generic title, "Factors affecting the germination, survival, and growth of . . . ," with the name of the species, and perhaps the forest site, being added to identify the specific rank.

Until recently such research was essentially a crude type of organismic plant physiology using large and difficult-to-handle organisms under poorly controlled environmental conditions. Although much of value was learned, the results were frequently biased by unmeasured and unexpected happenings such as frost, drought, and browsing. The typical experiment covered a span of years, and interpretation of the data was invariably difficult. As one scientist who spent many years on such research, I often categorized my job as involving, first, the salvage of experimental data of my predecessors, and second, the establishment of a new series of experiments on which my successors would, in turn, conduct salvage operations.

It was only with the development of controlled environmental growth chambers, popularized by the phytotron of Frits Went at the California Institute of Technology, that experiments with terrestrial organisms under controlled climatic conditions could be carried out on a replicable basis. As a result, autecology is now a reasonably precise science.

In contrast, synecology has long been treated on a philosophical rather than on an experimental basis. Its ancestry traces back directly to the naturalists of the eighteenth century. Indeed, many prominent ecologists today are essentially naturalists in tradition and in method as well as in literary skill. Thoreau acted in this tradition when he wrote about the succession of forest trees in New England in 1863. Darwin brought his powers of simple experimentation and deduction into play in his 1881 study of earthworm action, but still was in the tradition of the naturalist, if he was not indeed the epitome of the breed himself.

In the first half of the twentieth century, plant synecology developed along two major lines: (1) the American plant-succession school long dominated by Clements and (2) the European plant-sociological approach identified with Braun-Blanquet, Cajander, and Sukachev. The first was primarily deductive and the second highly subjective.

The American school had its origin with the Harvard geologist, William Morris Davis, who hypothesized that, assuming that climate were held constant, erosion would gradually reduce all eminences and fill up depressions so as to create a peneplane. Frederick E. Clements applied the same line of reasoning to vegetation and theorized that, holding climate constant and barring disturbances, such a peneplane would eventually be occupied by a climatic climax, an association of organisms best suited to the particular climate which was being held constant. It took us generations to learn that Clements' norm was highly atypical; that the climate is not fixed, but its changes affect the biota much more profoundly than does erosion; that fire, browsing, and windthrow are natural and to be expected in most parts of the world; and that there is nothing inherently more "natural" or "normal" about one type of biotic history than there is about any other.

In western and northern Europe, on the other hand, a number of national schools of plant sociologists developed concurrently, varying in their precepts and individual approaches, but united in the recognition that their environment lacked the undisturbed "virgin" standards of comparison still to be found in more recently civilized portions of the world and that they must start by classifying arbitrarily what they found around them. Their general approach was taxonomic—applying the Linnaean concepts to assemblages of organisms and utilizing subjective values such as abundance, cover, sociability, vitality, and periodicity. Despite the fact that genetic basis existed for the classification of plants and animals as communities, associations were classified in analogous fashion to individuals and were given highly subjective and arbitrary binomial names. As with the American school, the classic Europeans lacked a purely objective, experimental, and inductive approach. Neither group has ever been fully accepted by its peers in the more scientifically rigorous subfields of biology.

Only in the last two decades have the defects of the traditional schools been widely accepted, and the way has been opened for ecology to become a true experimental science on a nonteleological basis. Models based upon the assumption of the constancy of nature have become less valid as we have recognized the rapid rate of climatic variation and the immediacy of geologic change. Neither the Clementsian nor the Braun-Blanquet philosophy seems to have much applicability to an understanding of the forests of northern Michigan, for example, when we now know that the area was under continental glaciers ten thousand years ago; that major climatic shifts have taken place since that time; that major in-migration and out-migration of organisms have taken place, even within the last decade; and that fire, wind, ice, and browsing have been as influential in forming the biota before as after the advent of European man.

Concurrently, advances in technological and experimental design have opened the way for sound experimentation. Experimental design stemming from the pioneer work of R. A. Fisher and his successors in biostatistics has made it possible to design experiments with an increased number of variables interacting in a complex manner. The great breakthroughs in genetics have given us a better understanding of the internal controls and functioning of the individuals, species, and genera that make up our biotic ensembles. The present-day biotron permits the programing of controlled environments similar to those found in nature. Modern electronics has permitted automatic measurement of phenomena at many stations over a period of time and with a high degree of precision. Most importantly, the development of the whole concept of systems analysis and the construction of sophisticated and complex models, through the use of the computer, have made possible the integrated study of the data describing the natural ecological system.

In searching for a name to describe this new ecology, we have resurrected and enthroned Tansley's fortuitous coinage of 1935. Indeed, the term ecosystem is merely a contraction of ecological system, and how should such a system be studied if not by systems analysis and computer?

Quite possibly, ecosystem analysis will engender the next great advances in biological science. Certainly, we are developing the hypothetical models, the technology, and the hardware with which we may examine multivariate and highly interdependent systems involving a number of different organisms operating in a multitude of environments which vary in time and space. This approach is particularly suitable to the analysis of natural resource problems on an integrated and holistic basis. Our problems of management of our natural resources are clearly critical, whether we concern ourselves with our wildland or subdued land resources, or whether we concern ourselves with land, air, or water resources. We have every reason, therefore, to devote substantially increased effort to the application of ecosystem analysis to natural resource management problems. I trust that this volume will add impetus to that development.

Chapter II Historical Development of the Ecosystem Concept

JACK MAJOR

I. THE RICHNESS OF ECOLOGICAL TERMINOLOGY

Nominally the word ecosystem has a short history, dating from Tansley's introduction of the term in 1935. The idea itself is much older.

A. The Ecosystem

The "ecology" part of Tansley's idea dates formally from Haeckel (1866) as "nature's household" and the "system" part is fixed in English, but derived from Latin and ultimately from Greek, as a meaningful or useful agglomeration. It is so used in physical chemistry, biology, law, geology, literature, music, logic, and ordinary speech and thought according to Webster's dictionary.

B. Comparative Terminology

Men have evidently been much the same in habits of thought and irrationalities during all of recorded history, so it must be recorded somewhere that men of cultures other than our own had the explicit idea of an ecosystem. Ideas are expressed in language. Man as an animal has been at least a part of the wild natural scene and many of the terms in languages succinctly encapsulate the idea of a particular kind of ecosystem. The English terms carr, moss, fen, and heath have very precise meanings in terms of kinds of plants concerned, habitat factors, and resulting landscapes. Siberian terms for regionally extensive ecosystems, such as tundra and taiga, are now universally used. Steppe may be misused. Muskeg is a local Chippewa Indian term for a kind of ecosystem within the taiga (within most northern taigas). The *baldies* of the Uinta Mountains of Utah, at least, and the *goltsi* of the above-timberline summits of the central Siberian mountains are the same kind of ecosystem. Every Old World Mediterranean country has a term for a shrubby, aromatic, usually sclerophyllous plant formation formed by man from forest vegetation. The Spanish *chaparral* is now quite naturally applied in California in a widened sense. *Maquis* is the term for this formation that radiates from southern France; *shibliak* is the corresponding term in submediterranean Yugoslavia. Where soil losses have been extreme and much bare rock is exposed, the formation has degenerated to a *garrigue* in France, *tomillares* in Spain, and *phrygana* in Greece.

The names are different, but the aspect of the ecosystems, the dominant habitat factors, and often even many plants are the same. Particular grassland ecosystems may be prairie transposed from France to central North America, veld transposed from The Netherlands to South Africa, or pampas in Argentina from the local Quechua Indian language. Our own shortgrass and Palouse prairie are certainly ecosystem concepts. An undrained desert basin is a playa in the American Southwest, a *schott* in North Africa, and a *takyr* in Central Asia. Bajada is coming close to being an ecosystem term as the vegetation of bajadas is studied in Arizona and not remaining only a geomorphological term. Various words are in American use for particular kinds of native ecosystems: pocosin in the Southeast (cf. the German *Bruchwald*), flatwoods, bottomlands (cf. the German *Auenwald*), parkland in Saskatchewan (cf. Russian *lesostep*), motte, everglades, cross-timbers, shinnery, hammocks, and hog wallows (cf. Australian *gilgai*). Some of these are unique and this is one reason why the names are good ecosystem terms. Mallee, gibber, mulga, and spinifex are similarly such vivid, precise Australian ecosystem terms that it would be a shame to transplant them to different ecosystems no matter how superficially similar they were to each other.

South America does not lack similar terms: *cerrado* and *caatinga* in Brazil, and *paramo* and *pampas* in the Andes (see above). An amazing variety of terms has been developed, preserved, and is used for particular ecosystems in the ecologically diverse territory of the USSR. We could perhaps use some of these: *tugai* for desert floodplain forests; *takyr* for desert clay sinks; *bor* for pine (*Pinus sylvestris*) forests on poor, sandy soils; *kolki* for isolated aspen groves in the forest-steppe; *goltsi* for alpine tundra of rounded mountain summits above timberline; *liman* for a marsh at the mouth of a river; *tukulan* for drift sand areas in the wet *taiga* of Yakutia; and *alas* for depressed areas with lakes and bogs in the same area. The Scandinavian languages are rich in terms for other kinds of bog ecosystems of large and small extent. Troll (1950) lists many other local ecosystem terms in a variety of languages.

C. Antiquity of the Ecosystem Concept

It is fascinating to trace the impact of different natural ecosystems on men of various geographical areas, insofar as these impacts are reflected in their languages. The exercise leaves little doubt as to the great antiquity of the idea of the ecosystem as well as to its universality among mankind.

I must leave to classical as well as to Sumerian, Egyptian, Arabian, Chinese, and Mayan scholars the tracing of this idea in those older cultures. It will suffice in our context to trace some of the history of this idea in our own western civilization. Again, I lack the scholarly knowledge to begin with the Renaissance, but surely one could find the ecosystem in Leonardo da Vinci's work, although Boccacio was evidently more interested in personal relationships than in man-habitat-nature interactions. John Donne had the grace and insight to regard the human situation ecologically when he said, "for whom the bell tolls."

But we today regard the ecosystem idea as an essentially scientific one. We can look at plant ecology, of which range management is an application and a specialization, as it has developed up to our time.

D. Some Sequential Developments in Terminology

In a recent symposium in Stolzenau, Germany, Scamoni (1966) pointed out that Alexander von Humboldt, writing in 1807 on plant geography, said: "In the great chain of causes and effects no thing and no activity should be regarded in isolation." Haeckel surely echoed this idea some 60 years later. With specialization in science we find the same idea expressed in many of the biological sciences and their applications.

Möbius (1877) described the oyster beds in the North Sea off Kiel as a biocoenose and he specifically included in his description not only internal relationships among the organisms of the oyster bed but also their relationships to their environment.

Forbes, in Illinois (1887), described the lake as a microcosm; this is a tenet of limnology today, which has enriched general ecology with applications of the ecosystem concept such as Lindeman's ideas on trophic-dynamic aspects (1942) and with Thienemann's lifetime focus on life and environment (1956).

In 1889, Dokuchaiev, the founder of soil science, set up soil properties as a function of site factors in an equation similar to the equations of state established for physicochemical systems in general. Jenny (1941, 1961a,b) has profoundly refined Dokuchaiev's treatment. The modern tendency is to regard soils as a part of the ecosystem (Braun-Blanquet, 1964; Pearsall, 1950; Kubiena, 1953; Sukachev and Dylis, 1964; Crocker, 1952; Jenny, 1961b). This view is a direct consequence of Dokuchaiev's early formulation of the natural history and zonal arrangement of soils. Subsequently, these ideas were elaborated by several generations of very able Russian pedologists, geobotanists, and geographers. That soils are part of an ecosystem was an idea evidently tortuously arrived at by individual training and outlook; Glinka's textbook of soil science translated into English via German; and the general, international development of pedology. Jenny's tessera (1965) is a concrete, small sample of an ecosystem.

Morozov was a student of Dokuchaiev's and, as a forester, he developed the idea of the ecosystem in writing a classic textbook on silviculture. This was first circulated as lecture notes from 1902 to 1903; it was printed in 1912, with a seventh edition in 1949, and with German translations in 1928 and 1959. The present use of the terms habitat and site by foresters (Munns, 1950) includes the pertinent organisms (woody plants, anyway) and these are ecosystem concepts. Bakuzis' splendid outline for his course in forest synecology (1966) is probably the best available presentation of the ecosystem idea in many of its ramifications, but particularly in forestry. Certainly modern silviculture is simply an application of the ecosystem idea to forests.

E. Some Comparisons of Ecosystem Concepts

In 1914, Abolin, an associate of the late Professor Sukachev, in a journal on bogs, used the term *epigen* for our idea of the ecosystem. Later studies on bogs (Sjörs, 1963, 1948; Ruuhijärvi, 1960; Heinselman, 1963) are only examples from a very rich literature) are models of what ecosystem studies should be.

It is difficult to equate Negri's *ecoid* (1914) with our ecosystem concept, since he limited the organisms in his unit to one individual. Still, he did propose units of plant and environment, systems which are only relatively independent of each other. The writer has suggested elsewhere (Ponyatovskaya, 1961), and documented to an extent, the view that the individualistic approach to vegetation is a stage in the ontogeny of ecologists. Negri's neglected ideas find an agreeing echo in Maelzer's analysis (1965). The latter, in turn, neglected to confine environmental relationships to ecosystems to ecosystem properties and, therefore, became involved in the logical difficulty of relating properties of parts (individual organisms) of a whole (the ecosystem) to the functioning of the whole. Mason and Langenheim (1957) restricted themselves in analyzing "environment" to individual organisms, but operationally the operation must fail (Major, 1958, 1961), for autecology is obviously a contradiction in terms for plants which happen to occur in association with other plants. Competition between plants is certainly the most important fact in their phytosociological and ecological relationships (Braun-Blanquet, 1964; Harper, 1964; Major, 1958, 1961; Walter, 1960; Ellenberg, 1953, *et seq.*).

Another term equivalent to the ecosystem, which technically has priority, is the entomologist Friederichs' *holocoen* or *coen* (1927) which is frequently referred to in the German ecological literature (Scamoni, 1966), but should not be unknown in English (Friederichs, 1958). As a pedologist-phytosociologist and later university rector and president of the Swiss Schulrat, Pallmann (1948) spoke of the biochore. His student Etter (1954) later discussed the description of this ecosystem-under-another-name more fully. Present Swiss plant ecology is a splendid example of the acceptance and use of the ecosystem idea.

Man is part of the ecosystem, even economic man, but only some economics seems ecological. Still, if Adam Smith presented a deductive view of economics which stopped short of its logically inherent, deadly conclusions, Karl Marx observed and recorded accurately the realities of his time and made his picture ecological.

Geographers have used the term landscape for an idea very much like the ecosystem (Berg, 1958). They naturally often emphasize the chorological side of their object of study rather than the biological, but from Passarge's first suggestion (1913) right up to Troll (1950) and Polynov (1925, 1946), the relationship to ecosystem is clear. The plant geographer Troll's ecotope (1950), the soil scientist Polynov's elementary landscape (1925), Abolin's epimorph (1914), the Polish geographer Wodziczko's physiocoenose (1950, already suggested in 1932 according to Isachenko, 1956), the range scientist Larin's microlandscape (1926), the ecologist Ramenski's facies or epifacies (1938), Cain's natural area (1947), and Markus' nature complex (1926) all attempted to cut ecological units out of the landscape. Sukachev and Dylis relate the evolution of their ideas from Abolin's (1914). Many have used the term facies for our ecosystem idea (Sukachev, 1960; Isachenko, 1956). Facies is normally a geological term and Sukachev (1960) made an effort to restore that usage. At least Markus' concept did not differ from that of the ecosystem according to Bakuzis (1966); Polynov's, Wodziczko's, Ramenski's, and Cain's also clearly coincide. What roots did Markus' ideas, for example, strike? He was from Tartu (Dorpat) in Esthonia, but he published at least four papers in German and at least one in Russian.

Troll discusses other convergencies of ideas. Sjörs (1955) says the concept ". . . can be traced more or less obscurely behind the vegetation units of many early phytocoenologists (Sendtner, Drude, Norrlin, Warming, Schimper, Schröter, Flahault)." Many of these men were outstanding teachers whose students have developed modern plant ecology in its broadest sense. Perhaps some of their didactic success was related to the excitement generated in their students by their broad vision of the ecosystem or biogeocoenose concept. However, today it is almost impossible to trace this concept precisely with regard to its course of development, borrowings, and mutual and reciprocal influences. Too often these insights were achieved in isolation, illustrating the old adage that the wise man learns from the experience of others, others from their own. The history of such ideas must be written by those who lived and made it (Sukachev and Dylis, 1964).

II. RECENT CONSIDERATION OF THE ECOSYSTEM CONCEPT

We come to recent ecological positions. Evans, in 1956, proposed that the ecosystem was the basic unit for ecological study. I believe he expressed a general consensus. Most groups of American plant ecologists had used the ecosystem concept even if they did not use it under this name. Bakuzis (1966) quotes Livingston and Shreve (1921) as obviously using the idea. Not all ecological study, including not all applied ecological study in range management, forestry, and agronomy, explicitly defines the particular ecosystem on which it works, but most investigators recognize this as a deficiency.

A. Russian Concepts

In the USSR, their grand ecological tradition has included Sukachev's proposal of the term geocoenose in 1941 (according to Bakuzis, 1966; Sukachev and Dylis, 1964; Sukachev, 1942) and later biogeocoenose (Sukachev, 1942, 1944, 1945) to emphasize more adequately the biological nature of this fundamental ecological unit.

Because Russian geography, soil science, and plant ecology have always been extremely close-knit (Isachenko, 1956) (Dokuchaiev wrote his classic on natural zones in 1899 only 14 years after writing his classic on the Russian *chernozem*), the question immediately arises as to the relationship between biogeocoenose and landscape. Briefly, biogeocoenoses are structural parts of a landscape (Sukachev, 1960). The former occurs in an environmental habitat, the latter in a geographical locality (Major, 1958).

What is the relationship between ecosystem and biogeocoenose? Sukachev analyzed the problem (1960), but his treatment of the use of the term ecosystem was inadequate. Perhaps this is a fair summary: Ecosystem has the advantage of relating immediately to systems, their analysis by mathematical means, and their analogy to physicochemical systems. Biogeocoenose is more descriptive and, as defined by Sukachev (1960), also includes the transformations and exchanges of matter and energy about which ecologists are talking more and more. If the landscape is a geographical concept, the ecosystem is a functional one, and the biogeocoenose embodies relationships. The biological concept common to both ecosystem and biogeocoenose is the important point.

B. The Ecosystem and Natural Resource Management

We can bring this historical account of a concept up to date by illustrating its use in several problems currently of interest in range management, via plant ecology. As Heady aptly points out (1967), "We in range management have been involved with ecosystems as long as there has been a range management profession." Stoddart's outline (1966) for a beginning course in range management gives precision to Heady's statement. Range students receive formal training in applications of the ecosystem concept in their first professional course. And as Stoddart says, ". . . the outline has the interesting additional value of constituting a sort of definition of range management which will be of interest to members of the Society as well as to students. . . ." As a pastoralist husbandman, Abraham was an ecologist with other kinds of troubles. I hope that the necessary application of the ecosystem concept in applied ecology of whatever kind, including range management, has been obvious.

III. THE ECOSYSTEM RELATIVE TO SOME MODERN ECOLOGICAL IDEAS

The problems discussed below are some consequences of the spatial nature of ecosystems, their accurate delimitation by their vegetation, equilibrium, relationship to plant physiology, semantics of environment, productivity, mineral cycles and energy flow, systems theory, causal ecology, ecotypes within plant species, continuous nature of vegetation, range survey, multiple use of rangelands, and natural grazing areas. Discussions will not be complete. They will be biased and I hope they will add to what should be a continuing examination of our assumptions and logical positions.

An ecosystem occupies space, and therefore has a position on the earth's surface. This position can be described accurately by noting latitude, longitude, and altitude. A verbal description is likely to mislead many readers. A geographical location correlates with many factors of the environment about which the author may not be interested, but his readers should be.

The area occupied by an ecosystem has a particular vegetation. An animal community is associated. The latter is very often defined by the former. How do we describe vegetation? If we do not list the plant species occurring, with some measure of abundance, we have certainly failed. A physiognomic description is not sufficient. *Heracleum lanatum, Wyethia amplexicaulis,* and *Chenopodium album* are all widespread forbs in the aspen type. They may be rare and classed as "others" all too frequently. They may occur in contiguous ecosystems. They seldom occur together. They indicate very precisely very different kinds of ecosystems. They are more precise indicators for management than numbers describing precipitation, temperatures, evaporation, wind speed—even microclimatic measurements. They are more precise indicators than such soil parameters as depth, texture, cation exchange capacity, total N content, NO_3^- content, etc. They are easier to determine and their determination is more objective. If local floristic checklists to ease plant identification are not already available, build a fire under your botanical colleagues. Such incendiarism in the Forest Service, Bureau of Land Management, Soil Conservation Services, and universities could only be salubrious.

Soil data have an additional defect. If a soil sample is sent away to an analytical laboratory, what part of it is analyzed? Usually only the fraction passing a 2-mm sieve. In many range soils, well over half the solum will not be analyzed. The ecological and range literature is full of impossible figures on amounts of "available water" in soils, because such data have been used blindly.

If energy or mass units alone are used to describe an ecosystem, much of the information accumulated around our plant species names is lost. The ecosystem may be oversimplified and, in fact, unrecognizable.

One of the basic ideas related to the ecosystem is that such systems reach a kind of equilibrium suitable to an open system. A range area, grazed correctly of course, can be stable. We can all quote examples. Major Powell (1878) certainly had this idea when he wisely recommended how the arid lands of the West should be used. This kind of stability is an empirical fact of observation today. It has not always been universally recognized. Osgood (1929) quotes a resolution adopted by the Wyoming Stock Growers Association in 1879: "[Resolved] That in our opinion the question of whether grass will not disappear from the ranges with constant feeding is still unsettled, and that the stock business will not warrant the investment of so large a per cent of capital as one-sixth in what may, in a few years, be barren and worthless property." Range management has come a long way since 1879 both empirically and as a result of the development of ecological theory.

Many would tell us that "to understand range ecology we have to understand the physiology of the plant on the range." Empiricism hardly supports this generalization. As a matter of fact, plant physiology does not deal very much with ecosystems and it deals less with them daily. Plant competition is an unmentioned physiological fact.

Then, "The environment must be defined as organism-directed." This statement is a tautology and it leads to an abandonment of ecological work just as soon as the investigator forgets that the organism-directed measurements of environment he wants are correlated with the measurements it is possible for him to make. He can use plant reactions themselves as measures if he relates these to the ecosystem in which the measurements are made. The easiest, most objective way to do this is through floristic description of the vegetation of that ecosystem.

Productivity is a simplifying measure which can ignore the ecosystem idea and which then gives the same numbers for Death Valley as for the northernmost coast of Alaska. The widely held and often recurring idea that productivity increases with plant succession certainly does not hold in those ecosystems where soil fertility reaches a maximum and then declines.

Mineral cycles and energy-flow charts are often described for only a very large ecosystem, usually our planet. If not, if described for a specific, manageable ecosystem, they usually turn out not to be cycles at all, but storage takes place. Storage, however, is not detectable unless the elements of the cycle are quantitatively expressed within one carefully described and delimited ecosystem. Happily, Williams' energy flow diagram (1966) does apply to a delimited, well-described ecosystem. It and the examples discussed by Macfadyen (1964) are exemplary.

Systems theory needs some fundamental work on the forms of the relationships between major site variables and dependent ecosystem properties. Pure empiricism in this area has led to contradictory and unreasonable results. This is an area where theoretical advances would be most practical.

"Causal ecology" is simply unworkable in an ecosystem context; there are no direct, one-to-one relationships.

If the ecotype is the particulate unit of ecological study, then the habitat to which the genotype is responding must be defined. Historically, ecotypes and ecosystems have been only most crudely related. It is not altogether logical to breed forage grasses isolated from grazing. Notably, the competing plants within an association have not usually been considered as part of the environment of that association. Thus, "the individualistic nature of the plant association" is obviously true so far as spatial morphology and physiology go, but not in an ecosystem sense.

Since the ecosystem embodies the environment and change in the latter produces a change in the vegetation, then the spatially continuous nature of vegetation follows at once from the obviously continuous nature of many environmental parameters. It also has nothing to do with the real heuristic and predictive value of a vegetation classification.

IV. MANAGEMENT OF ECOSYSTEMS

It is good that range surveys are now including soil surveys. A soil may be the key to land management in that an ecosystem which has lost its soil through poor management is set back to time zero of a very long, primary succession. The vegetation, however, remains the most evident part of most ecosystems. It can be more objectively described than a soil. A range survey should therefore first concentrate on the description and classification of vegetation.

A watershed is an ecosystem subject to multiple use. The geomorphologist, the hydrologist, and the climatologist can all contribute to a watershed study, but their results are so bound up with other aspects of the ecosystem, such as soils and vegetation, that it might be wise to start with an integrated approach.

Much of the West is being grazed by domestic livestock today; once it was grazed by wild ungulates. The condition of the resulting ecosystems is partly an effect of this grazing factor acting within the context of the specific ecosystems concerned. For example, near Kluane Lake in the Yukon, Dall sheep heavily fertilize certain ridge noses where they bed down. Particular, nitrophilous, weedy plants occupy such sites and no "park preservation" should attempt to eliminate them. Similar, overgrazed sites are a part of any grazed landscape. Romell (1957) has described the lovely, wooded, grassy, bright-flowered fields of pastoral Sweden which easily and quickly disappear if the old land management practices, which included livestock husbandry, that formed them are not continued. I do not wish to be misunderstood. The overgrazing which almost destroyed both the deer herd and its habitat on the Kiabab Plateau, the present overuse by elk both on Yellowstone Park winter range and on summer ranges south of the park, the domestic sheep grazing that denuded the Wasatch Plateau of Utah, and the overgrazing that destroyed the bunchgrass vegetation of lowland California were all simply sloppy range management. They have no place in any continuing scheme of management. Society, however, has an important role to fill in ensuring that "natural areas" partly formed by grazing do not change to a condition which is unforseen and not desired simply because grazing is omitted as part of the ecosystem. If we put our own house in order, we will have acceptable advice to give to managers of natural areas.

We should obviously manage natural areas with intelligence and not exclusively by default (Stone, 1965). Grazing is continually used in many areas managed by the Nature Conservancy in England. In the Swiss National Park, management of grazing by reintroduced chamois and steinbock is a growing need. In North American national parks, game grazing has been continuous, even if often poorly managed. In any grassland national park, grazing by domestic livestock could be a compatible use. Longhorns in the Wichita Mountains Wildlife Refuge are an example.

ACKNOWLEDGMENT

H. F. Heady is thanked for presenting this paper at the symposium for the author who was incapacitated.

REFERENCES

Abolin, R. I. 1914. Tentative epigenological classification of bogs. *Bolotovedenie* **3** (cited from Sukachev and Dylis, 1964) (in Russian).

Bakuzis, E. V. 1966. "Forest Synecology" (Lecture notes). School of Forestry, University of Minnesota, St. Paul, Minnesota (mimeo.).

Berg, L. S. 1958. "Die geographischen Zonen der Sowjetunion," Vol. 1. Teubner, Leipzig. (Transl. from 3rd Russian ed. Ogiz, Moscow, 1947.)

Braun-Blanquet, J. 1964. "Pflanzensoziologie. Grundzüge der Vegetationskunde," 3rd ed. Springer, Vienna. 865 pp.

Cain, S. A. 1947. Characteristics of natural areas and factors in their development. *Ecol. Monographs* **17**, 185–200.

Crocker, R. L. 1952. Soil genesis and pedogenic factors. *Quart. Rev. Biol.* **27**, 139–168.

Dokuchaiev, V. V. 1889. Vol. 6. Akad. Nauk, Moscow; 1951. Collected Works, 1888–1900.

Ellenberg, H. 1953. Physiologisches und ökologisches Verhalten derselben Pflanzenarten. *Ber. Deut. Botan. Ges.* **63**, 24–31.

Etter, H. 1954. Grundsätzliche Betrachtungen zur Beschreibung und Kennzeichnung der Biochore. *Schweiz. Z. Forstw.* **2**, 11–14.

Evans, F. C. 1956. Ecosystem as the basic unit in ecology. *Science* **123**, 1127–1128.

Forbes, S. A. 1887. The lake as a microcosm. *Bull. Peoria Sci. Assoc.; Illinois Nat. Hist. Surv., Bull.* **15**, 537–550 (1925).

Friederichs, K. 1927. Grundsätzliches über die Lebenseinheiten höheren Ordnung und den ökologischen Einheitsfaktor. *Naturwissenschaften* **15**, 153–157 and 182–286.

Friederichs, K 1958. A definition of ecology and some thoughts about basic concepts. *Ecology* **39**, No. 1, 154–159.

Haeckel, E. 1866. "Generelle Morphologie der Organismen," 2 vols. Reimer, Berlin. 576 pp. and 462 pp. resp.

Harper, J. L. 1964. The individual in the population. *J. Ecol.* **52**, Suppl., 149–158.

Heady, H. F. 1967. Ecosystems—what we mean. *Paper Ann. Meeting Calif. Sect., Am. Soc. Range Management, Chico, Calif., 1967*, pp. 1–5.

Heinselman, M. L. 1963. Forest sites, bog processes, and peatland types in the glacial Lake Agassiz region, Minnesota. *Ecol. Monographs* **33**, 327–374.

Isachenko, A. G. 1956. Study of landscape in contemporary geobotany. *In* "Akademiku V. N. Sukachevu k 75-letiyu so dnia pozhdeniia," pp. 250–262. Akad. Nauk U.S.S.R., Moscow-Leningrad (in Russian).

Jenny, H. 1941. "Factors of Soil Formation." McGraw-Hill, New York.

Jenny, H. 1961a. Comparison of soil nitrogen and carbon in tropical and temperate regions. *Missouri Univ., Agr. Expt. Sta., Res. Bull.* **765**, 5–31.

Jenny, H. 1961b. Derivation of state factor equations of soils and ecosystems. *Soil Sci. Soc. Am., Proc.* **25**, 385–388.

Jenny, H. 1965. Bodenstickstoff und seine Abhängigkeit von Zustandsfaktoren. *Z. Pflanzenernaehr. Dueng. Bodenk.* **109**, 97–112.

Kubiena, W. L. 1953. "Bestimmungsbuch und Systematik der Böden Europas." Enke, Stuttgart.

Larin, I. V. 1926. Tentative definition of the soil's plant cover, parent material and agricultural economic value of pastures and other elements of the landscape in the central part of the Ural bay. *Tr. ob-va izucheniia Kazakhstana, Otd. Estestvozaniia i Geografii* **7**, No. 1; (cited from Sukachev and Dylis, 1964) (in Russian).

Lindeman, R. L. 1942. The trophic-dynamic aspect of ecology. *Ecology* **23**, 399–418.

Livingston, B. E., and F. Shreve. 1921. The distribution of vegetation in the United States as related to climatic conditions. *Carnegie Inst. Wash. Publ.* **284**, 1–585.

Macfadyen, A. 1964. Energy flow in ecosystems and its exploitation by grazing. *In* "Grazing in Terrestrial and Marine Environments" (D. J. Crisp, ed.), pp. 3–20. Blackwell, Oxford.

Maelzer, D. A. 1965. Environment, semantics and system theory in ecology. *J. Theoret. Biol.* **8**, 395–402.

Major, J. 1958. Plant ecology as a branch of botany. *Ecology* **39**, 352–363.

Major, J. 1961. Use in plant ecology of causation, physiology, and a definition of vegetation. *Ecology* **42**, 167–169.

Markus, E. 1926. Naturkomplekse. *Sitzber. Naturforsch.-Ges. Univ. Tartu* **32**, 79–94 (cited from Bakuzis, 1966).

Mason, H. L., and J. Langenheim. 1957. Language analysis and the concept of environment. *Ecology* **38**, 325–340.

Möbius, K. 1877. "Die Auster und die Austernwirtschaft." Wiegundt, Hempel & Parey, Berlin.

Morozov, G. F. 1949. "Studies on Forests," 7th ed. Goslesbumizdat, Moscow-Leningrad. 456 pp. (In Russian.)

Munns, E N 1950 "Forestry Terminology," rev. ed Soc. Am. Foresters, Washington, D.C.

Negri, G. 1914. Le unita ecologiche fondamentali in fitogeografia. *Atti Accad. Sci. Torino* **49**, 1–14.

Osgood, E. S. 1929. "The Day of the Cattleman." Pp. ix, 1–283. Univ. of Chicago Press, Chicago, Illinois (reprint).

Pallmann, H. 1948. "Bodenkunde und Pflanzensoziologie." Kultur- und Staatswissenschafliche Schr. Eid. Tech. Hochschule, Zürich No. 60. 23 pp.

Passarge, S. 1913. Physiogeographie und vergleichende Landschafts-geographie. *Mitt. Geograph. Ges. Hamburg* **27**, 119–151 (cited from Troll, 1950).

Pearsall, W. H. 1950. "Mountains and Moorlands. New Naturalist 11." Collins, London, 312 pp.

Polynov, B. B. 1925. Landscape and soil. *Priroda* Nos. 1/2, 74–84 (in Russian).

Polynov, B. B. 1946. The role of soil science in the study of landscapes. *Izv. Vses. Geog. Obshch.* **78**, 235–239 (in Russian).

Ponyatovskaya, V. M. 1961. On two trends in phytocoenology. *Vegetatio* **10**, 373–385; transl. from *Botan. Zh.* **44**, 402–407 (1959) (with notes by J. Major).

Powell, J. W. 1878. "Report on the Lands of the Arid Region of the United States, with A More Detailed Account of the Lands of Utah." Pp. xv, 1–195. U. S. Geogr. and Geol. Survey of the Rocky Mt. Region. Washington, D.C.

Ramenski, L. G. 1938. "Introduction to Combined Soil and Geobotanical Study of the Earth." 620 pp. Selkhozqiz, Moscow (cited from Isachenko, 1956, in Russian).

Romell, L. G. 1957. Man-made "nature" of the northern lands. *Intern. Union Conserv. Nature, Proc. Papers 6th Tech. Meeting, Edinburgh, 1956,* pp. 51–53. V.I.C.N.R., Brussels, Belgium.

Ruuhijärvi, R. 1960 Ueber die regionale Einteilung der nordfinnischen Moore. *Ann. Botan. Soc. Zool. Botan. Fennicæ "Vanamo"* **30**, 1–360; cf *Ecology* **44**, 221–222 (1963).

Scamoni, A. 1966. Biogeozönose—phytozönose. *In* "Biosoziologie. Bericht über das internationale Symposium in Stolzenau/Weser, 1960." (R. Tuxen, ed.), pp 14–22 Junk Publ., The Hague.

Sjörs, H. 1948. Myrvegetation i bergslagen. *Acta Phytogeograph. Suecia* **21**, 342 pp. Almquist & Wiksell, Uppsala.

Sjörs, H. 1955. Remarks on ecosystems. *Svensk Botan. Tidskr.* **49**, 155–169.

Sjörs, H. 1963. Bogs and fens on Attawapiskat River, northern Ontario. *Natl. Museum Can. Bull.* **186**, 45–133.

Stoddart, L. A. 1966. An introduction to range management. *J. Range Management* **19**, 133–134.

Stone, E. C. 1965. Preserving vegetation in parks and wilderness. *Science* **150**, 1261–1267.

Sukachev, V. N. 1941. On influence of the intensity of the struggle for existence between plants on their development. *Dokl. Akad. Nauk SSSR* **30**, 752–755 (in Russian).

Sukachev, V. N. 1942. The idea of development in phytocoenology. *Sovet. Botan.* 1942, Nos. 1/3, 5–17 (in Russian).

Sukachev, V. N. 1944 On the principles of genetic classification in biocoenology. *Zh. Obshch. Biol.* **5**, 213–217 (in Russian); transl. by F. Raney and R. Daubenmire, *Ecology* **39**, 364–367 (1958).

Sukachev, V. N. 1945. Biogeocoenology and phytocoenology. *Dokl. Akad. Nauk SSSR* **47**, No. 6, 447–449 (in Russian); in *English Do.*, 429–431.

Sukachev, V. N. 1960. Relationship of biogeocoenosis, ecosystem, and facies. *Soviet Soil Sci. (English Transl.)* No. 6, 579–584; *Pochvovedenie* No. 6, 1–10.

Sukachev, V. N., and N. V. Dylis. 1964. "Fundamentals of Forest Biogeocoenology." Nauka, Moscow, 575 pp. (In Russian.)

Tansley, A. G. 1935. The use and abuse of vegetational concepts and terms. *Ecology* **16**, 284–307.

Thienemann, A. F. 1956. "Leben and Umwelt." Rowohlts Deutsche Enzyklopädie 22, Hamburg, 153 pp.

Troll, C. 1950. Die geographische Landschaft und ihre Erforschung. *Studium Gen.* **3**, 163–181.

Walter, H. 1960. "Standortslehre. Einführung in die Phytologie," 2nd ed., Vol. III, Part I. Ulmer, Stuttgart. 566 pp.

Williams, W. A. 1966. Range improvement as related to net productivity, energy flow, and foliage configuration. *J. Range Management* **19**, 29–34.

Wodziczko, A. 1950. O biologie krajobrazu. *Przeglad Geograf. (Warszawa)* **22**, 295–301 (cited from Isachenko, 1956).

SECTION II EXAMPLES OF RESEARCH DEVELOPMENT AND RESEARCH RESULTS APPLYING ECOSYSTEM CONCEPTS

This section contains three chapters concerning field research projects in which the ecosystem concept is a central theme. Chapter III by Coupland, Zacharuk, and Paul details the procedures for designing intensive studies of grassland ecosystems using the Canadian Matador Project in the International Biological Program as a case example. The conduct of ecosystem research is, in part, shown by the nature of the authors' training, i.e., it must be interdisciplinary. Coupland has a broad background in grassland ecology. His undergraduate degree was in agriculture (plant science) in 1946 at the University of Manitoba. Subsequently, he studied under the late J. E. Weaver, the eminent grassland ecologist, and received his doctorate in 1949. He later became assistant professor and subsequently head of the Plant Ecology Department at the University of Saskatchewan at Saskatoon. His major research studies have included weed ecology, drought ecology, and soil-plant relationships. Coupland has traveled widely throughout grasslands of the world and has twice visited grasslands in the Soviet Union. He has been an editor or member of the editorial board of several technical journals and since 1966 has been active in the Canadian International Biological Program (IBP). At present, he is project director of the Canadian grassland studies. E. P. Paul, a soil scientist specializing in microbiology, obtained his B.S. and M.S. degrees at the University of Alberta and in 1958 received his Ph.D. from the University of Minnesota. Since 1959, he has been in the Soil Science Department at the University of Saskatchewan. His work involves the use of radioactive tracers to measure microbial transformations in the soil and the mechanisms of attack of soil humic acids by microorganisms as well as studies of symbiotic and nonsymbiotic nitrogen fixation. Recently, Paul has been active in the Canadian IBP grassland project and leads a group of investigators in the decomposer phases. R. Y. Zacharuk, the third author, represents the consumer trophic level, as he is a specialist in entomology. His undergraduate degree in 1950 and his master's degree in 1955 were obtained at the University of

Saskatchewan at Saskatoon where he specialized in entomology. His doctoral degree in 1961 at Glasgow University in Scotland was also in entomology. For 13 years following his undergraduate degree, Zacharuk was employed as a research scientist with a research branch of the Canadian Department of Agriculture on their Prairie Research Station. Since 1963, he has been professor and chairman of the Biology Department of the University of Saskatchewan at Regina. He was coordinator for consumer research in the Canadian IBP Matador Grasslands Project. Thus, the authors of Chapter III represent producer, consumer, and decomposer viewpoints toward the design of research studies and the establishment of research teams.

F. H. Bormann and G. E. Likens have been successful in applying ecosystem concepts to studying nutrient cycles on forests and watersheds in the Northeast. This team, too, represents different disciplines. Bormann received his undergraduate degree in botany from Rutgers University in 1948 and his doctorate from Duke University in 1952. Subsequently he has served as assistant professor at Emory University, professor of biology at Dartmouth College, and, since 1966, professor of forest ecology at the School of Forestry at Yale University. His major interests have been experimental ecology of plants, especially the pines, and ecosystem dynamics. Likens received his undergraduate training in biology at Manchester College in 1957; subsequently his master's and doctorate degrees were obtained, respectively, in 1959 and 1962 in zoology at the University of Wisconsin. His major interests are concerned with nutrient cycles from a biochemical approach and circulation and heat budgets of lakes and lake sediments. Their intriguing ecosystem study, described in Chapter IV, is another example of a team effort needed for comprehensive field research. Several other investigators, representing different agencies or institutions, have also been involved in this highly coordinated study.

A. M. Schultz brings to Chapter V the rationale, design, and some results of a long-term study of the arctic ecosystem. These studies, too, have involved persons of varied disciplines and have focused in recent years on nutrient cycling studies as an aid to the unraveling of the age-old problem of fluctuations of arctic microtine populations. Schultz has training in both plant and animal ecology. His bachelor's degree and master's degree were in zoology and animal ecology, respectively, at the University of Minnesota. He, too, obtained his Ph.D. under J. E. Weaver at the University of Nebraska. Since his doctorate degree, Schultz has held various positions in the School of Forestry at the University of California. His work and interests have covered grassland, brushland, forest, and tundra ecology. His present teaching is concerned primarily with ecology and the introduction of ecosystem concepts to students from a diversity of fields.

Chapter III Procedures for Study of Grassland Ecosystems

R. T. COUPLAND, R. Y. ZACHARUK, and E. A. PAUL*

I. INTRODUCTION

The concept of the ecosystem has been employed by biologists for several decades as a teaching technique to emphasize the interdependence of various strata of organisms within the same habitat and their relation to the environment (Odum, 1959). Intriguing hypotheses have been developed and theoretical models constructed to explain these relationships (Olson, 1964). However, research on ecosystems has not yet provided sufficient data to test these models.

The efficient management of renewable resources depends on a knowledge of the interrelations of organisms at various levels of activity and

* Issued as Canadian IBP No. CCIBP 16.

their relationships to abiotic subsystems. Most of our understanding of terrestrial ecosystems is limited to the relationship of a dominant plant species or vegetation to an edaphic situation, of a consumer to vegetation, or of one consumer to another. This knowledge has been applied to the management of systems in which such pairs of components exist, with the assumption that for each system the significant relationships have been studied. Perhaps we have been satisfied with this situation, because it was felt that research facilities were not available for more comprehensive studies.

The complexity of ecological systems dictates that a highly organized and integrated systematic approach be applied to their study. This chapter is devoted to a consideration of the procedures used and some of the problems that can be encountered in the organization and initiation of an integrated study of a grassland ecosystem.

II. AIMS OF AN ECOSYSTEM STUDY

The ecosystem approach is particularly well suited to the study of the biological basis of productivity that is being conducted under the International Biological Program (IBP). The objective in this instance is to develop a model of energy flow and nutrient transformations. Quantitative measurement is a preliminary aim, but will lead rapidly into the study of processes. Both are necessary for an understanding sufficient to synthesize a model that approaches reality and that can be tested by examining the effect of manipulation of components on the function of the system as a whole. Such studies are underway in a grassland at a field station established in 1967 at Matador, Saskatchewan, by the Canadian Committee on the International Biological Program (CCIBP) (Fig. 1). In the United States, in a program initially planned by the National Committee for the IBP, the concentrated effort is at the Pawnee Site in northeastern Colorado. Studies were initiated in 1968. This chapter is concerned primarily with the approach in the Matador study.

While it is of considerable interest to observe the magnitude of the pathways of energy and of nutrients through a single ecosystem, an important consideration in studying grassland is the comparative aspect. Interpretation of data relative to metabolism and nutrient transformations will require experimental manipulation and comparison of ecosystems that are basically similar. Such experiments are essential also in understanding processes. At first it will be necessary to manipulate one major factor at a time. In the Matador Project, manipulation initially will be through provision of different types of vegetative cover.

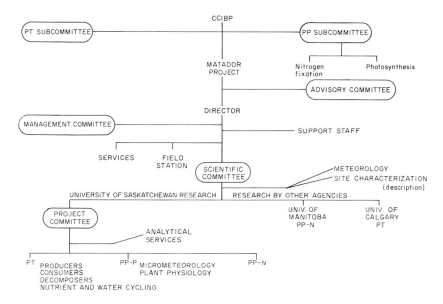

FIG. 1. An organization chart for the Matador Project, the grasslands study in the Canadian International Biological Program. CCIBP = Canadian Committee for the IBP; PT = Productivity Terrestrial; PP = Production Processes.

III. CHARACTERISTICS OF A STUDY SITE

The choice of a suitable location for such a study must be considered in relation to the objectives that have been defined. If only one ecosystem is to be studied, then emphasis must be given to homogeneity. If more than one ecosystem is involved, the comparisons that are to be made must be considered with respect to biotic or abiotic influences. If the project aims to compare ecosystems differing in biotic populations, then it is important that they be located in the same soil and climate. It is also important that each ecosystem be sufficiently extensive to permit the separation of biotic populations and environmental influences. The degree to which these principles can be followed depends upon the nature of the ecosystems involved. For example, a great difference in the height of vegetation may complicate the comparison of herbaceous and forest areas. Where cultivated and natural ecosystems are compared, the extent of the latter must be sufficient to provide a buffer zone that will assure isolation.

In the Matador study it was decided to compare cultivated and natural herbaceous ecosystems in an unforested region where tillage did not begin until the present century. The criteria that were established for site selection stressed the need for (1) a large enough area of relatively homogeneous native grassland that had not been modified much by agriculture and with a buffer zone of sufficient extent to provide a high degree of future protection; (2) absence of physiographical features that would affect environmental or biological characteristics of the site; (3) cultivated crops that had been grown on the same soil and physiographical type (as the native grassland) for at least 20 years within a reasonable distance of the natural area; and (4) reasonable proximity to a scientific establishment. The site that was selected is judged to fulfill these requirements and has the added advantages of being located in an area of soil highly suited for crop production and being representative of the most extensively cultivated situation in the region (Mitchell *et al.*, 1944). Initially 500 ha (1250 acres) of the natural system under study will be controlled. Another 2500–3000 ha (6200–7500 acres) of the same system, which protects the site from encroachment by tillage, is available for studies requiring larger sampling areas than are available on the controlled site. This allows for future manipulations.

The choice of site may be criticized because of emphasis on the natural ecosystem, when artificial ones also are to be studied. The agronomist may suggest that consideration be given to finding a convenient location for study of the arable ecosystems, with the expectation that a comparable natural system will be available nearby. Such a procedure will almost inevitably result in the choice of a natural system that has been avoided by tillage because of an unfavorable topographical or edaphic feature. The alternative is to seek out a natural system that survived within a tilled region because of some unusual historical development. The resulting choice then, of course, will be open to criticism as a relict not representative of any appreciable existing area of natural vegetation. However, it will represent conditions that previously existed in tilled land. The Matador site is the last known, large, surviving area of grassland that previously existed in 36% of the land that is now tilled in the brown soil zone of Saskatchewan. The choice of such a rare, natural system can be supported further in that, for economic reasons, it is less liable to future survival than more plentiful natural systems that have been avoided by tillage. Therefore, its study should not be delayed.

IV. ASSEMBLING THE RESEARCH TEAM

To assemble a sufficient number of competent and enthusiastic scientists knowledgeable of grassland from the required disciplines to conduct an ecosystem study is a major undertaking. It is almost certain that such workers are already engaged in full-time activities and are distributed among a number of university and government research groups. Even if funds were available to support them on a long-term project offering tenured positions, it is unlikely that a sufficient number of them could be attracted. The alternative is to encourage and facilitate their part-time participation.

In the Canadian IBP grassland study, we found that the direct approach, with a view to attracting specialists from various research groups before a plan was prepared, was not successful. In addition to being fully involved with their own programs, they were individually of the opinion that group research is not sufficiently rewarding to the individual or that adequate funds to support such a complex study would not be available. This approach, therefore, was abandoned. The alternative was found to lie in a request for assistance in the production of a hypothetical plan. The resulting discussions indicated not only the feasibility but also the applicability of the ecosystem approach to the investigator in his own field. When the project became a reality, the potential participants had the required background and enthusiasm to become thoroughly involved in the project.

The success of an ecosystem study will depend in large measure on the degree to which participants operate as a team. Individual initiative must be encouraged within a framework that includes sharing of basic data and collection of data, at least partially, as a service to others. The activities that are of greatest value to the group may be routine for the individual. However, opportunity also must be provided for related studies in which the individual may make more independent contributions.

Some full-time professional associates will be desirable in key positions to work in collaboration with the part-time participating specialists and to assure continuity of study. The need for support in different areas of activity will depend on the extent of participation of the specialist and on the degree of dependence of others on the data from that portion of the project. For example, meteorological measurements are of importance to almost every participant. They thereby warrant full-time professionals responsible to the project as a whole. Obviously coordination of activities is an important requirement at all levels and support staff will be needed for this purpose.

In a project of this type there will be a number of workers who wish to participate, but who are not attuned to working on field projects. Substitution of laboratory projects for essential field projects must not occur. Where laboratory studies are essential, these should not be divorced from field participation. This is a basic requirement to an interpretation of data collected in the real world. Another danger is that difficulties in recruitment may result in the decision to fit the project to the readily available scientific pool. It seems unlikely that a total ecosystem approach would be successful under these circumstances.

The total number of workers involved in an ecosystem study is likely to be considerable (see Fig. 1). In the Matador Project it is expected that the number will be between 80 and 100. Of these, about 30 will be professionals (usually Ph.D.'s), 20–30 will be graduate assistants, and the remainder will be summer assistants (4–5 months) and technical assistants.

V. PREPARING THE SCIENTIFIC PLAN

The plan must be produced by the people who will do the work. Moreover, participation in an international program, such as the IBP, requires that it must conform to agreed standards. We have been able to reconcile these matters in Canada by providing for planning within the project, but with the guidance of minimum standards that have been set up by IBP working groups. Planning, as well as research, must be coordinated; consequently, each participant should consider the plan of every other participant. In the Matador Project this is done through membership on the Scientific Committee (Fig. 1).

The participants will give much thought to the kind of research that should be conducted. In the beginning this will seem to include all disciplines of biological and environmental research. However, consideration of needs in relation to the theme (e.g., productivity) of a particular study will keep its scope within reason. The success of the activities related to this theme will depend largely on the extent to which erosion of their support can be avoided by inclusion of unrelated research.

Some difficulty will be experienced in planning because of differences of opinion concerning the intensity to which each aspect of the study should be conducted. Uniformity of treatment is vital. No guide can be given as to how uniformity in this respect can be achieved. It is a matter for frequent review by the participants throughout the period of study and for continuous consideration by the coordinators. Inevitably funds will become limiting, so pressure will always be present for uniform intensity of treatment. The decision concerning the number of ecosystems among which to share the funds may involve a consideration of whether all ecosystems will be studied in equal depth, but care should be taken to assure that all investigators apply the same priorities.

At all biotic levels in the ecosystem it will be possible to consider the activities of individual species only when these comprise a major proportion of the activities within a biological subsystem. This is most likely to be possible with the dominant producers and vertebrates, much less with the invertebrates, and probably not at all feasible with microorganisms. Consequently, this is not a feature upon which uniformity of treatment of different groups can be based.

Intensive planning by each interest group is necessary to consolidate and coordinate the plans of individual researchers, but integration of research into a master framework will only be possible by intensive intergroup discussions. The attendance in these sessions of chemists, data processors, and systems analyists will increase the effectiveness of the plan.

A particular problem with priorities may be experienced in attempting to keep the abiotic studies in balance with the biotic ones. Studies of the soil system have progressed much further than those of the total biotic system, and procedures are already well established for the former. Consequently, the soil workers are in a position to offer an initial plan that is complete; however, the biologists, who are working within a new framework, will find that additions to their plan are required as the study proceeds. As a result, if the full support is allocated at an early stage, treatment of the organismal portion of the system will suffer.

The project will seek to develop a model. Consideration in planning must be given from the outset to make certain that all significant information is collected to produce that model. It seems essential that the system be analyzed theoretically at an early stage to provide for this. However, opinions vary concerning the effectiveness of detailed system analysis before much preliminary data have been accumulated. At this time, opinions range from the desirability of developing a detailed theoretical model in the planning stage to that of working with a generalized model until some data have been collected. As knowledge of ecosystems increases, the feasibility of early analysis presumably would increase.

VI. ORGANIZING THE RESEARCH

In a productivity study it seems reasonable to organize the activities into four subsystems: producers, consumers, decomposers, and abiotic components. Superficially this seems to subdivide, respectively, green plants, animals, and microorganisms from environmental factors. However, this separation will be modified in application. Using microorganisms as an example, the decomposers will probably be investigated by the same personnel using the facilities that are employed in studies of nondecomposers (e.g., algae and nitrogen fixers) and parasites of producers and consumers. Similarly, while measurements of energy fixation and carbon assimilation by producers may be made by both biomass and photosynthesis approaches, these two activities will be less closely related than those of the physiologist and the micrometeorologist who will need to share equipment. Consequently, the organization of the activities at Matador (Fig. 1) consists of (1) biomass studies of producers by vegetationists; (2) consumer research by entomologists, ornithologists, and mammalogists; (3) microorganism research by bacteriologists, mycologists, and algologists; and (4) abiotic research by physicists, pedologists, micrometeorologists, chemists, and physiologists.

The following discussion of research procedures in a productivity study is concerned primarily with the organismal portions of the system. The methodology and concepts concerning the abiotic portion have been reasonably well developed elsewhere.

A. Producers

In a productivity study, research on producers must provide a reliable estimate of the amount of input (energy and carbon) into the system. This is important to interpretations of various other data. If methodology was sufficiently developed, gross measurements by physical or physiological techniques could be depended upon to furnish this estimate. At present, however, it seems necessary to use, as base measurements, net values obtained by biomass (harvest) studies. However, gross productivity measurements of CO_2 and energy by physicists using CO_2 flux and energy transfer techniques may be useful to the biomass determinator in considering the reliability of his data; these are considered together in a later section. The acquiring of accurate primary production data will provide a basis for comparison of the photosynthetic efficiency of various kinds of vegetative cover.

In the following discussion, the measurement of production by producers in terms of weight of plant tissue has been stressed, since past measurements of this parameter in herbaceous crops have only recently been shown to be deficient. Conversion of these measurements to calorific values is assumed, but other measures of quality also are desirable to provide reasonable comparisons between ecosystems. Carbon and nitrogen analyses usually will be minimum requirements. Westlake (1963) has discussed exhaustively the methods and standards to be applied and the means of presentation of comparative production data.

1. Stems and Leaves

The main source of information concerning the net productivity of herbaceous vegetation is an estimate of the standing crop of stems and leaves at the end of the growing season. This measure has serious short-comings. Let us consider in detail some of the problems that are experienced in measuring net productivity in herbaceous communities.

The peak standing crop does not always occur at the time when growth appears to have been completed. For example, Wiegert and Evans (1964) report that by clipping 20 times during one season the peak standing crop of aboveground parts in a Michigan old field was found to be 13% greater than the value at the time of cessation of growth (270 g/m² vs 238 g/m²).

The peak standing crop and the net primary production are identical only when the vegetation is composed of individuals that stop growing at a single instant in time (Wiegert and Evans, 1964). Thus, the peak standing crop might be expected to be a reasonable method of estimating the yield of an annual crop which germinates and develops uniformly or of a perennial grass sward consisting of one species, but it cannot be validly applied to a mixed stand of species (such as native grassland) that grow at different periods. The peak standing crop must be obtained for each species, if this method is to estimate accurately production in a mixed stand. Wiegert and Evans (1964) obtained a 26% higher estimate of peak standing crop in prairie by separation into only three categories (340 g/m² vs 270 g/m²). This value was 43% above the value (238 g/m²) that would have been obtained by a single clipping at maturity. These factors do not appear to have been considered in comparisons of the yield of native grassland with seeded, introduced grasses (Smoliak et al., 1967) and with annual crops (Ovington et al., 1963).

One of the major criticisms of the practice of equating total annual growth with peak standing crop of the current season is that mortality occurring during the sampling period is not measured. Because grass grows and dies continuously throughout the season, the error introduced by replacing annual growth with peak standing crop is greatest where grass (as opposed to forbs and shrubs) forms a high proportion of the cover. Wiegert and Evans (1964) used the following relationships to estimate mortality in Michigan:

(a) Change in standing crop of green material = growth − mortality
(b) ∴, Growth = change in standing crop of green material + mortality
(c) But mortality cannot be measured directly
(d) However, change in standing crop of dead material = mortality − material disappearing
(e) ∴, Mortality = material disappearing + change in dead standing crop

They measured change in standing crop of dead material (standing dead and litter) and rate of disappearance during 14 periods in 1959 and during 10 in 1960. Calculations indicated that growth was 2.4–2.5 times peak standing crop (on the basis of 10 and 8 samplings of green material) on upland and 4.5–4.9 times in swales. This disparity is probably much greater in prairie than in cropland (where one fast-growing species is harvested at the time of its maximum standing crop). It should be noted that, in a system that has reached equilibrium, the amount disappearing during the entire year should equal the net annual growth. In the present case, mean standing crop of dead material (195 g/m^2) × annual instantaneous rate of disappearance (1.72 g/g/year) = 335 g/m^2.

Repeated harvesting of the same plots is possible in stands of herbaceous plants, but it will provide a different result than if a sequence of plots is used. If the purpose is to measure the capacity of undisturbed vegetation to fix energy, then a sequence of plots must be used, with sampling becoming a greater problem. In order to avoid the possible differential effect, in subsequent years, of harvest at different seasons, the plot locations must be moved from year to year. This would seem to be the proper procedure to use in a project where the purpose is to compare the productivity of native grassland with a cereal crop or a seeded perennial hay crop. In ecosystems where harvest of grazing animals is the ultimate objective, comparisons are more legitimate if harvests are made at intervals on a sequence of plots that have been protected from grazing only for the period since the previous harvest, but the same plots may be sampled from year to year.

Measurements of primary productivity must be related to the proportion of shoots that are being removed by consumers. In a project that aims to compare the rate of production of native grassland and a cereal crop, it would seem logical that the native grass should receive the same amount of protection from herbivores during growth as does the crop. Since a considerable proportion of the energy fixed is removed from the cropland at harvest, consideration should be given also to the removal of a portion of the matured stems of the grassland. However, it would seem inadvisable to place the grassland under continuous grazing by domesticated livestock in such a comparison, for the resultant data would then involve a comparison between primary productivity in the crop and primary plus secondary productivity in the grassland.

2. UNDERGROUND PARTS

Another major shortcoming of many productivity studies is the failure to account for the photosynthates that are transferred to roots, rhizomes, and stem bases below the level of clipping. There is not great difficulty in sampling to obtain weights of these structures, but for perennial species it is difficult to make reliable estimates of annual increment. A minimum estimate has been obtained (Wiegert and Evans, 1964) by finding the difference in weight of underground parts between the high and low values for the year (143 g/m^2 for upland, 358 g/m^2 for swales). A more refined method has been to divide the soil profile into layers and to determine an increment for each layer, which may be different for different periods. Dahlman and Kucera (1965) obtained a value of 510 g/m^2 for Missouri prairie using 3 layers (452 g/m^2 when corrected for mineral matter) or 452 when the profile was taken in its entirety and not corrected, giving about 13% increase by profile division. Kucera *et al.* (1967) have attempted to separate underground parts of the current season from those of previous seasons and have obtained a similar value for production in the same prairie in the same year (179 g/m^2 of rhizomes and 369 of roots = 548 g/m^2 total). In both of these studies (Michigan and Missouri) the rate of decay of underground material approximated 25% per year. In an Indian study (Singh, 1967), biomass determinations were confined to the upper 30 cm of soil, being justified on the basis of the high proportion (95% or more) of the underground plant parts occurring near the surface.

B. Consumers

The studies of consumers should measure quantitatively and qualitatively: (1) the intake of primary production by herbivores (primary consumers); (2) the proportional transfer of this intake to carnivores (secondary consumers) and from all consumers to the environment as excreta, secreta, and dead organic materials; and (3) the losses from the ecosystem through consumer metabolism. A tremendous and costly task presents itself if the relations within the ecosystem of all consumer and trophic levels are to be considered. Establishment of priorities and levels of intensity for research will be necessary for each system.

Of primary importance is intensive research on herbivores, but this may have to be restricted to major or key species only. It may be necessary also to study intensively one or two key species of carnivores, if these may drastically affect intake of primary production through predation on herbivores. Also of primary importance will be an accurate estimation of standing crop of all consumer species within the ecosystem, minor as well as major, even if no initial, detailed research of their productivity or energetics is possible. Omnivorous animals, such as certain soil invertebrates, will present special problems in proportioning their activities as herbivores, carnivores, and decomposers. Portions of the studies on such animals will require close integration with the studies outlined in the following section on microorganisms. Thus, major animal groups that will require investigation in all grasslands are invertebrates (particularly selected insects and perhaps spiders), mammals, and birds.

A preliminary survey of the fauna in each ecosystem will be necessary to obtain a general idea of number and abundance of species in each group and at each trophic level. From this survey, the key species can be identified for initial, intensive study. Representative specimens of each species, perhaps three of each sex, should be preserved appropriately at this time for a reference collection. This survey undoubtedly will be concentrated on the study site. It is desirable, however, to include an appropriate zone surrounding the study site in an initial survey. Future problems involving wide-ranging species that visit the site sporadically, such as ungulates, and migrations to the site of animals with population centers nearby, such as grasshoppers, could thus be defined in the planning stage.

Faunal studies will be required above, at, and below the soil surface. Individual animal species will live in one, two, or all three layers at the same time, at different times, or in different life stages. In a systems study such as the Canadian one at Matador, where soil and aboveground invertebrates are being investigated by different groups of workers, careful integration and coordination of effort is imperative for completeness of coverage and accuracy. Further coordination of studies on soil animals, soil microorganisms, and primary productivity (root systems) is desirable and may even be necessary in such operations as sampling procedures and determinations of soil metabolism.

Primary productivity in grasslands could be appreciably affected by the feeding activities of sucking insects such as aphids and plant bugs. The effect, through removal of primary production by such consumers, may be minor in comparison with their effect on primary production processes. Some investigations on the total effect, while not without difficulty, should be initiated. Desroches (1958), Andrzejewska (1961), and Ricou and Duval (1964) evaluate the importance of, and give some methodology for, such studies in grasslands.

1. STANDING CROPS

Estimates of standing crops are required, not only for determinations of productivity in terms of biomass (weight/unit area), but also for relating to the system data from all subsequent studies on population dynamics and energetics (calories/unit area). Accuracy in the estimation of standing crops is, therefore, basic to the design of the model for the energetics of the system. Suitable quantitative sampling procedures and techniques for recovery of animals from samples must be devised for each ecosystem. Size, number, and pattern of samples must be considered in relation to type of soil, relief, vegetation, and other factors affecting animal distributions. The need for information on seasonal and annual fluctuations and age and stage structure of the populations also will require appropriate, periodic sampling programs. Recovery of the organisms during surface sampling or from soil samples could introduce further substantial errors in standing crop estimates, particularly in invertebrate studies. In the Matador Project the "quick-trap" method of Turnbull and Nicholls (1966) with vacuum suction recovery will be used for aboveground insects. Soil invertebrates will be sampled by a hydraulic core sampler, or the power-operated sampler of Burrage *et al.* (1963), with recovery to be devised for the heavy lacustrine soils using soil washing, temperature differential funnels, and/or hand-sorting techniques.

Researchers on consumers must consider the possible effect of sampling on reduction of populations, directly through removal of individuals, or indirectly through removal of breeding stock. Presumably this problem will increase with the size of the animal and its reduced numbers (the concentration of standing crop in a few large individuals). It may also arise in small organisms, such as wireworms or grasshoppers, if removal through intensive sampling coincides with a periodic concentration of breeding individuals. Such destructive sampling must be restricted to areas where other subsystems are not being studied.

2. POPULATION DYNAMICS AND ENERGETICS

In order to interpret data on standing crops in terms of rate of transformation of products of photosynthesis, detailed studies will need to be made on production through growth and reproduction, food ingestion, excretion, secretion, defecation, and metabolism. These would be expressed as caloric contents or losses. The energy budget for any consumer could be defined as:

$$I = Pg + Pr + E + M$$

where $I =$ caloric equivalent of material ingested, $Pg =$ production through growth, $Pr =$ production through reproduction (Pg and Pr expressed as caloric equivalent of protoplasm produced), $M =$ metabolic energy or heat loss through respiration and intraorganismal conversion, and $E =$ caloric equivalent of material egested through secretion, excretion, and defecation.

Assimilation (A), the caloric equivalent of the material retained by an organism, could be defined as $A = Pg + Pr + M$. Secondary productivity, the rate of production per unit time, would be: $(Pg + Pr)/T$. From the information derived from such studies, the assimilation/ingestion ratio (A/I) and the secondary production/assimilation ratio $(Pg + Pr)/A$ also could be calculated. The former is a measure of the efficiency with which a consumer extracts energy from the material ingested, and the latter measures the efficiency with which the energy assimilated by the consumer is transformed into products useful as energy sources for other organisms in the ecosystem.

The studies on growth and reproduction will provide a basis for conversion of the data on age and stage structure, derived through sampling for standing crop, to a clearer understanding of population dynamics of each species selected for detailed investigation. Measurements of ingestion will require studies on food preferences and rate of intake of different types of food by age, stage, and season. Caloric measurements on materials excreted, secreted, and defecated will indicate the energy that is ingested and not utilized by the consumer, but passed on for use by other organisms in the system. Energy lost to the ecosystem through consumer metabolism may be calculated by $M = I - Pg - Pr - E$. However, direct measurements of metabolic rates also should be made for greater accuracy in the calculation of energy budgets. While respirometry will give a measure of energy losses through respiration, adiabatic calorimetry will also be necessary if the energy loss through intracellular interconversions (e.g., fats to carbohydrates to proteins) are to be measured.

In most of the above studies, rearing in confinement will be necessary, but this should be coupled with field studies wherever possible. Such field studies can be accomplished through enclosures and/or exclosures in some instances. In other instances a taxon may be so active in comparison with other consumers, that food intake may be measured directly from clipping of herbage without manipulation of population. Radioactive tracer techniques are applicable to some food intake studies. It will be necessary to relate many of the above measurements to different temperature and activity regimes.

Budgetary, space, or manpower restrictions will limit studies on population dynamics and energetics of many minor species of consumers. Where such information is available in the literature for the species in question, or a related species, even if from a different ecosystem, it may be applied to the standing crop data that will be obtained to increase the accuracy of estimation of total energy budget for the system under study.

Many details on concepts and methodology for researchers on consumers in grassland ecosystems are available in the proceedings of the Symposium on Secondary Productivity held at Warsaw in 1966. Noteworthy are those by Macfadyen (1966), Petrusewicz (1966), and Wiegert and Evans (1967), among others.

3. PREDATION AND PARASITISM

The essential need to stress herbivore activities in a study of a grassland ecosystem will limit the resources available for research on carnivores. The importance of the latter in the system, however, should be under continuous surveillance. In some situations, herbivore and predator and/or parasite relationships may be so important that it will be necessary to establish a high priority for their more intensive study.

C. Microorganisms

Microorganisms, because of their small size and diversity, present a special challenge to the ecologist. Except for specialized studies, such as those of the rhizosphere (Macura, 1965; Rovira, 1965) and root nodule formation (Stewart, 1966), microorganisms have usually been studied with reference to a specific chemical reaction they carry out. The organisms that have received the most study do not constitute a significant portion of the microflora present in the soil–plant system at any one time (Burges, 1958). The role that the general soil population plays in energy transformations, in nutrient cycling, and in the breakdown of pollutants makes it imperative that microorganisms be studied in the detail accorded larger plants and animals.

The contributions that are expected of a microbiologist in a productivity study of grassland involve measurement of the degree of activity and functional role of microorganisms in energy flow and nutrient transformations. The methods developed for the study and the knowledge gained will be applicable to many other systems, for microorganisms tend to be ubiquitous in nature. Research in this area is complicated by the difficulty in the separation of the kinds of organisms present and in the determination of their biomass. Because changes in activity of the soil microflora take place after sampling, the samples must be utilized immediately, preferably in a field laboratory, as close to the point of sampling as possible.

1. INITIAL SURVEYS

The microbial population occurs in the soil, on the surfaces of, and within, the tissues of producers and consumers, and in the excreta of consumers. Microbial communities differ both qualitatively and quantitatively among these discontinuous microhabitats (Garrett, 1956; R. M. Jackson, 1965).

Sampling for microorganisms on a microhabitat basis is similar to that already described for soil animals, in that seasonal effects and vertical distribution will be measured using soil samples obtained with hydraulically operated soil samplers. In the case of microorganisms, sampling to the depth of root penetration is necessary.

The microhabitats to be considered are plant litter, rhizosphere, phyllo-
plane, animal excreta, living animals, and animal remains. Detailed characterization of the taxa constituting these populations is probably impossible, even with a fairly large group of specialists working in their own
field. However, concepts such as those utilized in numerical taxonomy
(Brisbane and Rovira, 1961) and biochemical ecology (Alexander, 1964),
plus the traditional classification procedures, should make general estimates feasible. The saprophytic fungi, actinomycetes, and bacteria, as
well as the parasitic fungi and bacteria, will have to be considered in
some detail. The protozoa and chemo- and photoautotrophs, if they are
not covered in the producer or consumer groups, should be considered.

The goal of the initial surveys will be to estimate the relative importance of these major groups in terms of numbers and biomass in order
to determine the microbiological homogeneity of the ecosystem. Synecological studies provide a means of assessing homogeneity and background information for studies of carbon and nitrogen turnover.

Photosynthesis and nitrogen fixation are being stressed in the current
IBP program. Nitrogen fixation and the input of available nitrogen into
the system by abiotic means, such as rainfall, are some of the major factors affecting both the quantity and the quality of primary productivity
in many grassland areas.

The extent of nitrogen fixation by asymbiotic and symbiotic microorganisms is to be measured under field and laboratory conditions. Iso-
topic labeling (^{15}N) work is slow and expensive. The acetylene reduction
technique, if used with appropriate standardization, may prove to be a
very useful tool (Stewart *et al.*, 1967), both in assessing the level of field
fixation and in laboratory studies. The sensitive ^{13}N technique also can
be used for laboratory investigations.

2. Biomass Measurements

Measurements of microbial biomass are essential to an understanding
of quantitative changes associated with the cycling of carbon and other
nutrients. Some techniques are available for estimating the standing crops
of microorganisms in soil samples, but few are available for assessing
microbial cell production and turnover in a particular microhabitat.
Methodology investigations must, therefore, continue to play an important role in the overall research design. Parkinson (1969) has recently reviewed the techniques available for estimating the biomass of
microorganisms.

Conversion of microscopic and plate counts to biomass can be made
only after chemical characterizations of the organisms. This, in turn,
necessitates culturing the organisms in a simulated soil environment.

Calorific and carbon and nitrogen measurements are of importance. Special components such as RNA, DNA, specific enzyme fluorescence, and diaminopimelic acid (el Shazily and Hungate, 1966) may provide effective complementary measurements (Rotman and Papermaster, 1966; Skujins, 1967).

3. DECOMPOSER CYCLE

Measurements of respiration in the field are necessary to interpret energy flow. The equation that was described above for consumers cannot be applied directly to microorganisms because of the diversity of the population and the disparity between the total biomass of microorganisms and the respiration they actually carry out, either under field or laboratory conditions (Clark, 1967).

Preliminary investigations at the Matador site have indicated a population of approximately 2×10^9 cells/g soil in the top 16 cm, with half of this amount being found at the 75-cm depth. Clark has stated that on the basis of 2 billion bacteria/g soil and measured bacterial respiration rates of 0.004 mg/hour/billion cells, the calculated activity of bacteria would be at least 10 times that normally found under field conditions. This indicates that a great majority of the soil organisms, although viable, must be present in a resting state. Calculation of the potential respiration of soil bacteria counted using the plate technique approximates field conditions much more closely than does the microscopic count.

Absorption systems incorporated into pits and infrared absorption spectrophotometers are applicable to field CO_2 measurements. Absorption techniques, such as alkali which removes the CO_2 from the system, may, however, alter the soil microbial activity and can yield artificial results if they act as a CO_2 sink. Paramagnetic and polarographic oxygen measuring systems are now available for field use. The polarographic technique is, however, very sensitive to temperature differentials and compensation techniques are required.

The roots of a grassland system not only produce large amounts of primary caloric materials, but also act as heterotrophs in gross soil CO_2 measurements. Therefore, it is essential that the decomposer cycle studies be closely associated with measurements of the growth rates of roots and of the amounts of organic materials excreted by the root systems (Harmsen and Jager, 1963; Rovira, 1962).

Tracers such as ^{14}C, ^{13}C, ^{32}P, ^{15}N, and ^{35}S utilized under canopies or incorporated into various substrates, give a measure of nutrient cycling and the metabolism of the products of primary producers (Paul et al., 1969). The humic materials contain the largest source of potential calories in the ecosystem and a measure of their turnover rate must be obtained. Mathematical models describing the turnover rate have been developed from data obtained using both tracer and nontracer techniques (Bartholomew and Kirkham, 1961; Jenkinson, 1965). These mathematical models can be readily made to contain the whole decomposer and consumer cycle in the soil. This information could be integrated relative to the aboveground consumers and the data from the primary producers and could be used to describe the overall turnover process.

D. Abiotic Studies

Abiotic studies pertinent to an ecosystem study of productivity must consider the environment in the soil and lower atmospheric layers in relation to the activity of producers, consumers, and microorganisms. To facilitate the common usage of equipment, investigations of photosynthesis, respiration, and transpiration must be integrated with micrometeorological investigations.

1. Photosynthesis

In recent years, physiological ecologists have become interested in the measurement of photosynthesis in the field. Elaborate, air-conditioned canopies have been devised to permit a reasonably accurate simulation of external conditions within the canopy (Musgrave and Moss, 1961; Eckardt, 1966). The utilization of CO_2 by producers is measured by the use of infrared gas analyzers, with corrections being made for respiration. A continuous record of photosynthesis throughout the growing period is most desirable. However, the nature of the gas sampling equipment used and the necessity for moving the canopy at frequent intervals will interfere with the continuous record. Perhaps sufficient data can be obtained by intensive measurements during representative periods. The objective of photosynthesis measurements is to obtain an estimate of gross primary productivity. This will serve as a basis for comparison with estimates that are obtained by biomass and micrometeorological methods.

2. Meteorology and Micrometeorology

Meteorological instrumentation should be set up to provide a continuous record of temperature, humidity, wind, rainfall, evaporation, and radiation. The purpose of the installation is twofold. First, it will provide participants with a detailed record of conditions that occurred near the sample plots at the time of sampling and in the intervals between sampling. Second, it will serve as a means of comparing the climate in the study area with that of standard weather stations where long-term records are available. To satisfy these requirements it will be necessary to employ standard instruments in addition to specialized ones. For example, a cup-counter anemometer is used as a standard to measure the daily run of wind, but in addition, a continuous record of speed and direction at a standard height also is required. The measurements of radiation will be the most sophisticated and costly.

Micrometeorological instrumentation should provide measurements of gradients of temperature, humidity, CO_2, and wind speed in the lower atmosphere; radiation components and spectral composition; temperatures and heat flux in soil; and fluxes of heat and moisture above the soil surface. The application of Bowen ratio, aerodynamic, and eddy correlation methods will provide measures of the activity of organisms in the ecosystem (Lemon, 1960, 1965, 1966; Tanner, 1963). Because of the high cost of setting up data acquisition facilities for this purpose, it may be desirable to install them in portable shelters.

3. Water Cycling

The amount of water available to organisms is reflected in the floristic composition of producers and in the magnitude of primary production. This relationship is exaggerated in native grasslands that owe their existence to a dry, warm climate. It is important, therefore, to consider the physics of soil water and the efficiency of its use by producers. The main objective is to examine the moisture flux in the soil, the air, and the plant and to determine its relationship to growth. It will be necessary to characterize plant environment as it changes with time. Soil parameters that are of consequence include rate of gaseous diffusion, porosity, resistance to penetration, moisture retention, and permeability. Where laboratory measurements of these variables are made, they must be related to moisture contents in the field.

To obtain data on the rate of water use by plants, a lysimeter should be provided. Additional characterization of transpiration in the producers can be obtained by measuring leaf water potentials at various times of the day and during the growing season, in as many species as practicable. Such measurements can be related to soil and atmospheric water fluxes.

4. Soil Nutrients

The fluxes of mineral elements, especially nitrogen and phosphorus, in an ecosystem are closely related to the flow of carbon in that system. Measurement of these nutrients, therefore, yields information on the quality of primary production. In addition, it can aid in the interpretation of the extent and relevance of the energy transformations involved. Determination of the total amounts of nutrients present is relatively straightforward (M. L. Jackson, 1958; Black *et al.*, 1965). However, these nutrients occur in a number of different forms, both in the organisms and in the soil (Bear, 1965). In an intensive study, the various forms must be measured. In addition, nitrogen undergoes a complex series of biological cycles, some of which can lead to losses from the system, making the determination of a balance sheet difficult (Bartholomew and Clark, 1965). The extent to which nutrient cycling is studied will depend on the resources available.

An essential feature of analyses, especially where the systems approach is used, is a requirement for consistency of measurements of all components. To measure phosphorus in some components, nitrogen in others, and dry matter in still others could result in a great deal of meaningless data.

The other aspect of nutrient cycling pertains to the possibility that one or more of the nutrients could be in short supply. They would then be one of the abiotic factors limiting either primary production or the energy and carbon transformations by decomposers and consumers. Analyses of grassland soils have shown that nitrogen, phosphorus, potassium, copper, zinc, molybdenum, boron, and sulfur fluctuate considerably. Some of these, as well as a few others, such as selenium, magnesium, and sodium, can be present in such concentrations that toxicity factors limit production.

The initial soil nutrient assessment can best be correlated with a detailed soil survey. This indicates the variability of the site and yields basic information required in establishing the experimental design. It is also useful for interpreting the data at a later stage. The scope of the routine survey depends on the site and on the aims of the research. The routine survey analyses are pH, conductivity (soluble salts), cation exchange capacity, exchangeable bases, mechanical analysis, total carbon, inorganic carbon, and total nitrogen (Black *et al.,* 1965; Stelly, 1967).

The manipulation of comparative sites requires an assessment of (1) available nutrient levels in the soil, and (2) the need for extra nutrients to achieve optimum growth. The losses through leaching, erosion, and volatilization and the input of nutrients into the system by rainfall, excreta, or soil management will affect the interpretation relative to nutrient cycling and abiotic factors controlling plant growth (Fried and Broeshart, 1967).

VII. CONCLUSION

This discussion of procedures involved in a productivity study of a grassland ecosystem has been shaped by the experience of the authors in the planning and initiation of such a project. Many lessons are still to be learned concerning the processing and interpretation of data on an integrated basis and the most effective means of presenting the results. Plans are being prepared concurrently by a number of IBP groups and will almost immediately become available to supplement the above treatment. An example is the handbook, "Methods for Estimating the Primary Production of Grasslands, Arid Lands and Dwarf Shrublands," by R. E. Hughes, C. Milner, R. O. Slayter, and G. H. Giningham sponsored by Section PT.

Hopefully, a large number of ecosystem studies established in many countries during the IBP will result in detailed understanding of the organization of systems which are being depended upon for sustained production of food crops. They will undoubtedly result in important advances in techniques of study.

REFERENCES

Alexander, M. 1964. Biochemical ecology of soil microorganisms. *Ann. Rev. Microbiol.* **18**, 217–252.

Andrzejewska, L. 1961. The course of reduction in experimental *Homoptera* concentrations. *Bull. Acad. Polon. Sci., Ser. Sci. Biol.* **9**, 173–178.

Bartholomew, W. V., and F. E. Clark. 1965. "Soil Nitrogen," Agron. Ser. No. 10. Am. Soc. Agron., Inc., Madison, Wisconsin. 615 pp.

Bartholomew, W. V., and D. Kirkham. 1961. Mathematical descriptions and interpretations of culture induced soil nitrogen changes. *Trans. 7th Intern. Congr. Soil Sci., Madison Wisc., 1960.* Vol. 3, pp. 471–477. Elsevier, Amsterdam.

Bear, F. E. 1965. "Chemistry of the Soil," 2nd ed. Reinhold, New York. 515 pp.

Black, C. A., D. P. Evans, J. L. White, L. E. Ensminger, and F. E. Clark. 1965. "Methods of Soil Analysis," Agron. Ser. No. 9. Am. Soc. Agron., Inc., Madison, Wisconsin. 1572 pp.

Brisbane, P. G., and A. D. Rovira. 1961. A comparison of methods for classifying rhizosphere bacteria. *J. Gen. Microbiol.* **26**, 379–392.

Burges, A. 1958. "Microorganisms in the Soil." Hutchinson, London. 188 pp.

Burrage, R. H., R. O. Vibert, and M. N. MacLeod. 1963. Note on a power operated soil sampler for field wireworm studies. *Can. J. Plant Sci.* **43**, 242–243.

Clark, F. E. 1967. Bacteria in soil. *In* "Soil Biology" (N. A. Burges and F. Raw, eds.), pp. 15–49. Academic Press, New York.

Dahlman, R. C., and C. L. Kucera. 1965. Root productivity and turnover in native prairie. *Ecology* **46**, 84–89.

Desroches, R. 1958. Le Ray-grass anglais est-il encore le roi de la prairie francaise? *Bull. Engrais. Suppl.,* 12–15.

Eckardt, F. E. 1966. Le principe de la soufflerie climatisée, appliqué a l'étude des échanges gazeux de la couverture végétale. *Ecol. Plant* **1**, 369–400.

el Shazily, K., and R. E. Hungate. 1966. Methods for measuring diaminopimelic acid in total rumen content and its application to the estimation of total bacterial growth. *Appl. Microbiol.* **14**, 27–30.

Fried, M., and H. Broeshart. 1967. "The Soil-Plant System in Relation to Inorganic Nutrition." Academic Press, New York. 358 pp.

Garrett, S. D. 1956. "Biology of the Root Infesting Fungi." Cambridge Univ. Press, London and New York. 283 pp.

Harmsen, G. W., and G. Jager. 1963. Determination of the quantity of carbon and nitrogen in the rhizosphere of young plants. *In* "Soil Organisms" (J. Doeksen and S. Van der Drift, eds.), pp. 243–251. North-Holland Publ., Amsterdam.

Jackson, M. L. 1958. "Soil Chemical Analysis." Prentice-Hall, Englewood Cliffs, New Jersey. 498 pp.

Jackson, R. M. 1965. Antibiosis and fungistasis of soil microorganisms. *In* "Soil Borne Plant Pathogens" (K. F. Baker and W. C. Snyder, eds.), pp. 363–373. Univ. of California Press, Berkeley, California.

Jenkinson, D. S. 1965. Studies on the decomposition of plant material in soil. I. Losses of carbon from C14 labelled rye-grass incubated with soil in the field. *Soil Sci.* **16**, 104–115.

Kucera, C. L., R. C. Dahlman, and M. L. Koelling. 1967. Total net productivity and turnover on an energy basis for tallgrass prairie. *Ecology* **48**, 536–541.

Lemon, E. R. 1960. Photosynthesis under field conditions. II. An aerodynamic method for determining the turbulent carbon dioxide exchange between the atmosphere and a corn field. *Agron. J.* **52**, 697–702.

Lemon, E. R. 1965. Micrometeorology and the physiology of plants in their natural environment. *In* "Plant Physiology" (F. C. Steward, ed.), Vol. 4A, pp. 203–227. Academic Press, New York.

Lemon, E. R. 1966. Energy conversion and water use efficiency in plants. *In* "Plant Environment and Efficient Water Use" (W. H. Pierre *et al.*, eds.), pp. 24–48. Am. Soc. Argon., Inc., Madison, Wisconsin.

Macfadyen, A. 1967. Methods of investigation of productivity of invertebrates in terrestrial ecosystems. *Symp. Principles Methodol. Secondary Productivity Terrestrial Ecosystems, Warsaw, 1966* pp. 383–412. Panstwowe Wydawnictwo Naukowe, Warsaw.

Macura, J. 1965. Interrelations between microorganisms and plants, in the rhizosphere. Plant microbes relationships. *Proc. Symp. Relationships Between Soil Microorganisms Plant Roots, Prague, 1963* pp. 26–33.

Mitchell, J., H. C. Moss, and J. S. Clayton. 1944. "Soil Survey of Southern Saskatchewan," Soil Surv. Rept. No. 12. University of Saskatchewan, Saskatoon, Saskatchewan. 259 pp.

Musgrave, R. B., and D. N. Moss. 1961. Photosynthesis under field conditions. I. A portable, closed system for determining net assimilation and respiration of corn. *Crop Sci.* **1**, 37–41.

Odum, E. P. 1959. "Fundamentals of Ecology," 2nd ed. Saunders, Philadelphia, Pennsylvania. 546 pp.

Olson, J. S. 1964. Gross and net production of terrestrial vegetation. *J. Ecol.* Jubilee Suppl., 99–118.

Ovington, J. D., D. Heitkamp, and D. B. Lawrence. 1963. Plant biomass and productivity of prairie, savanna, oakwood and maize field ecosystems in central Minnesota. *Ecology* **44**, 52–63.

Parkinson, D. 1969. Heterotrophic microorganisms. *Symp. Methods Study Soil Ecol., Paris, 1967* (in press).

Paul, E. A., V. O. Biederbeck, and N. S. Rosha. 1969. The use of C14 in investigating the metabolism of soil organisms. *Symp. Methods Study Soil Ecol., Paris, 1967* (in press).

Petrusewicz, K. 1967. Concepts in studies on the secondary productivity of terrestrial ecosystems. *Symp. Principles Methodol. Secondary Productivity Terrestrial Ecosystems, Warsaw, 1966* pp. 17–49. Panstwowe Wydawnictwo Naukowe, Warsaw.

Ricou, G., and E. Duval. 1964. Contribution a l'étude de l'action des cicadelles pur quelques graminées de prairies. *Compt. Rend Acad. Agr. France*, 472–476.

Rotman, B., and B. W. Papermaster. 1966. Membrane properties of living mammalian cells as studied by enzymatic hydrolysis of fluorogenic esters. *Proc. Natl. Acad. Sci. U.S.* **55**, 134–141.

Rovira, A. D. 1962. Plant-root exudates in relation to the rhizosphere microflora. *Soils Fertilizers* **25**, 167–172.

Rovira, A. D. 1965. Interactions between plant roots and soil microorganisms. *Ann. Rev. Microbiol.* 19, 241–266.

Singh, J. S. 1967. Seasonal variation in composition, plant biomass and net community production in the grasslands of Varanasi. Ph.D. thesis, Dept. of Botany, Banaras Hindu University, Varanasi, India.

Skujins, J. J. 1967. Enzymes in soil. *In* "Soil Biochemistry" (A. D. McLaren and G. H. Peterson, eds.), pp. 371–414. Marcel Dekker, New York.

Smoliak, S., A. Johnston, and L. E. Lutwick. 1967. Productivity and durability of crested wheatgrass in southeastern Alberta. *Can. J. Plant Sci.* **47**, 539–548.

Stelly, M. 1967. "Soil Testing and Plant Analysis." Soil Sci. Soc. Am., Inc., Madison, Wisconsin.

Stewart, W. D. P. 1966. "Nitrogen Fixation in Plants." Oxford Univ. Press (Athlone); London and New York. 168 pp.

Stewart, W. D. P., G. P. Fitzgerald, and R. H. Burris. 1967. *In situ* studies on N_2 fixation using the acetylene reduction technique. *Proc. Natl. Acad. Sci. U.S.* **58**, 2071–2078.

Tanner, C. B. 1963. "Basic Instrumentation and Measurements for Plant Environment and Micrometeorology," Soil Bull. 6. Dept. of Soil Sci., University of Wisconsin, Madison, Wisconsin. 354 pp.

Turnbull, A. L., and C. F. Nicholls. 1966. A "quick-trap" for area sampling of arthropods in grassland communities. *J. Econ. Entomol.* **59**, 1100–1104.

Westlake, D. F. 1963. Comparisons of plant productivity. *Biol. Rev.* **38**, 385–425.

Wiegert, R. G., and F. C. Evans. 1964. Primary production and the disappearance of dead vegetation on an old field in southeastern Michigan. *Ecology* **45**, 49–63.

Wiegert, R. G., and F. E. Evans. 1967. Investigations of secondary productivity in grasslands. *Symp. Principles Methodol. Secondary Productivity Terrestrial Ecosystems, Warsaw, 1966* pp. 499–518. Panstwowe Wydawnictwo Naukowe, Warsaw.

Chapter IV The Watershed-Ecosystem Concept and Studies of Nutrient Cycles

F. H. BORMANN and G. E. LIKENS

I. INTRODUCTION

The operation of a factory is somewhat analogous to the utilization of land for agricultural purposes, from intensive agriculture to forest management. Materials and energy flow into the factory. In a variety of steps they pass through the processing divisions of the plant where they are subject to chemical and physical reconstitution; energy is dissipated at each step, and at some points materials may be recycled to an earlier point in the process. Finally, finished products and a variety of wastes emerge from the plant. Under early systems of management, factories were operated with minimal knowledge of flux rates among component parts of the plant and between the plant and its environment. However, modern conditions of commerce and competiton and the recognition of new cost factors, such as pollution abatement, have forced the evolution of increasingly sophisticated systems of management. For individual plants, it has become necessary to understand, in detail, input and output relationships and every operation of the plant itself including the complex of interactions among its component parts. Operations analysis has served to elucidate these relationships, to detect weak links, and to suggest alternative linkages. As a consequence, it has been possible to maximize output with a concomitant reduction of costs.

Managers of agricultural, semiwild, and wild lands seek similar goals of maximization of output and reduction of costs. Also they have the added responsibility of protecting the producing capital of their lands for use by future generations. Often, management entails not one end product, but several, such as wood, water, wildlife, and recreation. Given the basic abiotic and biotic complexity of land, the phenomena of succession and retrogression, a multiplicity of managerial goals, and a desire for more efficient use of the land, it is obvious that some theoretical framework upon which we can assemble and interrelate these diverse components is a necessity. In one form or another, foresters and range managers have long been aware of this need (Lutz, 1957, 1963; Costello, 1957).

A. Ecosystem Concept

The ecosystem concept provides this framework. The ecosystem is a basic functional unit of nature which includes both organisms and their nonliving environment, each interacting with the other and influencing each other's properties, and both necessary for the maintenance and development of the system (Odum, 1963). An ecosystem, then, may be visualized as a series of components, such as species populations, organic debris, available nutrients, primary and secondary minerals, and atmospheric gases, linked together by food webs, nutrient flow, and energy flow. Knowledge of these components and the links involved leads to an understanding of the interrelationships within the systems and of the ramifications of any manipulation applied at any point in the system.

B. Input–Output Relationships

Boundaries of an ecosystem are most often defined to meet the pragmatic needs of the investigator (Evans, 1956). But once this definition is made, the ecosystem may be visualized as being connected to the surrounding biosphere by a systems of inputs and outputs. Inputs and outputs may be in the form of radiant energy, water, gases, chemicals, or organic materials moved through the ecosystem boundary by meteorological, geological, or biological processes (Bormann and Likens, 1967). Knowledge of input–output relationships is necessary if we are to understand fully (1) energy and nutrient relationships of the individual ecosystem, (2) the comparative behavior of ecosystems, (3) the effect of geological processes such as erosion and deposition, mass wasting and weathering on ecosystem dynamics, (4) the effects of meteorological variations on ecosystem behavior, (5) the relationship of the individual ecosystem to worldwide biospherical cycles, and (6) the effects of managerial practices on the structure and function of individual ecosystems and on other ecosystems intimately linked to the manipulated system.

Measurement of these critical input–output relationships presents difficulties particularly in studies of nutrient cycling. Nutrient cycles are strongly geared, at many points, to the hydrological cycle. As a consequence, measurement of nutrient input and output requires simultaneous measurement of hydrological input and output. Estimation of nutrients entering or leaving an ecosystem by subsurface seepage or sheet or rill flow presents a difficult problem. In many systems, the problem is further complicated because solutes leave by way of deep seepage and eventually appear in another drainage system. This loss may be very significant; it is virtually impossible to measure.

C. The Small Watershed Approach to Ecosystem Studies

About eight years ago we recognized that in some ecosystems the nutrient cycle, hydrological-cycle interaction can be turned to good advantage in the study of the nutrient cycles and other basic parameters of the ecosystem. This is particularly so if an ecosystem meets these specifications. (1) The ecosystem is a watershed; (2) the watershed is underlain by a tight bedrock or other impermeable base, such as permafrost; and (3) the watershed has a uniform geology. Given these conditions, for chemical elements without a gaseous form at biological temperatures, it is possible to construct nutrient budgets showing input, output, and net loss or gain from the system. Moreover, these data provide estimates of weathering and erosion.

If the ecosystem were a watershed, input would be limited to meteorological (bulk precipitation and dry fallout) and biological (moved by animals or man) origins. Geological input (alluvium and colluvium) need not be considered because there would be no transfer between adjacent watersheds.

In humid regions where the contribution of dust is small, meteorological input can be measured from a combination of hydrological and precipitation-chemistry parameters. From periodic measurements of the elements contained in precipitation and from continuous measurements of precipitation entering a watershed of known area, one may calculate the temporal input of an element in terms of grams per hectare.

FIG. 1. A weir showing the V-notch, record house, and ponding basin (courtesy of the Northeastern Forest Experiment Station; from Bormann and Likens, 1967).

Losses from this watershed ecosystem would be limited to geological and biological output. Given an impermeable base, geological output (losses due to erosion) would consist of dissolved and particulate matter in either stream water or seepage water moving downhill above the impermeable base.

Geological output can be estimated from hydrological and chemical measurements. A weir, anchored to the bedrock (Fig. 1), will force all drainage water from the watershed to flow over the notch, where the volume and rate of flow can be measured. These data, in combination with periodic measures of dissolved and particulate matter in the outflowing water, provide an estimate of geological output which may be expressed as grams of an element lost per hectare of watershed.

The nutrient budget for a single element in the watershed ecosystem then becomes: (meteorological input + biological input) less (geological output + biological output) = net loss or gain. This equation may be further simplified if the ecosystem meets a fourth specification—if it is part of a much larger, more or less homogeneous, biotic unit. Biological output would tend to balance biological input if the ecosystem contained no special attraction or deterrent for animal populations moving at random through the larger vegetation system, randomly acquiring or discharging nutrients. On this assumption, the nutrient budget for a single system would become: (meteorological input per hectare) less (geological output per hectare) = net gain or loss per hectare. This fundamental relationship provides basic data for an integrated study of ecosystem dynamics.

II. THE HUBBARD BROOK ECOSYSTEM STUDY

A. Undisturbed Forest Ecosystems

Several years ago we began a study of northern hardwood forest ecosystems utilizing the small watershed approach. This work is being done through the cooperation of the Northeastern Forest Experiment Station at the Hubbard Brook Experimental Forest in the White Mountains of central New Hampshire. This is the major installation of the U.S. Forest Service in New England for measuring aspects of the hydrological cycle as they occur in small forested watersheds. Summarizing data from the Hubbard Brook Ecosystem Study, we would like to illustrate the utility of the small watershed approach in the study of the structure, function, and management of terrestrial ecosystems.

Six watersheds, tributary to Hubbard Brook and 12–43 ha in size, are being studied (Fig. 2). Each watershed is forested by a well-developed, second-growth stand composed primarily of *Acer saccharum* Marsh. (sugar maple), *Fagus grandifolia* Ehrh. (beech), and *Betula alleghaniensis* Britt (yellow birch). Some of the largest trees were cut 52 years ago but since that time there has been no disturbance by cutting or fire.

The watersheds are underlain by impermeable gneiss of the Littleton Formation. Consequently, losses of water and nutrients are confined to materials carried out of the watersheds by small first- and second-order streams. Losses by deep seepage may be eliminated from consideration.

Precipitation entering these watersheds is measured by a network of gaging stations (Fig. 2) and drainage water leaving the watershed is measured by a V-notch weir (Fig. 1) or a combination flume and weir anchored on the bedrock across the stream draining each watershed. Since there is no deep seepage, evapotranspiration is calculated by subtracting hydrological output from hydrological input.

1. HYDROLOGICAL PARAMETERS

The average monthly water budget for the period 1955–1967 (Table 1) indicates that precipitation is distributed rather evenly throughout the

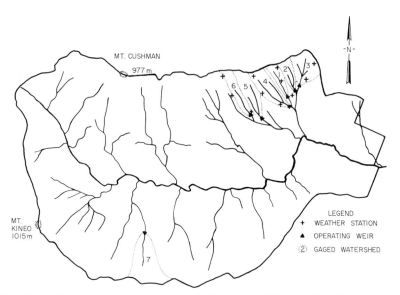

Fɪɢ. 2. Outline map of the Hubbard Brook Experimental Forest, West Thornton, New Hampshire, showing the various drainage streams that are tributary to Hubbard Brook (from Likens *et al.,* 1967).

year whereas runoff is not. Most of the runoff (57%) occurs during the snowmelt period of March, April, and May. In fact, 35% of the total runoff occurs in April. In contrast, only 0.7% of the yearly runoff occurs in August. The annual budget for 1964–1965 shows an extreme drought year; 1966–1967 shows a very wet year; 1963–1964 a relatively dry year; and 1965–1966 shows a moderately wet year (Table 1).

A detailed comparison of watersheds 1, 2, and 3 has shown that the hydrology (precipitation and streamflow characteristics) is nearly identical (Hart, 1966). The similarity in hydrological factors between these watersheds adds still further evidence for the assumption that deep seepage is negligible. Average distribution of water loss from the Hubbard Brook Experimental Forest (42% evapotranspiration, 58% runoff during 1955–1967) differs from that obtained by similar methods at other major catchment installations. For instance, at Wagonwheel Gap in Colorado the evapotranspiration was 71% and runoff 29%; at Coshocton in Ohio evapotranspiration was 72% and runoff 28%; and at Coweeta in North Carolina evapotranspiration was 48% and runoff 52% (Bates and Henry, 1928; Dreibelbis and Post, 1941; Dils, 1957). Differences in climate, vegetation, and geological structure of the watersheds undoubtedly account for these variations in pattern of water loss from various ecosystems. However, some of these calculated values for evapotranspiration (precipitation minus runoff) probably do not adequately take into account deep seepage, groundwater, unknown watershed boundaries, etc. At Hubbard Brook we feel confident that these factors are negligible in the calculation of evapotranspiration.

2. CHEMICAL PARAMETERS

To measure chemical parameters (dissolved substances) of the ecosystems, weekly water samples of input (rain and snow) and output (stream water) have been collected and analyzed for Ca^{2+}, Mg^{2+}, K^+, Na^+, Al^{3+}, NH_4^+, SO_4^{2-}, NO_3^-, Cl^-, HCO_3^-, and SiO_2. Data on these ions, their methods of analyses, and their behavioral patterns have been presented by Likens *et al.* (1967) and Fisher *et al.* (1968). Discussion here will be limited to Ca^{2+}, Mg^{2+}, Na^+, and K^+ during 1963–1965. Precipitation chemistry, stream water chemistry, and nutrient budgets are considered in the following paragraphs.

TABLE 1

AVERAGE RUNOFF (R) AND PRECIPITATION (P) FOR WATERSHEDS 1–6 OF THE HUBBARD BROOK EXPERIMENTAL FOREST, EXPRESSED IN CENTIMETERS OF WATER PER UNIT AREA[a]

Month	1955–1967		1963–1964		1964–1965		1965–1966		1966–1967	
	R	P	R	P	R	P	R	P	R	P
June	1.8	8.7	0.8	6.6	0.4	5.7	1.4	11.1	3.7	11.0
July	1.1	10.8	0.2	7.3	0.9	12.2	0.4	9.6	0.7	9.7
August	0.5	10.3	0.3	14.8	1.5	15.2	0.1	8.7	2.5	19.2
September	0.9	9.1	0.1	3.9	0.1	3.6	3.8	19.7	2.3	11.8
October	5.1	11.7	0.1	3.1	0.5	5.4	7.7	12.5	4.8	8.4
November	8.5	14.1	8.8	20.9	5.3	11.6	9.1	14.6	12.8	14.3
December	6.0	10.0	5.9	7.4	7.9	11.4	3.1	7.6	5.6	10.2
January	3.7	9.7	6.0	12.2	2.8	6.3	3.7	6.7	2.6	5.8
February	2.6	8.8	2.1	4.4	3.2	10.1	2.3	6.2	1.7	10.7
March	7.2	8.2	10.0	14.3	4.4	2.7	12.3	12.0	4.8	5.5
April	24.9	10.7	25.6	11.1	18.1	8.3	18.5	6.6	26.6	14.0
May	8.1	9.1	7.8	11.1	3.6	2.4	10.9	9.3	12.6	11.5
Annual	70.4	121.2	67.7	117.1	48.7	94.9	73.3	124.6	80.7	132.1
P–R (evaporation and transpiration)	50.8		49.4		46.2		51.3		51.4	

[a] Data are based on a water year from June 1–May 31. Data from watershed 2 (cutover) after June 1, 1965, are not included.

The precipitation values reported here represent bulk precipitation, i.e., a mixture of rain and dry fallout (Whitehead and Feth, 1964). The contribution of Ca^{2+}, Mg^{2+}, K^+, and Na^+ from dry fallout, or water-soluble aerosols, is apparently not large as we have not observed any marked differences in these cation concentrations between samples obtained in collectors that were continuously open and those that opened only during rainstorms.

The concentration of the four cations in weekly precipitation samples is variable. For example, during 1963–1965, the values in milligrams per liter ranged from 0.02 to 3.9 for Ca^{2+}, 0.007 to 1.09 for Mg^{2+}, 0.008 to 3.83 for K^+, and 0.006 to 2.52 for Na^+ (Likens *et al.*, 1967; Juang and Johnson, 1967).

Weighted average cation concentrations in mg/liter [Σ_{1-52} weekly precipitation in liters $\times$ weekly concentration in mg/liter $\div$ total liters/year = weighted concentration in mg/liter] are given for water years June 1, 1963–May 31, 1964, and June 1, 1964–May 31, 1965, in Table 2. Our results compare reasonably well with those of Junge and Werby (1958) for the central New Hampshire area.

The yearly average concentration in 1964–1965 for both magnesium and sodium was about double that of 1963–1964, whereas average calcium and potassium values remained about the same for both years (Table 2). Sodium and magnesium are abundant in seawater, in contrast to Ca^{2+} and K^+, and their increase in 1964–1965 may have been due to a greater intrusion of maritime air into our region.

These results suggest that shifts in precipitation chemistry might play some role in altering patterns of productivity and nutrient cycling in terrestrial ecosystems, particularly where a shift is prolonged over a period of years and where the terrestrial ecosystem is heavily dependent on precipitation as a source of nutrients.

C. M. Johnson and Needham (1966) have shown a strong inverse re-

TABLE 2

WEIGHTED AVERAGE CATION CONTENT OF PRECIPITATION
SAMPLES COLLECTED WITHIN THE HUBBARD BROOK
EXPERIMENTAL FOREST[a]

Time period	Cations (mg/liter)			
	Ca	Mg	K	Na
June 1, 1963–May 31, 1964	0.26	0.06	0.21	0.09
June 1, 1964–May 31, 1965	0.30	0.12	0.19	0.22

[a] From Likens *et al.* (1967).

lationship between Ca^{2+}, Mg^{2+}, and Na^+ concentrations and stream discharge in mountain streams of California. Our data do not clearly show this relationship; in fact, the weighted averages of cationic concentration in drainage water were remarkably constant even though the discharge of water from the watersheds ranged over four orders of magnitude (Fig. 3). The relative constancy of concentration was particularly striking in the spring and summer, since about 37% of the yearly runoff occurred in April and less than 0.2% in September (Table 1; Fig. 4).

This chemical stability in the Hubbard Brook system is very likely due to the mature and highly permeable podzolic soils. Because of the microtopography, loose soil structure, and absence of frozen ground during the winter, virtually all of the drainage water must pass through the soil. The intimate contact afforded by this passage, plus the relatively large buffering capacity of the soil materials (relative to the quantities lost to the percolating water) and the relatively small range of temperatures within the soil mass (10°C at a depth of 91 cm) apparently succeed in buffering the chemical composition of the transient waters. Therefore, concentrations of calcium and magnesium remain relatively constant. The negative correlation between concentration and stream discharge for sodium might be explained on the basis of dilution since there is relatively little sodium available on the exchange complex. The potassium relationship is somewhat more complicated. Here, preferential base exchange and biological activity undoubtedly play major roles. The latter is inferred by the low concentration of K^+ measured in stream water during periods of active plant growth and the higher concentration associated with dormancy (Fig. 5).

A nutrient budget of dissolved cations for the Hubbard Brook ecosystem was determined from the difference between the meteorological input per hectare and the geological output per hectare (Bormann and Likens, 1967). Since we assumed that dust in this system is negligible, the cationic input was calculated from the product of the cationic concentration (mg/liter) and the volume (liters) of water as precipitation. These calculations were greatly simplified, as there was no significant difference in the quantity or cation quality of precipitation collected throughout the watershed area (Likens et al., 1967). Output was calculated as the product of the volume (liters) of water draining from the system and its cationic concentration (mg/liter). All of the input and output values were calculated for a common area (hectare), making individual watersheds more readily comparable. Calculations were made on a weekly basis and the results summed for the wateryear June 1 to May 31. Data covering six watersheds and four water years, based on over 5000 chemical determinations (N. M. Johnson et al., 1968) are presented in Table 3.

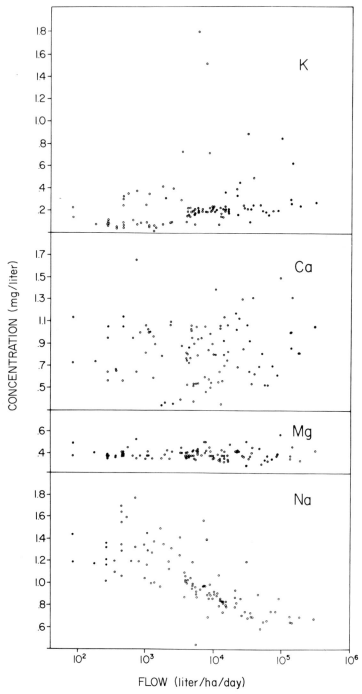

Fig. 3. Relationship between cation concentration and volume of flow for stream water in watershed 4 from 1963 to 1965 (from Likens *et al.*, 1967).

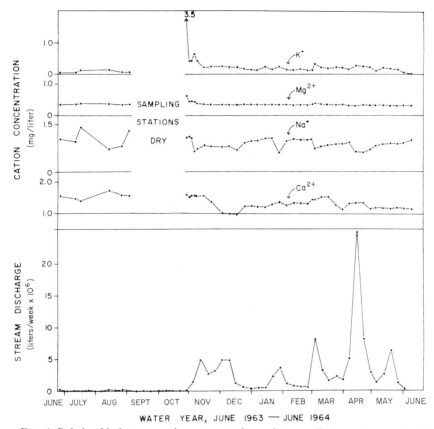

FIG. 4. Relationship between cation concentration and stream discharge in watershed 2 (from Likens *et al.*, 1967).

Since there was not a strong correlation between cationic concentration and discharge in general, the total output was more strongly dependent on volume of stream flow than on ionic concentration. This is shown graphically in Fig. 6 where average gross output for six watersheds is plotted against average annual runoff. The output of sodium, magnesium, and potassium shows a fairly close relationship to the annual runoff, while calcium shows more scatter. This may be due, in part, to the fact that it was necessary to correct for chemical interference in the 1963–1965 calcium values (Likens *et al.*, 1967; Johnson *et al.*, 1968).

Because of the strong correlation between output and runoff, the total export of cations during summer periods of low flow was small, whereas major export occurred during periods of high runoff in the late autumn and in the spring snowmelt period. Conversely, cation input was more or less randomly scattered throughout the year.

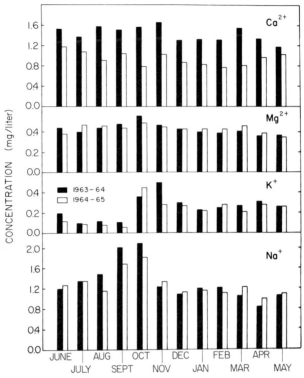

Fɪɢ. 5. Weighted monthly average concentrations of Ca^{2+}, Mg^{2+}, K^+, and Na^+ in drainage water from watershed 4 (from Likens *et al.*, 1967).

Data for four water years, 1963–1967, indicate an average of 2.6, 1.5, 0.7, and 1.4 kg/hectare of calcium, sodium, magnesium, and potassium, respectively, entered the ecosystems, while 10.6, 6.1, 2.5, and 1.5 kg/ha were flushed out of the system in solution in stream water (Table 3). Comparisons of data for individual years indicate considerable difference between years. These differences result from variations in the hydrological cycle and changes in the concentrations of ions in precipitation and stream water (Likens *et al.*, 1967).

Net losses of calcium, sodium, and magnesium occurred in every year spanning a range of wet, dry, and average years. The potassium budget, on the other hand, was just about balanced in each year, suggesting that K^+ is accumulating in the ecosystem relative to the other cations; several factors may be accountable. Part of the K^+ may be retained in illitic clays developing in the ecosystem, or, if the biomass of the ecosystem is increasing, K^+ may be retained in the developing biomass in proportionally greater amounts than the other cations.

TABLE 3

ANNUAL BUDGETS FOR CALCIUM, SODIUM, POTASSIUM, AND MAGNESIUM FOR SIX SMALL WATERSHEDS OF THE HUBBARD BROOK EXPERIMENTAL FOREST[a,b]

Water year (June–May)	Calcium (kg/ha)			Sodium (kg/ha)			Potassium (kg/ha)			Magnesium (kg/ha)		
	Ppt.	Runoff	Net	Ppt.	Runoff	Net	Ppt.	Runoff	Net	Ppt.	Runoff	Net
1963–1964	+3.0	−12.8[c]	−9.8	+1.0	−5.9	−4.9	+2.5	−1.8	+0.7	+0.7	−2.5	−1.9
1964–1965	+2.8	−6.3[c]	−3.5	+2.1	−4.5	−2.4	+1.8	−1.1	+0.7	+1.1	−1.8	−0.7
1965–1966	+2.1	−11.5[d]	−9.4	+1.7	−6.9	−5.2	+0.6	−1.4	−0.8	+0.6	−2.9	−2.2
1966–1967	+2.3	−12.3	−10.0	+1.2	−7.3	−6.1	+0.6	−1.7	−1.1	+0.4	−3.1	−2.7
4-year average	+2.6	−10.6	−8.0	+1.5	−6.1	−4.6	+1.4	−1.5	−0.1	+0.7	−2.5	−1.8

[a] Data represent input in precipitation, output in stream water, and net loss or gain of chemicals.

[b] Modified from N. M. Johnson et al. (1968).

[c] These values have been increased by a factor of 1.6 to compensate for chemical interferences during analysis.

[d] The values for June–December, 1965, have been increased by a factor of 1.6 to compensate for chemical interferences during analysis.

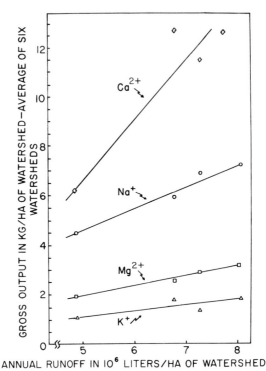

FIG. 6. Relationship between gross output of calcium, magnesium, potassium, and sodium and annual runoff of water in six tributary streams to Hubbard Brook.

3. PARTICULATE MATTER

The output losses discussed above are based on dissolved substances swept out of the system in stream water and do not indicate losses of organic and inorganic particulate matter swept out of the ecosystem by stream water erosion and transportation (Fig. 7). To measure particulate matter losses, we (1) periodically collect the debris settled in the ponding basin behind the weir, (2) weekly sample the suspended debris by collecting larger particles washed over the V-notch with a 1-mm mesh net, and (3) collect finer debris by passing a water sample through a 0.45-μ filter (millipore). The sampling design for the measurement of total losses in stream water is shown in Fig. 8.

Output in stream water in terms of organic and inorganic particulate matter has been measured over a 2-year period (Bormann *et al.*, 1969). These data along with data on dissolved substances (Likens *et al.*, 1967; N. M. Johnson *et al.*, 1968; Fisher *et al.*, 1968) constitute a total view of output relationships of our undisturbed forest ecosystems.

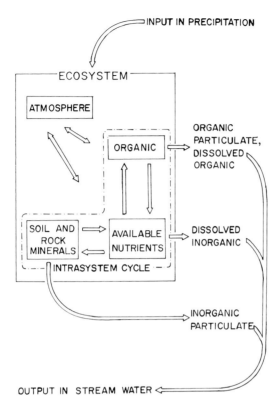

FIG. 7. Nutrient relationships for a terrestrial watershed ecosystem. Sites of accumulation, major pathways, and origins of chemical losses in stream water are shown.

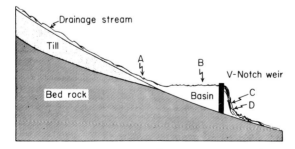

FIG. 8. Sampling design for measurement of output components in stream water of a terrestrial watershed ecosystem. A, water sample for dissolved substances; B, sediment load dropped in basin; C, water sample for millipore filtration; D, net sample. Total losses = dissolved substances (A) + particulate matter (B + C + D).

TABLE 4

AVERAGE ANNUAL GROSS LOSSES OF INDIVIDUAL ELEMENTS AS DISSOLVED
SUBSTANCE AND AS PARTICULATE MATTER[a]

Element	Kilograms per hectare				Percent of total		
	Organic part. matter	Inorganic part. matter	Dissolved substance	Grand total	Organic part. matter	Inorganic part. matter	Dissolved substance
Aluminum	d	1.00[b]	2.60	3.60	d	27.7	72.3
Calcium	0.07	0.18	9.85	10.10	0.7	1.8	97.5
Iron	0.01	0.49	c	0.50	2.0	98.0	c
Magnesium	0.06	0.11	2.80	2.97	2.0	3.7	94.3
Nitrogen	0.12	c	1.90	2.02	5.9	c	94.1
Potassium	0.01	0.32	1.50	1.83	0.5	17.5	82.0
Sodium	0.00	0.19	6.55	6.74	0.0	2.8	97.2
Silicon	d	3.69[b]	15.92	19.61	d	18.8	81.2
Sulfur	0.01	0.01[b]	16.42	16.44	0.1	0.1	99.8

[a] Based on 2 years of data.
[b] Rock and till percent used in calculation of 0.105 mm fraction.
[c] Not measured but very small.
[d] Not measured.

Data (Bormann *et al.*, 1969) on solution losses and particulate matter losses from watershed 6 for 2 years are given in Table 4. Particulate matter losses constitute a relatively small part of the total for most elements.

Although one would expect in an ecosystem of this type that solution losses would exceed particulate matter losses, the magnitude of the differences was unexpected—particularly in view of the fact that this ecosystem has a strong grade with an average of 26% (many slopes of 40%, and a few are great as 70%).

The relatively minor particulate matter losses from such a precariously placed ecosystem attest to the great stability of the system. There is no question that this stability is in large part a function of the biological fraction.of the system. The biota and organic debris, as it influences the absorption of energy of falling rain and downslope movement of water, plays a major role in determining chemical fractions lost either in solution or as particulate matter. The location of this ecosystem indicates that environmental manipulations causing markedly reduced vegetation coverage or severe disturbance of soil surface conditions would favor increased movement of surface water and decreased movement of subsurface water and, consequently, would tend to shift the balance of nutrient losses toward particulate matter losses.

4. WEATHERING

Using the small watershed technique, weathering, or the rate at which an element bound in primary or secondary minerals is made available, can be estimated from net losses of that element as calculated by the nutrient budget method (Bormann and Likens, 1967; Johnson *et al.*, 1968). Restricting consideration to elements without a gaseous phase at ecosystem temperatures, atoms of an element in an ecosystem may be located in (1) soil and rock minerals, (2) the biota and organic debris, and (3) the pool of available nutrients (Fig. 7). Exchanges between these three compartments constitute the intrasystem cycle of that element. Cycling is most intense between compartments (1) and (2) as large quantities of ions are taken up annually by the vegetation and released by direct leaching or by a stepwise decomposition in the food chain. Ions are continually released to the intrasystem cycle by weathering of soil and rock minerals within the system. Although some ions may be reconstituted as secondary minerals within the soil and rock compartment, generally, there is a net loss of ions from the compartment. New substrate for weathering is introduced into the ecosystem as it is lowered in place by erosion, as roots extend the boundary of the system by growth into previously unoccupied portions of the regolith, or as the result of burrowing animals or other mechanisms that introduce new primary or secondary minerals into the system.

If the ecosystem is in a state of dynamic equilibrium as the presence of a climax plant community would suggest, ionic levels in the intrasystem cycle must remain about the same over the course of years. Thus, for a climax system, with its dynamic stability, net ion losses (output minus input) must be balanced by equivalent additions derived from weathering of soil and rock minerals. For successional systems, where nutrients are still accumulating within the system, net losses provide an underestimate of weathering rates. True weathering rates for successional systems would be calculated by adding to annual net output losses the net quantities of an element accumulated annually in the organic and available nutrient compartments.

Based on species composition and vegetation structure, the northern hardwood ecosystem at Hubbard Brook may be considered as an immature climax. Its modest basal area, ca 100 ft²/acre, indicates that the system is still acquiring biomass; hence weathering rates calculated by the nutrient budget method are conservative estimates.

N. M. Johnson *et al.* (1968) have estimated the amount of bouldery till that weathers each year in the Hubbard Brook Ecosystem. Based on net losses of cations and a knowledge of the bulk chemical composition of the bouldery till, it is estimated that 800 ± 70 kg/ha of bouldery till undergo decomposition each year. Since deglaciation about 14,000 years ago, 11×10^6 kg/ha of rocky till should have been completely weathered.

5. CALCIUM CYCLE

As previously mentioned, we now have for the undisturbed northern hardwood ecosystems of Hubbard Brook estimates of chemical input in precipitation, output in stream water, and the rates at which ions are generated by the weathering of minerals within the system. To complete the picture of nutrient cycling according to the model shown in Fig. 7, it is necessary to obtain the nutrient content of the vegetation, litter, and the available nutrient in the soil, as well as estimates of annual rates of nutrient uptake by the vegetation and nutrient release from the biota. The measurement of these parameters is underway through the cooperation of R. H. Whittaker, G. Voigt, and R. Pierce. To gain some idea of these parameters in our system, we took Ovington's data (1962) for the calcium content of trees, litter, and exchangeable calcium in the soil of a beech forest in West England and incorporated them in our model (Fig. 9). Some 203 kg/ha are localized in the trees and litter, while 365 kg/ha represent exchangeable calcium in the soil. This gives a total of 568 kg of Ca/ha in organic matter and as available nutrient. Using unadjusted calcium data for 1963–1964, annual input is 3 kg/ha, output is 8 kg/ha, and 5 kg/ha are generated by weathering. (These values have subsequently been raised by 1.6 to account for chemical interference in measurement of Ca^{2+} concentration.) Data (Table 3) collected subsequent to the construction of Fig. 9 indicate that average annual input of calcium is 2.6 kg/ha, output is 10.6 kg/ha, and 8 kg/ha are generated by weathering. These data will be incorporated in a new diagram as soon as we have our own data on calcium in trees and litter and as exchangeable calcium in the soil. In the meantime, data presented in Fig. 9 will serve the purposes of this discussion.

Data in Fig. 9 suggest a remarkable ability of these undisturbed forests to hold and circulate nutrients, for the net annual loss of 5 kg/ha represents only about 1% of the calcium circulating within the system. However, if percentage loss calculations were based on the actual amounts of calcium circulated annually (annual uptake and release) rather than total calcium, percentage losses would be much higher and perhaps less remarkable.

6. INPUT–OUTPUT RELATIONSHIPS

The individual Hubbard Brook Ecosystem is part of a matrix of ecosystems with which the individual system has communication through its set of inputs and outputs. These relationships are shown in Fig. 7 for

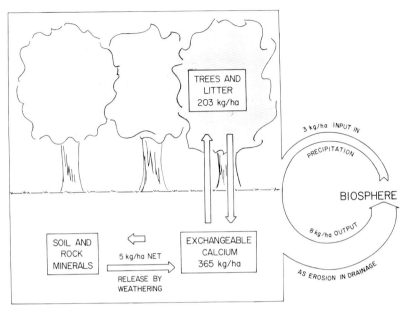

FIG. 9. Estimated parameters for the calcium cycle in an undisturbed northern hardwood ecosystem in central New Hampshire. Data on trees, litter, and exchangeable calcium from Ovington (1962) (from Bormann and Likens, 1967).

nutrient cycles without a gaseous phase at ecosystem temperatures in terms of quantities measurable with the small watershed approach to ecosystem dynamics. Sources of output within the individual compartments of the ecosystem are also shown.

At Hubbard Brook we have rate measurements of all of these quantities: nutrient input in precipitation, nutrient output as dissolved inorganics, organic particulate matter, and inorganic particulate matter, as well as a few data on nutrient output as dissolved organic matter in stream water. As mentioned in the previous section, we also have data on the net release of elements by weathering of minerals within the system.

These data provide not only the key to understanding relationships between ecosystems but also the role of individual ecosystems in important earth processes. Some of these uses are listed below.

(a) Input and output data reveal the role of these small ecosystems in the larger biogeochemical cycles of the earth.

(b) The major geological processes acting in the fluvial denudation of a landscape are weathering, erosion and transportation. Our data provide quantification of these processes.

(c) Transportation of eroded materials out of an ecosystem may follow either of two pathways of output: as dissolved substances in surface or subsurface waters or as particulate matter in surface waters. The importance of both of these pathways can be determined. In our undisturbed forest systems, the bulk of the nutrients are lost as dissolved inorganic substances in subsurface waters.

(d) Manipulation of the ecosystem can result in basic changes of the hydrological pattern. The proportions of water leaving via surface or subsurface drainage may be shifted. This, in turn, will affect the nature of nutrient output—shifts toward increases in surface runoff causing increases in particulate losses. Our data provide estimates of these conditions in both undisturbed and manipulated ecosystems (to be discussed later) and thus permit an evaluation, in quantitative terms, of the effects of manipulation.

(e) The kind and amount of output in stream water from these small terrestrial ecosystems has a strong influence on the structure and function of aquatic ecosystems downstream. Data on levels of dissolved nutrients and of dissolved and particulate organic matter have importance in any consideration of primary and secondary productivity in the downstream ecosystem. Although output of inorganic particulate matter is presently inconsequential at Hubbard Brook, manipulations of the terrestrial ecosystem could lead to greatly increased inorganic particulate output. In turn, this could result in configurational shifts in the downstream stream bed with consequent effects on linked aquatic and streamside terrestrial ecosystems.

(f) Data on the quantitative and temporal aspects of water draining from these small ecosystems plus information on the organic and inorganic characteristics of the water provide direct information on quantity and quality of water exiting from the ecosystem. This information is of fundamental importance in land-use planning.

(g) Finally, implicit in the foregoing is the idea that our study on undisturbed northern hardwood forest ecosystems is providing essential baseline data on the behavior of a natural ecosystem. It is information of this kind that will help us to evaluate the efficiency of human manipulations of natural ecosystems.

B. A Manipulated Forest Ecosystem

The small watershed approach provides a means by which we can conduct experiments at the ecosystem level. Using a watershed ecosystem calibrated in terms of hydrological-nutrient cycling parameters it is possible to impose treatments and to determine treatment effects either by comparison with control watersheds or with predicted behavior had the watershed not been treated. This is, of course, the long-standing method employed by the U.S. Forest Service in its study of the hydrological parameters of forests. This approach makes it possible to evaluate managerial practices in terms of the whole ecosystem rather than isolated parts and to test the effects of various land-management practices (cutting, controlled burning, grazing, etc.) or to determine the effect of potential environmental pollutants (pesticides, herbicides, etc.) on the behavior of nutrients, water, and energy in the system.

1. CLEAR-CUTTING EXPERIMENT

In 1965, the forest of one watershed was clear-cut in an experiment designed (1) to determine the effect of clear-cutting on streamflow, (2) to examine some of the fundamental chemical relationships of the forest ecosystem, and (3) to evaluate the effects of forest manipulation on nutrient relationships and eutrophication of stream water (Bormann *et al.*, 1968).

The experiment was begun in the winter of 1965–1966 when the forest biomass of one watershed, 15.6 ha, was completely leveled by the U.S. Forest Service. All trees, saplings, and shrubs were cut, dropped in place, and limbed so that no slash was more than 1.5 m above ground. No products were removed from the forest and great care was exercised to prevent disturbance of the soil surface thereby minimizing soil erosion. The following summer, June 23, 1966, regrowth of the vegetation was inhibited by an aerial application of 28 kg/ha of the herbicide, Bromacil. Bromacil is a substituted uracil, $C_9H_{13}BrN_2O_2$. Approximately 80% of the applied mixture was Bromacil, while 20% was largely inert carrier. This use of an herbicide in an ecosystem is analogous to the use of a genetic block to elucidate a biochemical pathway.

Stream water samples were collected weekly and analyzed, as they had been for a 2-year period preceding the treatment. Similar measurements on adjacent undisturbed watersheds provided comparative information.

2. EFFECTS ON THE NUTRIENT AND HYDROLOGICAL CYCLES

The treatment had a pronounced effect on runoff. Beginning in May of 1966, runoff from the cutover watershed began to increase over values expected had the watershed not been cut. The cumulative runoff value for 1966 exceeded the expected value by 40%. The greatest difference occurred during the months of June–September, when actual runoff values were 418% greater than expected values. The difference is directly attributable to the removal of the transpiring surface and probably reflects

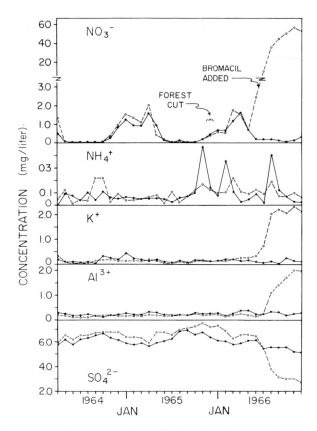

F<small>IG</small>. 10. Average monthly concentrations of selected cations and anions in stream water draining from an undisturbed forest ecosystem (solid line) and a forest ecosystem cutover (dashed line) in the winter of 1965–1966 (Bormann *et al.*, 1968).

wetter conditions within the soil profile, at least a few centimeters below the soil-air interface.

The striking loss of nitrate nitrogen in stream water (Fig. 10) suggests that alteration of normal patterns of nitrogen flow played a major role in nutrient loss from the cutover ecosystem. A comparison of nitrate concentrations in stream water draining watershed 6 (undisturbed) and watershed 2 (cutover) indicates a similar pattern of losses prior to cutting, and through May of 1966 (Fig. 10). Beginning on June 7, 1966, 16 days before the herbicide application, nitrate concentrations in watershed 2 show a precipitous rise, while the undisturbed ecosystem shows the normal late spring decline. The increase in nitrate concentrations is a clear indication of the occurrence of nitrification in the cutover ecosystem. Since an NH_4^+

substrate is required, the occurrence of nitrification also indicates that soil C/N ratios were favorable to the production of NH_4^+ in excess of heterotrophic needs sometime prior to June 7.

The net loss of nitrate from the cutover system must be credited to the destruction of a large portion of the vegetation by the clear-cutting operation the previous winter. Allison (1955) has documented similar losses of nitrate from uncropped fields or fields with poorly established crops.

It seems likely that some of these conclusions hold for the undisturbed ecosystem, i.e., that sometime prior to June 7 C/N ratios were favorable for the flow of ammonium either to higher plants or to the nitrification process. The low levels of NH_4^+ and NO_3^- in the drainage water of the undisturbed ecosystem (W-6) may attest to the efficiency of the oxidation of NH_4^+ to NO_3^- and the efficiency of the vegetation in utilizing NO_3^-. However, Nye and Greenland (1960) state that growing, acidifying vegetation represses nitrification; thus the vegetation may be drawing directly on the NH_4^+ pool and little nitrate may be produced within the undisturbed ecosystem. In this case, it must be assumed that cutting drastically altered conditions controlling the nitrification process.

The action of the herbicide in the cutover watershed seems to be one of reinforcing the already well-established trend of NO_3^- loss induced by cutting alone. This is probably effected through the destruction of the remaining vegetation, herbaceous plants, and root sprouts by the herbicide. Even in the event of rapid transformations of all the nitrogen in the Bromacil this source could at best contribute 5% of the nitrogen lost as nitrate.

During 1966, the cutover system showed a net loss of 52.8 kg/ha of N as compared to a net gain of 4.5 kg/ha for the undisturbed system (Table 5). Assuming the cutover system would have normally gained 4.5 kg/ha, the adjusted net loss from the cutover system would be about 57 kg/ha. The annual nitrogen turnover in undisturbed systems is approximately 60 kg/ha, based on an equilibrium system in which annual leaf fall is about 3200 kg/ha (Hart *et al.*, 1962) and annual loss of roots is about 800 kg/ha. Consequently, an amount of elemental nitrogen equivalent to the annual turnover was lost in the first year following cutting.*

Nye and Greenland (1960) state that a high level of nitrate ion in the soil solution implies a corresponding concentration of nutrient cations and ready leaching. This is precisely what is seen in the cutover ecosystem. Increases in Ca^{2+}, Mg^{2+}, Na^+, and K^+ concentrations in the stream water occur almost simultaneously with the increase in nitrate. This is followed about 1 month later by a sharp rise in the concentrations of Al^{3+}. Sulfate, on the other hand, shows a sharp drop in concentration coincident with nitrate rise (Fig. 10).

* See Likens *et al.*, 1969, for more recent data.

TABLE 5

PARTIAL NITROGEN BUDGETS FOR WATERSHEDS 2 AND 6[a,b,c]

Year	Input in precipitation			Output in stream water			Net gains (+) or losses (−)		
	(NH$_4$-N)	(NO$_3$-N)	Total N	(NH$_4$-N)	(NO$_3$-N)	Total N	(NH$_4$-N)	(NO$_3$-N)	Total N
				Watershed 6 (13.2 ha, undisturbed)					
1965	2.1	2.8	4.9	0.6	1.0	1.6	+1.5	+1.8	+3.3
1966	2.0	4.5	6.5	0.5	1.5	2.0	+1.5	+3.0	+4.5
				Watershed 2 (15.6 ha, cut in winter 1965–1966)					
1965	2.1	2.8	4.9	0.4	1.3	1.7	+1.7	+1.5	+3.2
1966	2.0	4.5	6.5	1.2	58.1	59.3	+0.8	−53.6	−52.8

[a] All data are in kilograms of elemental nitrogen per hectare of watershed.
[b] Gains by biological fixation or losses by volatilization are not included.
[c] Modified from Bormann et al. (1968).

Net losses of the cations Ca^{2+}, Mg^{2+}, Na^+, and K^+ are approximately 9, 8, 3, and 20 times greater, respectively, than those for five undisturbed ecosystems from June, 1966, to June, 1967.

These results indicate that this ecosystem has limited capacity to retain nutrients when the bulk of the vegetation is removed. The accelerated rate of nutrient loss is related to the cessation of nutrient uptake by plants and to the larger quantities of drainage water passing through the ecosystem. Accelerated losses may also be related to an increased mineralization rate resulting from changes in the physical environment, e.g., temperature, or an increase in substrate available for mineralization.

However, the effect of the vegetation on the process of nitrification cannot be overlooked. In the cutover ecosystem, the increased loss of cations is correlated with the increased loss of nitrate anions. Consequently, if the intact vegetation inhibits the process of nitrification (Rice, 1964) and removal of the vegetation promotes nitrification, release from inhibition may account for major nutrient losses from the cutover ecosystem.

A study in the summer of 1967 (Smith et al., 1968) has indicated that populations of nitrifying bacteria (*Nitrosomonas* and *Nitrobacter* groups) have increased by 18-fold and 34-fold on the cutover watershed. This result, coupled with the timing of the initial acceleration of NO_3^- loss, indicates that release of nitrifying organisms from inhibition by the vegetation must be given serious consideration as a major factor in nutrient loss following clear-cutting.

Results of the particulate matter study indicate a basic change in the pattern of losses. It seems likely that losses of particulate organic matter will soon undergo a sharp decline as a result of the virtual elimination of primary production of organic matter within the ecosystem and the progressive removal of a large quantity of organic matter stored in the stream bed. Loss of organic material from the stream bed has been accelerated because (1) of an increase in amount of flow and in flood peaks resulting from the elimination of transpiration, and (2) of accelerated heterotrophic activity in the stream bed due to an increase in water temperatures during the growing season. Increased heterotrophic activity results in weakening of the organic dams and in their accelerated removal from the system.

Conversely, the relative output of inorganic particulate matter is increasing in watershed 2. This results because of the greater erosive capacity of the now augmented streamflow and because several biological barriers to stream bank erosion have been removed. The extensive network of fine roots that tended to stabilize the bank is now dead as a result of cutting and herbicide treatment, and the dead leaves that tended to plaster overexposed banks are now gone. It seems probable that within a year or two, continued denudation of watershed two will see an exponential increase in output of inorganic particulate matter.

3. IMPLICATIONS FOR FOREST MANAGEMENT

These results suggest several conclusions important for environmental management.

(a) Clear-cutting tends to reduce the nutrient capital of a forest ecosystem (1) by reducing transpiration and hence increasing the amount of water passing through the system, (2) by simultaneously reducing root surfaces able to remove nutrients from the leaching waters, (3) by the removal of nutrients in forest products, (4) by adding to the organic substrate available for immediate mineralization, and (5) in some instances by producing a microclimate more favorable to rapid mineralization. These effects may be important in other types of forest harvesting depending on the proportion of the forest cut and removed and the nature of the local ecosystem. Nutrient loss may be greatly accelerated in cutover forests where the soil microbiology leads to an increase of dissolved nitrate in leaching waters.

(b) Management of forest ecosystems can make significant contributions to the eutrophication of stream water. Nitrate concentrations in the small stream draining from the cutover ecosystem have exceeded established pollution levels (10 ppm) (Rainwater and Thatcher, 1960) for over a year and algal blooms have appeared during the summer months.

(c) These conclusions are very similar to those reported by Nye and Greenland (1960) for tropical ecosystems and probably have a wide application to forest ecosystems in general.

III. GENERAL CONCLUSIONS

1. The ecosystem concept provides a theoretical framework for the study and management of our natural resources.

2. Knowledge of input-output parameters is necessary to understand nutrient and energy relationships of the individual ecosystem; however, quantitative measurement of these parameters is often difficult or impossible.

3. The small watershed approach, utilizing measured parameters of hydrological and chemical input, output, and net change, is a powerful tool for the study of biogeochemical relationships of individual ecosystems. It has been applied to a northern hardwood forest ecosystem, and has yielded basic information on nutrient budgets, erosion, and weather-

ing. Concomitant studies of biomass, nutrient uptake and release, exchangeable nutrients, and other parameters are also in progress. With these data, the interrelationships among the biota, the hydrological cycle, nutrient cycles, and energy flow are being quantified.

4. Finally, the small watershed ecosystem is being used to evaluate the complex of ecological effects of a forest management procedure (clear-cutting) on the northern hardwood forest.

ACKNOWLEDGMENTS

Financial support for this work was provided by NSF Grants GS 1144, GB 4169, GB 6757, and GB 6742. This is contribution No. 8 of the Hubbard Brook Ecosystem Study, published as a contribution to the U.S. Program of the International Hydrological Decade, and the International Biological Program. Most of the results reported herein were obtained by team effort. We acknowledge, with full credit, the contributions of Noye M. Johnson, Robert S. Pierce, and Donald W. Fisher to these studies.

REFERENCES

Allison, F. E. 1955. The enigma of soil nitrogen balance sheets. *Advan. Agron.* **7,** 213–250.

Bates, C. G., and A. J. Henry. 1928. Forest and streamflow at Wagon Wheel Gap, Colorado. *Monthly Weather Rev.* Suppl. 30, 1–79.

Bormann, F. H., and G. E. Likens. 1967. Nutrient cycling. *Science* **155,** 424–429.

Bormann, F. H., G. E. Likens, D. W. Fisher, and R. S. Pierce. 1968. Nutrient loss accelerated by clear-cutting of a forest ecosystem. *Science* **159,** 882–884.

Bormann, F. H., G. E. Likens, and J. S. Eaton. 1969. Biotic regulation of particulate and solution losses from a forest ecosystem. *BioSci.* **19**(7), 600–610.

Costello, D. F. 1957. Application of ecology to range management. *Ecology* **38,** 49–53.

Dils, R. E. 1957. A guide to the Coweeta Hydrologic Laboratory. *Southeastern Forest Expt. Sta., Misc. Publ.* 40 pp.

Dreibelbis, F. R., and F. A. Post. 1941. An inventory of soil water relationships on woodland, pasture and cultivated soils. *Soil Sci. Soc. Am. Proc.* **6,** 462–473.

Evans, F. C. 1956. Ecosystem as the basic unit in ecology. *Science* **123,** 1127–1128.

Fisher, D. W., A. W. Gambell, G. E. Likens, and F. H. Bormann. 1968. Atmospheric contributions to water quality of streams in the Hubbard Brook Experimental Forest, New Hampshire. *Water Resources Res.* **4,** 1115–1126.

Hart, G. E., Jr. 1966. Streamflow characteristics of small forested watersheds in the White Mountains of New Hampshire. Ph.D. thesis, University of Michigan, Ann Arbor, Michigan. 141 pp.

Hart, G. E., Jr., R. E. Leonard, and R. S. Pierce. 1962. Leaf fall, humus depth, and soil frost in a northern hardwood forest. *Northeastern Forest Expt. Sta., Forest Res. Note* No. 131.

Johnson, C. M., and P. R. Needham. 1966. Ionic composition of Sagehen Creek, California, following an adjacent fire. *Ecology* **47,** 636–639.

Johnson, N. M., G. E. Likens, F. H. Bormann, and R. S. Pierce. 1968. Rate of chemical weathering of silicate minerals in New Hampshire. *Geochim. Cosmochim. Acta* **32,** 531–545.

Juang, F. H. T., and N. M. Johnson. 1967. Cycling of chloride through a forested watershed in New England. *J. Geophys. Res.* **72**(22), 5641–5647.

Junge, C. E., and R. T. Werby. 1958. The concentrations of chloride, sodium, potassium, calcium and sulphate in rainwater over the United States. *J. Meteorol.* **15**, 417–425.

Likens, G. E., F. H. Bormann, N. M. Johnson, and R. S. Pierce. 1967. The calcium, magnesium, potassium, and sodium budgets for a small forested ecosystem. *Ecology* **48**, 772–785.

Likens, G. E., F. H. Bormann, and N. M. Johnson, 1969. Nitrification: Importance to nutrient losses from a cutover forested ecosystem. *Science* **163**, 1205–1206.

Lutz, H. J. 1957. Applications of ecology in forest management. *Ecology* **38**, 46–49.

Lutz, H. J. 1963. Forest ecosystems: Their maintenance, amelioration, and deterioration. *J. Forestry* **61**, 563–569.

Nye, R. H., and D. J. Greenland. 1960. The soil under shifting cultivation. *Commonwealth Bur. Soil Sci. (At. Brit.), Tech. Commun.* **51**.

Odum, E. P. 1963. "Ecology." Holt, New York.

Ovington, J. D. 1962. Quantitative ecology and the woodland ecosystem concept. *Advan. Ecol. Res.* **1**, 103–192.

Rainwater, F. H., and L. L. Thatcher. 1960. Methods for collection and analysis of water samples. *Geol. Surv. Water-Supply Paper* 1454.

Rice, E. L. 1964. Inhibition of nitrogen-fixing and nitrifying bacteria by seed plants. *Ecology* **45**, 824–837.

Smith, W. H., F. H. Bormann, and G. E. Likens. 1968. Response of chemoautotrophic nitrifiers to forest cutting. *Soil Sci.* **106**, 471–473.

Whitehead, H. C., and J. H. Feth. 1964. Chemical composition of rain, dry fallout, and bulk precipitation at Menlo Park, California, 1957–1959. *J. Geophys. Res.* **69**, 3319–3333.

Chapter V A Study of an Ecosystem: The Arctic Tundra

ARNOLD M. SCHULTZ

I. INTRODUCTION

A few years ago some ecologists predicted that the ecosystem bubble would soon burst, after which investigators would "go back to tried and true methods" for probing nature. Others have said there is nothing new in the ecosystem concept at all except a fancy name and another language for students to learn. Like any other science, ecology has had its share of fads, some of which turned out to be nothing but old ideas dressed in new semantics.

The concept of system is indeed very old but not so the area we have come to call loosely systems analysis. The way in which modern systems analysis has been applied to the ecosystem—just since 1962—is truly fantastic. It comes closer to a breakthrough than any other event in the history of ecology.

We now have a conceptual tool which allows us to look at big chunks of nature as integrated systems (Schultz, 1967). Also we now have the technical tools to handle the information obtained in this framework. Ecologists no longer fear complexity. These innovations have come along not any too soon. The alarm of "Silent Spring" is still ringing in our ears. Finally we realize that nature is not as piecemeal as science is.

Where does one begin, to study an ecosystem? The first part of this chapter explains some concepts fundamental to ecosystem study. In a discussion of models, it tells how the author decided to look at the tundra. The second part gives our first-hand experience in studying a tundra ecosystem in northern Alaska. In a discussion of results, it tells what we decided to look for.

II. THE REAL SYSTEM AND THE MODEL

A. The Language of Systems

The concept of system dates back to the very dawn of thought although its language is quite modern. From the beginning man has perceived only wholes; his penchant for taking them apart is rather a recent development but his ability to put the parts together again has scarcely developed at all. Our language of systems derives from these three ways of looking at things: taking things apart (analysis), assembling parts into wholes (synthesis), and seeing things only as wholes.

Therefore a system is a whole thing; it has three kinds of components (Fig. 1).

The elements of a system are the physical objects, often thought to be the "real" parts. In an ecosystem the elements are space–time units in that they occupy some volume in space for a certain length of time. Rain-

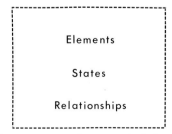

Elements

States

Relationships

FIG. 1. The components of a system including boundary.

drops, sand grains, and mosquito larvae are examples of elements in a system. Each element has a set of properties or states, e.g., number, size, temperature, color, age, or value. Between two or more elements or between two or more states there are relationships which can be expressed as mathematical functions or less formally, with plain English verbs.

A system can now be defined as a set of elements together with relations among the elements and among their states (Hall and Fagen, 1956).

The term "set" implies that the components can be bounded. In the diagram the boundary is permeable to indicate an open system into which elements can enter from outside.

We might think that the elements, being physical entities of some kind such as nitrogen ions or living organisms, are the real and important components while the others are mere abstractions. In system thinking, however, we put more emphasis on the state. In a thermostatic control system, for example, it is not the air in the room but its temperature that is important; nor is it the hardware of the furnace but its state of being off or on, low or high that we consider. The elements of most interest to us in a system are capable of taking on two or more alternative states. In other words, the element is a variable and over time its difference in state is what we observe, measure, and record. We cannot record the element itself.

There would be no need to invoke the systems concept were it not for the crucial component relationship. A system becomes a whole thing only because its elements and states are connected together in some way. Thus, by understanding the linkages we see how the whole system works.

We like to use this very simple model of a system (Fig. 1) as a reminder of what is real, what is abstraction, and why we have the systems concept in the first place.

How different is this approach from the one used in science before systems analysis? Scientists have always studied systems but they have shied away from complex ones. Early physicists learned that the mathematics needed to describe the attractions of more than two bodies at a time were beyond their powers of calculation or too time-consuming to carry out. Ecology is the science of relationships; yet ecologists, though awed by the complexity of nature, have long used methods which treat one factor at a time. How can one unravel the many interrelationships in a species-rich temperate forest, for example, or for that matter, in the arctic tundra?

B. Homomorphic Models

It is improbable that any ecosystem will ever be studied in its full complexity. In some systems there may be thousands of kinds of organisms and billions of interacting individuals. If we had spectroscopic X-ray

. . . .   Atoms, Molecules
Cells, etc.

(a) (b) (c) (d) (e) (f) (g) (h) (i) (j) (k) (l) (m) (n) (o) (p)   Individuals

(a + b) (c + d) (e + f) (g + h) (i + j) (k + l) (m + n) (o + p)   Species, Genera, etc.

(a + b + c + d) (e + f + g + h) (i + j + k + l) (m + n + o + p)   Trophic Levels

(a + b + c + d + e + f + g + h) (i + j + k + l + m + n + o + p)   Living & Non-living
Biomass

(a + b + c + d + e + f + g + h + i + j + k + l + m + n + o + p)   Entire System

FIG. 2. Homomorphic models showing six levels of discrimination of system parts.

eyes we could see yet finer subdivisions: cells, molecules, atoms, perhaps even electrons. Obviously, to study the many states and relations in such a complex system would be too much for our best computers. The system must be simplified.

In the past ecologists would select a certain few parts of a larger system for study, for example, a species population or a relationship such as plant succession or competition. This kind of simplification falls short of studying the whole system; the other parts and the other relations that occur are completely neglected.

A complex system can be simplified by making a homomorphic model of it. Here the system remains intact. Its parts are discriminated at some level that can be handled conveniently. Figure 2 shows the ABC's of homomorphic models.

In the row second from the top are sixteen distinct individuals. Not everyone can see them, however, so my statement is only a point of view. Suppose the letters *a* to *p* represent grass plants in a dense sward. You would not be able to distinguish individuals at all. The next row down shows the units combined by twos into superunits. We can think of these as populations of taxonomic groups (species, genera, families) or of physiognomic ones (herbs, shrubs, trees). All of the finer units are still present but undiscriminated at the next higher level and so on through the hierarchy—trophic levels, then living and nonliving biomass, until the entire system emerges as a Gestalt and there is no discrimination of parts whatsoever.

Homomorphic models are not designed just for lazy taxonomists. There is a very practical reason for looking at systems this way. Let us look at the top and bottom lines of Fig. 2. At the top, every possible state has been distinguished but the sheer bulk of information is so overwhelming that we can make little use of it. At the other end, all the states are fused into one grand but platitudinous expression and all you can say for it is "There it is!" In between are a number of handy simplifications. On the fine end some realism is retained. On the coarse end we gain generality. We now have a set of models which allows us to coordinate all the discoveries made by specialists on the separate parts of the system. Any part can be handled as a black box coupled into the system at the appropriate level of organization.

C. The Tundra as a Simple Ecosystem

The history of scientific investigation of arctic tundra follows a pattern different from that of other regions of the world. Because of the sparse indigenous human population, pressure for agricultural research, as we know it in lower latitudes, has not occurred. Basic studies from diverse disciplines and with broad objectives were initiated; they did not assume the single-minded purpose of maximizing crop yields.

Many ecologists were lured to the tundra because it was supposed to be simple. Here in the arctic could be found a paucity of growth forms and species, shallow soil, a short growing season, an extreme climate, and essentially no disturbance of the landscape by humans. If ever the total processes of nature could be put in order, it would be done for the tundra. Let us see how simple the tundra really is.

I can cite numerous detailed descriptions of coastal tundra in the vicinity of Point Barrow, Alaska. These include my own investigations and those of my students which began in 1958 (Schultz, 1964; Pieper, 1964; Van Cleve, 1967) and the excellent work on microtines and their predators by F. A. Pitelka and others begun in 1952 (Pitelka *et al.*, 1955; Pitelka, 1958; Maher, 1960). Intensive studies on soils, meteorological phenomena, bacteriology, and other aspects of the tundra have been going on at Point Barrow since 1950.

It is not my intent here to redescribe the tundra. Rather I shall point out the proximity between the real ecosystem and its simplified model.

It would be possible to print on two pages of this book a list of all the species of plants, animals, and microorganisms known to occur in the massive ecosystem under study. Perhaps it would go on one page. The list could be reduced to about ten species, and still include 90% or more or the biomass in each major group. It would include sphagnum moss among the lower plants, several grasses and sedges among the higher plants, the brown lemming and pomarine jaeger among herbivores and carnivores, respectively. For quantitative studies of energy and major nutrients, an analysis of samples from these few predominant groups would give essentially the same results as would a total ecosystem study. To put it another way, the properties or states of the trophic levels are at least 90% predictable from just one or two of their component parts.

We have already seen how the complexity of a system is determined by the number of distinguishable parts. At the species population level of discrimination, the tundra is fairly simple compared to other ecosystems of the world. But there is another determinant of complexity: the number of recognizable states which the parts can assume. Some examples are depth of active soil layer, exchangeable calcium in the soil, phosphorus level of forage plants, population density of lemmings, jaegers, and owls, and decomposition rate of organic matter. The number of states depends entirely on the yardstick and stopwatch used to measure. The investigator can make it as simple or complex as he wishes.

We must come to the conclusion that the easiest way to analyze the tundra or any other ecosystem is to lump all populations of organisms into trophic levels and the resources (atmosphere and soil) into convenient compartments. For each compartment a sample, in exactly the proportion in which the various populations occur, can be digested for nutrients and bombed for energy. Egler (1964) calls this the "meatgrinder" approach. A wealth of information inside each box is conveniently ignored. Our primary interest is directed to relationships between boxes. How much phosphorus flows from one tropic level to another? In this framework, one system is as simple as another—tundra or temperate zone grassland.

D. Isomorphic Models

We can now return to the third component of systems, relationships. The homomorphic model does not help us here; it is concerned only with the power of resolution used for the elements and their states. We need an isomorphic model for studying relationships.

An isomorphic model is a map. A road map of New Mexico is a model of the real geographical area of the state. It shows among other things the distance and direction between Albuquerque and Santa Fe. By inference from the kind of highway, it also shows the rate of traffic flow between the two cities. So a map can be a flow chart.

Figure 3 is a map of a homomorphic model of an ecosystem—two kinds of models combined. This one has been proposed by the subcommittee on Terrestrial Productivity for the International Biological Program as a tentative model for studying all terrestrial ecosystems. The arrows indicate paths of nutrient or energy flow from one box to another. Research on ecosystems can be standardized and routinized by using

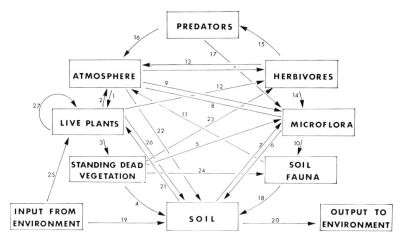

FIG. 3. Isomorphic model mapping the relations between ecosystem compartments.

such a scheme. One can use the numbers on the arrows as a checklist to see which relations of Fig. 3 one has forgotten to measure.

Let us consider what kinds of relationships there are and how they are measured.

Arrow No. 12 represents a process with a familiar name: grazing. The model suggests that something travels physically along the path from *live plants* to *herbivores*. We measure this something in units of chemical compounds, biomass, or energy. It is measured as a rate—so many pounds of dry matter per unit time. Rarely is the transfer measured directly. The *live plants* box is weighed at time one (T_1) and again at time two (T_2). The difference in state between T_1 and T_2 is read as a gain or loss. A series of exclosures and enclosures will indicate what proportion has been gained or lost via pathways No. 1, 25, 2, 3, 12, or 21. The same is done for *herbivores* and every other box in the model. We should note here that enclosures and exclosures are experimental tools not to be preempted by range managers. The bacteriologists' agar plates and the radiobiologists' isotope tracers are based on the same principle as a fenced plot.

Isomorphic models can be constructed for many kinds of relationships. A correlation matrix is such a model; the coefficients show spatial or phenetic distances. Regression coefficients and ratios may be used to express causation, conjunction, or succession of events. Finally, a model can be set up with vernacular expressions like "boy meets girl," or more appropriately, as in André Voisin's definition of grazing, "cow meets grass." You will recognize pathway 15 as owl eats lemming, 5 as bacterium rots straw, and 20 as ocean takes soil. I am not being trite. If we could clearly establish verbally expressible relationships between all of the parts we would indeed understand ecosystems very well and could develop valuable theorems about them. The numbers would then be superfluous.

E. System Boundary and Environment

Earlier in this chapter an ecosystem was described as a space-time unit—a volume that exists in time. But you cannot see an ecosystem in the same way that you can see, for instance, a lemming. The skin clearly marks the boundary of the animal. It is simply a matter of perception and all observers would agree as to what is lemming and what is not lemming.

Not all ecosystems have a skin you can touch. Defining the boundaries of his ecosystem is one of the biggest problems an ecologist has to face. Some insight into the problem can be obtained by considering the nature of boundary for a generalized system (Fig. 4).

The system has within it all the elements the observer is interested in (the set, by definition). These elements have a certain density or concentration. Outside, the elements are different either in kind or in concentration. The observer is not interested in these to the same degree. If they had been of the same kind of elements or had the same density as inside, he would have included them in the system. Thus, the environment (e) (outside) is different from the system (s) itself. It is not entirely true that the observer is disinterested in the environment; he cares about what effect it has on the system. There is pressure for some of the elements to cross over the boundary.

Think of the boundary as a membrane with a certain thickness Δx. It has a texture which determines how easily any of the elements can flow across it. This can be thought of as a permeability factor (m). Since there are different densities of elements inside and outside, there must be a concentration gradient across the boundary. If Δx is narrow, the gradient will

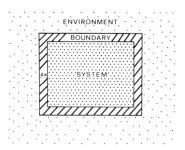

Fig. 4. The boundary between system and environment.

be steep. The flow of elements across the membrane is governed by the following three factors (Jenny, 1961):

$$\text{Flow} = -\left(\frac{\text{concentration}_e - \text{concentration}_s}{\Delta x}\right) m$$

This can be illustrated with an exclosure in a pasture. The fence, of course, represents the boundary. The exclosed plot has no animals, but outside are 10 steers per acre. Some of these are pressing against the fence. The gradient is sharp—from 10 to 0 over a distance the diameter of the wire. If the fence is strong ($m = 0$), the flow will be zero no matter what the steer pressure is.

In most ecosystems there is no actual membrane separating system from environment. The boundary is imaginary and is located at the convenience of the observer. One choice is to set the boundary in a zone where there is no gradient (where concentration$_e$ − concentration$_s = 0$). Now any crucial variations in density are trapped within the system and must be measured as state transformations. The other approach is to use a natural boundary (where $m \rightarrow 0$) such as the shore line of an island. Here the problem is that the permeability coefficient may be low for only one kind of element, and as we have already seen, a complex may have a thousand kinds.

F. The Role of the Observer

One of the tenets of systems research is that the observer is always a part of the system. Refer back to the four diagrams and picture the role of the observer. He is an element of the system, with definite properties and unique relations to the other elements and states (Fig. 1). The observer decides the level of discrimination to be used in the study (Fig. 2). He selects from the large number of possible relations just those he wants to measure (Fig. 3). He fixes the boundary of his system according to his resources and his interests (Fig. 4). Apart from the observer there can be no unique ecosystem. No one can go to Point Barrow and see the same system that I see. It follows that I cannot possibly describe to you the real or absolute ecosystem, only my model of it.

Within the system he has circumscribed, the observer looks for time-invariant relationships. During the investigation he records the activity of the system. This includes haphazard events which may happen only once, activities which occur frequently, and those which occur every time. These represent, respectively, the temporary, the hypothetical, and the permanent or real behavior of the system. In graphical form, the three kinds of activity can be shown as scatter diagram, prediction curve, and equation for an absolute law.

It is within this framework that I present some of the results of my research.

III. THE TUNDRA AS A HOMEOSTATIC SYSTEM

A. Cyclic Phenomena

Many of the activities of the tundra ecosystem under study are cyclic, with a periodicity of three or four years, but with varying amplitudes between cycles. We can think of a cycle as a series of transformations of state (Ashby, 1963). Thus, if a subsystem (compartment) has four clearly recognizable alternative states, a, b, c, and d, and the transformation always goes a → b → c → d → a → b, etc., then the sequence of states is a cycle. This can be shown kinematically:

$$a \rightarrow b$$
$$\uparrow \quad \downarrow$$
$$d \leftarrow c$$

or, when put on a time scale, as a sine wave:

and so on.

At Point Barrow we have good records of yearly lemming population densities, starting from 1946 (Fig. 5). Lemming peaks occurred in 1946, 1949, 1953, 1956, 1960, and 1965. Neither the amplitude nor the wavelength of these cycles is always the same. Yet there are some striking similarities. Using the systems language given above with a 1-year time interval, generally we can recognize four states: a, high density; b, very low density; c, low density; and d, medium density. The states can be named by reading the histogram, without any knowledge of lemming population dynamics or life histories. Sometimes c is missing and once c and d are transposed. Always the high year was immediately followed by a very low year. But even during the low years, there were found local "pockets" supporting a denser population; for example, on the outskirts of the Eskimo village of Barrow, the fluctuations in lemming numbers have never been as pronounced as on the open tundra. Also of significance, some areas are out of phase with Point Barrow. At a point 100 miles east of Barrow, the population peaked in 1957, a year after the Point Barrow high. By 1960, it was in phase again.

What can be said about lemming population cycles? By Ashby's criterion, we have definitely observed cycles; but an engineer would say,

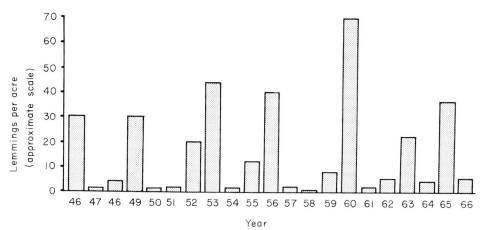

FIG. 5. Lemming population cycles at Point Barrow, Alaska, from 1946 to 1966.

if he saw the waves on an oscilloscope, that there was a lot of noise on the channel.

We have data on standing crop of plants (tops only), starting from 1958. Clippings were made in the vicinity of traplines used for the lemming census. Ninety percent of the dry matter was contributed by three species (*Dupontia fischeri, Eriophorum angustifolium,* and *Carex aquatilis*); these same species constitute the bulk of the lemming diet. When a graph is constructed for standing crop each year at phenologically equivalent dates, a cyclic pattern appears. Moreover, the pattern is in synchrony with that of the lemming population cycle, noise and all. A short lag develops between the two curves as the high point approaches. I do not want to explain the facts at this time, but I should say, in passing, that the high correlation between forage yield and lemming stocking rate would not come as a surprise to a range management audience.

Samples of the material clipped for production records were analyzed for total nitrogen, phosphorus, calcium, and several other elements. Concentrations of nutrients in the herbage, at any given phenological stage (i.e., date) increased through the year corresponding to the peak lemming year, then dropped to low values the year after, only to increase again. Figure 6 shows the activity for phosphorus superimposed on the histogram of lemming population density. Within a season, nitrogen, phosphorus, and potassium decrease percentagewise as grasses mature, while calcium increases. The line in Fig. 6 should not be construed as a continuous increase in phosphorus level from 1957 through early 1960.

Calcium, potassium, and nitrogen show the same trends as phosphorus. Due to greater plot-to-plot variability, the nitrogen data are not as sig-

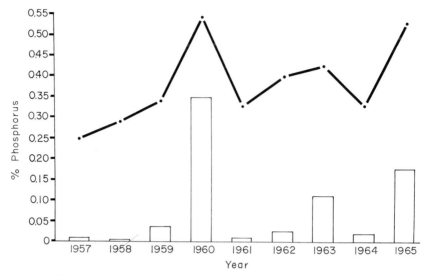

Fɪɢ. 6. Phosphorus levels in forage at Point Barrow, Alaska. Bars represent relative lemming numbers.

nificant as are those of the other three elements, but the trend is nevertheless the same. Magnesium and sodium show no relationship to lemming numbers at all, nor were the data cyclic.

Still another activity studied was decomposition of organic matter on the soil surface. This, too, turned out to be a cyclic phenomenon, and correlation with the activities already mentioned is high.

I have given a rough sketch of the behavior of the tundra, as discovered by survey techniques. The observations seem to fit closely a hypothesis of synchronous cycles. But at this stage, the results could be spurious. The close fit might result from artifacts of sampling. The transformation sequence a → b → c and/or d → a might occur frequently just by chance.

The next step is experimental: to introduce a disturbance at any one of the compartments and watch for reverberations throughout the system.

B. Experiments in Stressing the System

If the fluctuations in herbage production and nutrient level are related to immediate grazing history, then the cyclic aspect should disappear when grazing is eliminated. A simple exclosure, in effect, removes the herbivore from the system.

In 1950, a series of exclosed plots was established, alongside paired plots open to normal grazing (Thompson, 1955). Records kept for 13 years show cyclic variation on the outside paired plots, while the fenced plots show a constant decline. Since 1958, percentages of phosphorus, calcium, potassium, and nitrogen in the herbage from the grazed plots show the same marked cycles that occur elsewhere on the tundra (see Fig. 6). By comparison, year-to-year fluctuations inside the exclosures are slight and not cyclic.

With regard to decomposition of litter, the outside plots responded as did the tundra on the whole; inside the exclosures, decomposition rates were low and constantly decreasing.

An unexpected bonus came from the exclosure experiment. It gave an opportunity to assess the effect of lemming activity on the depth of thaw. By comparing, at the time of maximum thaw, soil depths inside and outside exclosures, I could separate the lemming-caused (within-system) effects from the summer temperature (environmental) effects. The results were most interesting. During a peak lemming year, the thickness of the active soil layer was maximum and it gradually diminished to the shallowest point the year before the next peak.

A second experiment was to stabilize artificially the fluctuating nutrient levels in the soil. This was done by fertilizing annually 6 acres of tundra with nitrogen, phosphorus, potassium, and calcium. Heavy applications were made to make sure that the variations in native soil nutrients were completely masked. What effect would this kind of disturbance have on primary and secondary production?

Net primary production, for the 4 years studied, was stabilized at a level 3–4 times that of the control plot. Annual variation was obliterated. Herbage quality was also stabilized. Protein levels, for example, were 4–5 times those of the vegetation of the control plot. Percent of calcium and phosphorus in the green tissue at equivalent dates remained high and constant in the four years.

The first fertilization was applied in 1961. No animals were seen either on fertilized or on control plots in 1961 or 1962. In 1963, animals were abundant all over the tundra (see Fig. 5 or 6), while in 1964, they were generally sparse. Immediately after the snowmelt in 1964, 30 winter nests per acre were counted on the fertilized area, none on the control plot, and less than 1 per 10 acres on the tundra in general. However, jaegers had found this 6-acre pantry and picked it clean. The few survivors observed at the time of the winter nest survey were large and fat. In 1965, a year of high lemming density all over the tundra, lemmings were abundant both on the fertilized and unfertilized control plot.

C. Hypotheses of Ecosystem Cycling

Only a fraction of the information collected so far can be presented in this paper. For the sake of brevity, I have shorn away all evidence of

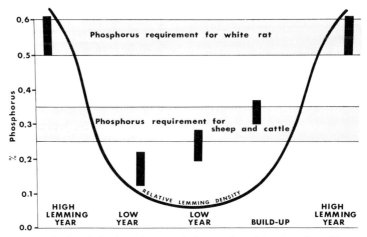

Fig. 7. Hypothesis relating lemming populations and nutritional quality of forage. Lemming density curve is generalized, as are the black bars showing phosphorus in forage.

"noise" and shown, so to speak, only the slick regression lines. These represent the hypothetical behavior of the system compartments.

With the evidence at hand, let us develop a more general hypothesis to explain the synchronous cycles apparent in our ecosystem. Tentatively it might be called the nutritional threshold hypothesis (Fig. 7) but such a name places undue emphasis on just one part of the system.

Let us review a generalized 4-year cycle.

1. Early in summer of the high lemming year, the forage is calcium- and phosphorus-rich. Because of high production and consumption, much of the available calcium and phosphorus is tied up in organic matter. The soil has thawed down deep into the mineral layer because grazing, burrowing, and nest-building has altered the albedo and insulation of the surface.

2. The next year, not only is forage production low, but also the percent of calcium and phosphorus in the diet is below that which would be required for lactation by sheep or cattle. Nutrients in organic matter have not yet been released by way of decomposition.

3. The following year production is up, the plants are recovering from the severe grazing two years earlier, and dead grass from the previous year insulates the soil surface. At the same time, decomposition of that dead material is speeding up. Forage quality is still quite low. Whether there is enough calcium and phosphorus in the diet to support lemming reproduction and lactation depends on how closely the species resembles domestic livestock, on the one hand, or the laboratory rat, on the other, in mineral requirements.

4. In the fourth year plants have fully recovered from grazing. Forage species accumulate minerals in their stem bases. Freezing action concentrates solutes in upper soil layers. Dead grass from several years has accumulated and plant cover is high; soil surface is well insulated and the thawed layer is very shallow. Decomposition rate is high. Calcium and phosphorus (and also potassium and nitrogen) content of forage is satisfactory for reproduction. There is enough food to support a large population of herbivores.

Next, the sequence is repeated.

Not until a nutritional threshold has been reached can a large lemming population build up. But the population does not keep getting bigger and bigger. This would be disastrous to the vegetation. So a deferred-rotation grazing scheme is built into the system. No grazing at all would also be disastrous to the vegetation and to the soil as well. Predators play a role at the time of herbivore decline. Indeed, all parts of the system play a role. It is a homeostatically controlled system.

This is only a hypothesis. It can be tested in the framework of the ecosystem concept: First, by showing that all parts bear some relationship to all others; second, by experimentally stressing the system to see how it adapts to disturbance; third, by opening the black box and studying its physiology—that is, explanation of a phenomenon at a lower level of organization.

D. Contemporary Hypotheses on Cycling

Needless to say, the nutritional threshold hypothesis is at variance with several prominent hypotheses that have been advanced in recent years. The hypotheses of Christian and Chitty minimize the role played by energy and nutrition in controlling animal populations. The stress hypothesis of Christian (1950) associates population declines with shock disease and changes in adrenal–pituitary functions. The increase in adrenal activity at high population densities lowers reproduction and raises mortality. The hypothesis involving genetic behavior (Chitty, 1960) suggests that when animal numbers fluctuate, the populations change in quality. This is brought about through selection resulting from mutual antagonisms at high breeding densities.

All hypotheses concerning animal population cycles have in common the notion of feedback. There are two kinds of feedback, negative and positive. The kind generally involved in control mechanisms is negative or deviation-counteracting while the "vicious circle" kind is positive or deviation-amplifying (Maruyama, 1963). Most ecosystems have both kinds. We can think of loops running through a series of compartments

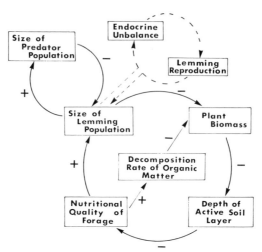

FIG. 8. Feedback-loop model showing homeostatic controls in the arctic tundra ecosystem.

so that the state of each compartment either counteracts (−) or amplifies (+) the change of state of the next (Fig. 8).

It is probable, in fact common, that any given element will be stationed on several loops. It may be checked via one loop and amplified via another. Consider the herbivore compartment of any ecosystem. Amount of forage, its quality, and availability of space are all positive; predators and pathogens are negatively related. In some cases, a density control mechanism operates within the compartment itself—as described by the stress theory or the genetic selection theory. This is simply an additional loop in the system. There is no reason why a control system should have but one governor.

The idea of one cause–one effect is left over from the nineteenth century when physics dominated science. The whole notion of causality is under question in the ecosystem framework. Does it make sense to say that high primary production causes a rich organic soil and a rich organic soil causes high production? This kind of reasoning leads up a blind alley. We are dealing with the different dependent properties of the same system. Only things outside the system can cause something to happen inside. For the same reason, we cannot say that the lemmings are the driving force, any more than the vegetation, the soil, or the microflora, in making the ecosystem tick.

ACKNOWLEDGMENTS

Work on the coastal tundra near Point Barrow, Alaska, was done under grants from the Arctic Institute of North America, the National Science Foundation, and the Office of

Naval Research. R. D. Pieper, who participated in some of these studies, is thanked for presenting this paper at the symposium while the author was in Great Britain.

REFERENCES

Ashby, W. R. 1963. "An Introduction to Cybernetics." Wiley, New York. 295 pp.

Chitty, D. 1960. Population processes in the vole and their relevance to general theory. *Can. J. Zool.* **38**, 99–113.

Christian, J. J. 1950. The adreno-pituitary system and population cycles in mammals. *J. Mammal.* **31**, 247–260.

Egler, F. E. 1964. Pesticides—in our ecosystem. *Am. Scientist* **52**, 110–136.

Hall, A. D., and R. E. Fagen. 1956. Definition of system. *Gen. Systems Yearbook* **1**, 18–28.

Jenny, H. 1961. Derivation of state factor equations of soils and ecosystems. *Soil Sci. Soc. Am. Proc.* **25**, 385–388.

Maher, W. J. 1960. The relationship of the nesting density and breeding success of the pomarine jaeger to the population level of the brown lemming at Barrow, Alaska. *Alaskan Sci. Conf., Proc.* **11**, 24–25.

Maruyama, M. 1963. The second cybernetics: Deviation-amplifying mutual causal processes. *Am. Scientist* **51**, 164–179.

Pieper, R. D. 1964. Production and chemical composition of arctic tundra vegetation and their relation to the lemming cycle. Ph.D. thesis, University of California, Berkeley, California.

Pitelka, F. A. 1958. Some characteristics of microtine cycles in the arctic. *Ann. Biol. Colloq.* **18**, 73–78.

Pitelka, F. A., P. Q. Tomich, and G. W. Treichel. 1955. Ecological relations of jaegers and owls as lemming predators near Barrow, Alaska. *Ecol. Monographs* **25**, 85–117.

Schultz, A. M. 1964. The nutrient-recovery hypothesis for arctic microtine cycles. II. Ecosystem variables in relation to arctic microtine cycles. *In* "Grazing in Terrestrial and Marine Environments" (D. J. Crisp, ed.), pp. 57–68. Blackwell, Oxford.

Schultz, A. M. 1967. The ecosystem as a conceptual tool in the management of natural resources. *In* "Natural Resources: Quality and Quantity" (S. V. Ciriancy-Wantrup and J. J. Parsons, eds.), pp. 139–161. Univ. of California Press, Berkeley, California.

Thompson, D. Q. 1955. The role of food and cover in population fluctuations of the brown lemming at Point Barrow, Alaska. *Trans. 20th N. Am. Wildlife Conf.*, pp. 166–175.

Van Cleve, K. 1967. Nutrient loss from organic matter placed in soil in different geographic regions. Ph.D. thesis, University of California, Berkeley, California.

SECTION III ECOSYSTEM CONCEPTS IN NATURAL RESOURCE MANAGEMENT FIELDS

In this section (Chapters VI–IX), five of the natural resource management fields are reviewed and discussed with respect to the applications and implications of ecosystem concepts. In Chapter VI, J. K. Lewis describes the field of range management in an ecosystem framework. In this field management simultaneously attempts to maximize both primary and secondary productivity. Lewis has broad background in animal husbandry, animal nutrition, and range ecology. His B.S. degree was received at Colorado A&M College and his master's in animal nutrition in 1951 at Montana State College. Then he joined the faculty in the Animal Science Department at South Dakota State University, where since 1950 he has been in charge of range management and range nutrition investigations. He has also taken advanced work in range management and supporting fields at Texas A&M University. Lewis' chapter presents a thorough consideration of application of ecosystem concepts to management problems, including rangeland classification, and an integration of classic concepts of progression and regression in plant succession.

In Chapter VII Bakuzis reviews the field of forestry in an ecosystem perspective. Like Major, the author of Chapter II, Bakuzis has command of several European languages and this is reflected in his exhaustive review and incorporation of the European literature concerning forest ecosystems. Bakuzis completed his first two degrees in Europe; his master's degree was in forest engineering in 1935 at the University of Latvia. Subsequently he studied agronomy in Latvia and forestry in Germany at Hamburg University. He completed his Ph.D. in forestry in 1959 at the University of Minnesota. His background includes research and practice in forestry as well as teaching in both European and American universities; since 1951 he has been at the University of Minnesota where he is now a professor. His major interests include forest ecology and silviculture.

Chapter VIII, by F. H. Wagner, concentrates on ecosystem applications and implications, both in practice and in theory, in fishery and wildlife biology. Thus, ecosystem concepts are applied to, and drawn from, studies of a wide variety of organisms including both freshwater and marine aquatic organisms and upland game and big game as terrestrial organisms. Wagner's undergraduate degree in 1949 was obtained at Southern Methodist University in zoology and botany. Both his advanced degrees were in wildlife management at the University of Wisconsin, in 1953 and 1961, respectively. Wagner has practical ranching experience in the Trans-Pecos area of Texas and has worked in wildlife management on desert game ranges in Nevada and in upland game management in Wisconsin. For the past nine years he has been in the Department of Wildlife Resources at Utah State University where he is now a professor. His major interests are in population dynamics, with emphasis on limiting factors and regulatory mechanisms, and in energy flow with emphasis on interspecific competition and predation. His work has covered a variety of organisms with classic studies on the ringneck pheasant in Wisconsin.

The fields of range, forest, and fishery and wildlife management are concerned primarily with optimizing the yield of a biological product from wildland ecosystems. Watershed management, on the other hand, has as a major concern the understanding of water flow through the ecosystem and its yield to man. Chapter IX, by C. F. Cooper, emphasizes problems of the field of watershed management and suggests procedures of modeling watersheds to include multiple-use objectives. Cooper received his undergraduate degree in 1951 in forestry at the University of Minnesota. Subsequently he was employed as a forester and range conservationist by the Bureau of Land Management in the Southwest before returning to graduate school and obtaining his master's degree in 1957 in range management at the University of Arizona. His doctorate was obtained in 1958 at Duke University in plant ecology. Following this he had a teaching position at Humboldt State College for two years and then was a research hydrologist with the Agriculture Research Service, USDA, before taking a position in the University of Michigan in 1964 where he is now an associate professor of natural resources ecology. Cooper's major interests are in ecological hydrology, snow hydrology, and computer simulation of ecosystem behavior.

Chapter VI   Range Management Viewed in the Ecosystem Framework

JAMES K. LEWIS

I. INTRODUCTION

Range management as a science appears to have had its origin on the grazing lands of North America, although it has been enriched by the inclusion of concepts and practices developed or expanded in other lands. The early history of range management has been intensively reviewed by Campbell *et al.* (1944) and in less detail by others including Chapline (1951), Sampson (1955), and Costello (1964a). The early explorers and the botanists who accompanied them sometimes made notes concerning the nature of the grazing lands through which they passed. However, no systematic effort was made to study these lands until the users became alarmed because of widespread deterioration. The first studies of grazing problems were made by the Division of Agrostology which was organized by the U.S. Department of Agriculture in 1895.

The first grazing experiment was organized by Jared Smith at Channing and Abilene, Texas, in 1898. In "Grazing Problems in the Southwest and How to Meet Them," J. G. Smith (1899) described range deterioration in terms of (1) reduction in the grazing capacity; (2) ". . . disappearance of the best grass . . ."; (3) increase in short-growing grasses; (4) ". . . rapid spread of prickly and thorny shrubs in the south and of the mesquite bean on the table lands and higher prairies"; (5) invasion of ". . . a vast number of rampant weeds which are not eaten by any grazing animal"; (6) ". . . ground is trampled and compacted . . ."; (7) ". . . decrease in fertility (of the land) through exposure of the surface layers to the sun and air"; (8) ". . . less of the rainfall absorbed . . ."; (9) ". . . destructive action of torrential rains"; (10) rapid increase of prairie dogs and jack rabbits. A progress report of the grazing study was also given. Eight treatments were compared, four of them under cattle use with a grazing season terminating in October. The treatments were (1) deferred until June 1; (2) deferred until June 1, plus cut with disc harrow; (3) grazed until June 1, then rested; (4) alternate grazing with cattle moved every 2 weeks; (5) complete rest; (6) rest, plus harrowing; (7) rest, plus discing; (8) rest, plus cultural treatments, including furrowing at intervals, seeding with native and exotic forage species, and transplanting native turf-forming grasses. Eighty acres were devoted to each of the eight treatments. At the end of the first year, treatment (2) was judged to be the most successful. This study was continued at Abilene, but apparently not at Channing and was reported by H. L. Bentley (1902).

J. G. Smith (1899) also discussed prickly pear control; attributed the mesquite problem to protection from fire in part; discussed the effect of shading by the mesquite canopy on grass species composition and nutritional value; recommended water development, ". . . so that cattle will never have to travel more than a couple of miles to water"; production of hay from natural or artificial meadows; production of cultivated forage crops on "naturally well-watered" or irrigated lands; and forage preservation by making "stack silage."

Clements (1920, pp. 310–330) after an extensive literature review outlined a ". . . complete system (of range improvement) based upon investigation as well as practice." After listing the essential features of his system, Clements commented, "Practically all of these have been regarded as more or less essential to range improvement since the first proposals of Smith, 25 years ago, and the present treatment assumes only to correlate them more closely and to work some of them out in greater detail." Dyksterhuis (1955) presented Clements' system with explanatory comments as follows:

The essential factors were presented as seven processes, namely:

(1) *Proper stocking;* to be determined by actual trial accompanied by measurement of the result,

(2) *Rotation or deferred grazing;* under which Clements included all methods of alternate grazing and rest, whether both occurred in one year or more,

(3) *Control of rodents, poisonous plants, weeds, etc.;* and here the importance of natural succession is stressed, along with direct measures by man,

(4) *Manipulation of the range;* including use of fire, irrigation, fertilization, cultivation, cutting, sowing and planting,

(5) *Development of feed and forage for droughts and winter;* to permit better utilization of the range and against the chance that weather may be abnormal,

(6) *Development of water;* to permit more even utilization of the range; and

(7) *Herd management;* under which is included all features which relate to the handling of livestock such as fencing or herding methods that can contribute to the improvement or prevent deterioration of the range.

Dyksterhuis further commented that these practices " . . seem equally appropriate today."

By 1928, eight federal and one state experiment station had been established to study range vegetation and range livestock. Monumental contributions had been made by Sampson, Jardine, Clements, Shantz, Shreve, and others. Range management was being taught in nine colleges with a range management curriculum in three of them (Sampson, 1954) and three textbooks on different phases of range management had been published (Sampson, 1923, 1924, 1928). The passage of the McSweeney-McNary Forest Research Act in 1928 provided additional funds for federal range and forest research and ushered in a period of expanded research activity lasting until World War II. Renner *et al.* (1938) listed 8274 references in "A Selected Bibliography on Management of Western Ranges, Livestock, and Wildlife." The great drought of the thirties with attendant dust storms intensified interest in proper land use and good range management. Legislation providing for improved management and conservation of the range resource was passed and federal agencies created to administer the new regulations. After World War II range management boomed with the rest of the economy. The American Society of Range Management was formed and began publishing the *Journal of Range Management.* Doctorates in range management were offered by several schools and the pace of range research was quickened by federal and state experiment stations. Lag time between research and application was reduced by the extension service and federal action agencies. Range deterioration was halted and upward trends initiated on many ranches and public grazing areas. Range management philosophy was disseminated worldwide.

Yet 70 years after the publication of the first progress report of the first grazing study (J. G. Smith, 1899), how much progress have we made? We have corrected some mistakes; we have verified many hypotheses and validated many concepts, but as a quantitative science range management is in its infancy. In comparison with other fields such as aviation, space technology, electronics, medicine, animal nutrition, or agronomy, we have accomplished little. Why have we lagged so far behind? There are at least four reasons.

First, resources devoted to range research are meager compared with other fields (Box, 1967). For example, in the land grant universities and the U.S. Department of Agriculture in 1965 only 146 scientist man-years were devoted to range research (including the areas of conservation, 85; protection, 44; and efficiency of production, 17), as compared with timber and forest products (1004), dairy cattle (601), small fruit and tree nuts (527), beef cattle (514), poultry (469), wheat (304), corn (298), swine (259), ornamentals and turf (245), citrus fruit (242), and sheep (203). Furthermore, the projected growth shows that only 180 man-years will be needed in 1977 compared with 1550 in the area of improving biological efficiency of field crops (Agricultural Research Institute, 1966). The situation is not as bleak as indicated by these figures since range was defined rather narrowly, and some research pertinent to range ecosystems was included in other categories. The situation is improving slowly as research programs of the U.S. Forest and Range Experiment Stations, the Agricultural Research Service, the state experiment stations, and others continue their range research. Nevertheless, expansion is urgently needed.

Second, range research has not been able to adequately study range ecosystems because of their complexity. In the range environment, the large number of uncontrolled variables often results in inconclusive experimental results. However, when organisms are removed from the range environment for study, experimental results may not be applicable to range ecosystems. Too often in field studies, researchers have studied the range by parts without consideration of the whole. The range has been approached from the standpoint of vegetation, or livestock, or soil, and almost never from the standpoint of the entire ecosystem. Oversimplification has been the rule. Costello (1957) warned: "Simplification of methods of vegetation measurement is a fetish which should be abandoned or at least compromised in favor of sounder ecology. Nature is not simple. And the proper interpretation of her processes is not simple." Yet Costello spoke in regard to the vegetation alone. Because of the heterogeneity of range ecosystems and the small amount of resources devoted to range research, properties of range ecosystems have seldom been sampled with adequate precision. Likewise, because of the great expense involved, range research has often been done with an inadequate number of replications. Range pastures are often more variable than grazing animals, yet researchers who would not consider basing their recommendations on fewer than five animals treated alike have been forced to make recommendations based on only one or two pastures treated alike.

Third, methodology for studying range ecosystems has been inadequate. A high degree of human control over range ecosystems is usually either not possible or not economical. If a high degree of human control is economical the land is usually cultivated and ceases to be range. Consequently, range must be manipulated by extensive methods which are ecological in nature rather than by intensive methods which are agronomic in nature. Yet quantitative ecology is relatively new and is still actively developing its methodology. The ecosystem concept, which provides a framework for the consideration of the full complexity of ranges, the methodology of quantitative ecology, and the development of digital computers now in their third generation will permit us to study range ecosystems on a more complex level (Van Dyne, 1966). For examples and suggestions on the use of mathematical models in the understanding of grassland ecosystems, see Van Dyne (1968). More researchers adequately trained in the new methodology and adequately supported are urgently needed. More attention and more funds must be channeled to range research.

Fourth, the principles and techniques which have been discovered have not been fully applied. The National Inventory of Soil and Water Conservation Needs (U.S. Department of Agriculture, 1962) indicated that the ranges of the United States were producing only about half of their potential. The reasons for the slow diffusion from research to application are somewhat obscure. However, contributing factors have been (1) the lack of man power and resources devoted to the task, (2) the slow, often unspectacular, response to recommended range improvement practices, and (3) the cost-price squeeze which has made many land owners reluctant to invest in range improvement programs.

In the United States, the situation is improving. The American Society of Range Management (ASRM) was organized in 1948 and now has a membership of about 4200 members organized in 20 sections including Mexico and East Africa (Canadian members are included in those sections that border Canada). The ASRM has been rather actively engaged in education and in application of range management principles. Our universities are teaching more range management courses to more range management students than ever before. In 1968, 212 seniors were enrolled in the colleges and universities affiliated with the Range Management Education Council. In addition 122 students were engaged in a master's degree program and 64 in a Ph.D. program (Tueller, 1968). Some of these students will be ranchers, many of them technicians or administrators, some of them teachers, research workers, and extension specialists. There are now 25 schools that offer a B.S. degree in range management or equivalent; 23 schools award the M.S. degree and 14 the Ph.D. Practically all of the western states now have range specialists on the State Agricultural Extension Staff. More range-trained men are being hired as county agents, as work-unit conservationists with the Soil Conservation Service, and as range conservationists with various federal agencies. Banks as well as public and private lending agencies working in the range area are hiring more range-trained people.

However, in the developing countries of the world, ". . . deterioration is continuing at an alarming rate, especially in arid and semi-arid regions. . . . The shocking fact is that badly needed animal production from natural grazing lands has been cut in half as a penalty charge against mismanagement while destruction continues. A hungering world gets hungrier while a valuable natural resource erodes away for lack of rational programs and proper management" (R. E. Williams *et al.*, 1968). A beginning has been made with the use of technical advisors and the training of native students, but the trend of range regression has not been reversed except locally (Chapline *et al.*, 1966). Pearse (1966) stated that ". . . what has held back the expansion of range management abroad is the lack of a doctrine of range management such as we know it. Sustained progress demands faith in the knowledge that management of natural resources based on sound ecological principles will lead to fullest development and productivity." Very high priority must be given to the training of cadres of professional range management specialists in each country, ". . . who must utilize and adapt their own training, knowledge, and experience for and to the ecological and sociological conditions that prevail, and who must formulate management systems that are acceptable to the graziers of the country" (Johnston, 1966). The need is critical and must not be ignored.

II. RANGE MANAGEMENT: PERSPECTIVE AND DEFINITIONS

A. Development of the Philosophy of Range Management

Early workers used the concept of range management but did not define it. Sampson (1914) expressed the view that "Ideal range management would mean the utilization of the forage crop in a way to maintain the lands at their highest state of productiveness and at the same time afford the greatest possible return to the stock industry."

Jardine and Anderson (1919) stated that ". . . grazing on the National Forests is regulated with the object of using the grazing resources to the fullest extent possible consistent with the protection, development, and use of other resources." He continues, "As the National Forests were established primarily for the protection and development of the Forests and the protection of the watersheds, great pains must be taken to harmonize grazing with these primary purposes. Also . . . more and more care must be exercised to see that the wildlife of the Forests is not unduly restricted. . . . The recreational features of the National Forests, too, are of increasing importance, and increased attention is necessary to harmonize grazing use with recreational use." Thus, the concept of multiple use of the forest was stated quite clearly very early.

By the time of World War II, Stoddart and Smith (1943) gave the following definition: "Range management is the science and art of planning and directing range use so as to obtain maximum sustained livestock production consistent with conservation of the range resource." Definitions given by Sampson (1952, 1954), Stoddart and Smith (1955), Hanson (1962), and the American Society of Range Management (1964) are essentially the same.

In recent years there has been a trend toward broadening of the concept of range management. Dyksterhuis (1955) defined range as "native pasture on natural grazing land," and range management as, "economic improvement or maintenance of natural pastures for the production of animals and animal products." He continues, "This does not preclude wildlife management, or watershed management, or application of some forestry practices on range areas. Rather, other disciplines, which may be practiced on range, are thus also left some autonomy in developing their principles." Range is here recognized as the resource base for wildlife, watershed, and other uses as well as grazing.

Costello (1957) implied that wildlife should be included when he wrote, "Range management essentially is habitat management for sustained optimum production of forage for grazing animals." In an editorial Davis

(1961) made a plea for the study of the range resource as a complex in which wildlife is an integral part. Julander (1962) stated, "Since range is the *basic resource,* both deer and livestock are greatly affected by management of the range" (italics the author's). Hedrick (1966) in an editorial, "What Is Range Management?" stressed the importance of technology as well as science and art and defined range management as ". . . the manipulation of the soil, plant, and animal complex used by grazing animals. This management is based on the best scientific information available on these complexes which occur largely on uncultivated land where native plants are predominant, and where other natural resource values—watershed, forestry, wildlife, recreation, etc. may be important." In another recent editorial, Poulton (1967) stressed the range resource ". . . as the food and cover base for both wildlife and domestic livestock populations and as the core of watershed protection and quality on millions of acres." Stoddart (1967) added ". . . or game production . . ." to the definition given in Stoddart and Smith (1943).

Heady (1967) in a series of lectures on range management delivered at the University of Queensland in 1966, presented ". . . Range Management as a land management discipline which depends upon basic sciences; limits its activities to uncultivated lands in subhumid, semi-arid and arid regions; centers its activities on grazing animals and forage; and is concerned with the production of animal products, water, timber, wildlife and recreation which are useful to mankind."

In concluding his resume of "Range Education Needs for the Next 20 Years," Leinweber (1967) said ". . . if we consider range management as the total diverse use of the resource and range science as the body of information which bulwarks the application of principles, then we as educators have rather clearcut objectives in furthering the science."

At the keynote session of the twentieth annual meeting of the American Society of Range Management, Pechanec (1967) asked, "What role do we want this Society to play? The broad role of professional concern about rangelands and all the goods and services they may provide people—Or a narrower role dealing primarily with forage production and livestock grazing, with only peripheral concern for other products and values?" He further said, "I hope we can accept the broader role of professional concern in all matters relating to the conservation and use of rangelands for all goods and services they may provide for the American people." Pechanec concluded, "If we don't move forcefully in this direction, there are certainly others far less knowledgeable regarding rangelands who will." While there is still heterogeneity among range men with reference to their identity, there is growing unanimity of opinion that range management is range ecosystem management and that the objectives of management should consider all of the products and services which the ecosystem can supply. That civil service qualification standards for the range conservation series reflect this understanding is shown by the description of work which includes ". . . managing the natural resources of rangelands and related grazing lands . . ." (Civil Service Commission, 1967).

B. Classification of Range Ecosystems

Dyksterhuis (1958) emphasized the importance of the ecosystem concept in range evaluation when he stated, "The first principle to be recognized is that range is an ecosystem, involving the accumulation, circulation, and transformation of energy and matter through such biological processes as photosynthesis, herbivory, and decomposition, with the non-living part involving precipitation, erosion, and deposition, reacting to the living part, and with coactions between organisms."

Ecosystems have most often been classified by single factors such as climate, soils, vegetation, or animal communities, or by plant and animal communities. Whittaker (1962) reviewed the very extensive literature on the classification of natural communities. Eyre (1963) classified and described the soils and vegetation of the world. Kuchler (1964) provided a map of the vegetation of the conterminous United States. Knapp (1965) presented a detailed classification and description of the vegetation of North America. Penfound (1967) presented a physiognomonic classification of the vegetation in the same area. Classification of habitats has proceeded more slowly because of the difficulties involved in relating communities with their environment. The range site concept has been used extensively and satisfactorily in the United States (Dyksterhuis, 1958; Soil Conservation Service, 1967). However, in areas where the relationships between habitats and communities are not well understood, this method of classification can be used only in a limited way (Pratt et al., 1966). A world site classification that will be useful for a variety of purposes is urgently needed and hopefully will be obtained (Nicholson, 1966, p. 111).

For the broad delineation of terrestrial ecosystems, the classification shown in Table 1 is suggested. This scheme is based on the distinction between those ecosystems which are (1) naturally forested or unforested; (2) intensively managed with a high degree of environmental control, using agronomic techniques, or extensively managed with little environmental control, using ecological principles; (3) cultivated or uncultivated; (4) used or not used for urban, suburban, or industrial development. Although these criteria differ sharply in their modal condition, each of them

TABLE 1

A UTILITARIAN CLASSIFICATION OF
TERRESTRIAL ECOSYSTEMS

Glacial: Ice caps
Forest: Natural forests, including temporary openings
Range
   Natural pastures
    Biomes
     Deserts
     Grasslands
     Shrublands
     Savannas (including open, noncommercial forests)
     Tundra
    Long-enduring pastures of primary succession
     Grazable marshes
     Shrubs and grass successional to forest
   Derived pastures extensively managed
    Derived from forests
    Derived from natural pastures
     Native vegetation
     Introduced vegetation
Cultivated
  Wood lots
  Croplands
  Sown pastures
  Derived pastures intensively managed
Urban–suburban: Residential–industrial areas, airports,
   etc.

---

forms a continuum so that classification must be arbitrary in the transitional zones.

Freshwater ecosystems are considered as part of the matrix of the particular terrestrial ecosystem in which they occur. Open, noncommercial forests are grouped with savannas as natural pastures. However, economic conditions determine what is or is not a commercial forest. Forest openings which are successional to forest occur frequently due to fire, clearing, insects, etc. These openings may be an important grazing resource, but are classified with forest ecosystems because of their temporary nature. However, distinguishing between forest openings and natural pastures at a given time may require careful consideration of soil, climate, and weather influences.

Pastures derived from forest ecosystems where succession to forest is prevented by the activities of man are classified as derived pastures. If these pastures are managed extensively by ecological principles with little environmental control, they are classified as range ecosystems. Conversely, if they are intensively managed with fertilization, plant control, and other agronomic techniques, they are classified as cultivated ecosystems. Obviously, the transition area is difficult to classify even though the modal conditions are distinct. Love (1961) made no distinction between cultivated and range ecosystems, nor between range and grazable forest ecosystems. Dyksterhuis (1955), however, stated, "Range obviously cannot encompass both forest sites and natural pasture sites, and still have unique and universal principles of management." Although principles of management do differ on forest and range sites, it seems advisable to include extensively managed pastures derived from forest ecosystems as a kind of range because (1) the objectives of management include grazing; (2) the pasture is permanent unless the management objectives change; and (3) the methods of management are ecological rather than agronomic.

Intensively managed tame or sown pastures are classified under cultivated ecosystems, even though reestablishment may be required infrequently, because the management is more agronomic than ecological. However, extensively managed plantings of well-adapted introduced species that will persist for many decades without cultivation or agronomic treatment have been considered as a kind of range. This classification is considered to be necessary because of common usage in the western United States where vast acreages of crested wheatgrass are regarded as range. However, such areas are not natural pasture. This treatment of extensively managed sown pastures follows the Range Term Glossary (American Society of Range Management, 1964) where range is defined as "all land producing native forage for animal consumption and lands that are *revegetated naturally or artificially to provide a forage cover that is managed like native vegetation.* Generally considered as land that is not cultivated" (italics the author's). Range seedings made with only one or two climax dominants are regarded as range. However, they should not be considered as natural pastures but as derived pastures because of the simplified flora. Many grazing lands will be difficult to classify and heated controversy may develop concerning proper classification. The most complicated problems of management usually occur in these transition areas. Different kinds of ecosystems may occur intermingled on the same ranch or on the same administrative unit. Usually, however, one kind of ecosystem is clearly more important than the others.

The proposed classification is consistent with the statement of Watts (1951), former Chief, U.S. Forest Service, "The range lands of the world occupy more than half of the earth's entire land surface. They include grasslands on all continents that are too dry, too rough, or too rocky to produce cultivated crops. They include open forests and savannas where much grazable vegetation occurs under scattered tree growth. They include desert shrub types, mountain meadows, and alpine grasslands near or above the timberline. They include the tundras of the far north." The suggested classification is also consistent with the American Society of Range Management Committee on International Biological Program (R. E. Williams *et al.*, 1968): "About one-third of the sphere is land or 34 billion acres. Approximately 10% is farmed, 28% is in forest which is grazed by domestic animals and wildlife part of the year, another 15% is covered with icecaps or fresh water, leaving 47% of the global land area too steep, shallow, sandy, arid, wet, cold, or saline for crops and suitable for grazing by domestic livestock and game animals frequently or occasionally." Urban-suburban ecosystems are not mentioned. However, in the United States, cities occupy 3.6% of the land and the urban area is expanding at the rate of over 1 million acres per year (Dickinson, 1966, pp. 471 and 606; for a more conservative estimate, see Clawson *et al.*, 1960, p. 110).

C. Potential Goods and Services from Range Ecosystems

A partial list of the goods and services which can be supplied by range ecosystems is shown in Table 2. While the major use of range ecosystems is usually grazing because of their very nature, other uses may be important. For example, water yield is an important value of much of the alpine tundra, the mountain grasslands, and many derived pastures. Water quality is an important value, especially in water-deficient areas. Ranchers are finding more and more opportunity to market wildlife and recreational values as well as grazing. On public lands of the United States wildlife and recreational values face increasing demand because of the pressure

TABLE 2

SOME POTENTIAL GOODS AND SERVICES
SUPPLIED BY RANGE ECOSYSTEMS

Grazing and/or habitat
  Livestock
  Wildlife
Water
Recreation
Minerals
Beauty
Preservation of a healthful environment
Preservation of natural or seminatural ecosystems for scientific study
Preservation of endangered species
Preservation of germ plasm for domestication or breeding
Timber (small value)

of a growing population with more leisure time, greater mobility, and greater affluence. In most cases on range ecosystems, optimum range management for livestock grazing will be good range management for most other uses as well. However, some uses, such as the preservation of some endangered species and attempted restoration of ecosystems of the past (Allen, 1966; National Park Service, 1968) will require livestock exclusion.

The number and proportion of uses will depend upon the capabilities of the ecosystem, the level of technology, economic demands, and social pressures. The last three of these factors will change with time, resulting in a change in the number and proportion of goods and services yielded by the range ecosystem. However, a change in the kind or proportion of uses shown in Table 2 does not change the basic nature of the ecosystem. It is still a range ecosystem.

D. Definitions

Having classified range ecosystems we can now define range management. Range management is the management of a renewable resource composed mainly of one or more range ecosystems for the optimum, sustained yield of the optimum combination of goods and services. Management means decision-making in the presence of uncertainty and involves the manipulation of one or more of the dependent and/or the controlling factors. Composed mainly implies that range ecosystems may be mingled with other kinds of ecosystems, such as forests or cultivated lands, which also require manipulation by the manager. Range ecosystems are natural pastures or derived pastures managed extensively on the basis of ecological principles. The optimum combination of goods and services is determined by the capabilities of the ecosystem, the level of technology, economic demands, and social pressures. The objectives may include any of the values which the ecosystem is capable of producing. Management for optimum yield requires a selection of alternatives to maximize values and minimize costs or negative values. Sustained yield requires continuous energy flow with orderly cycling of matter. The restrictions imposed by the word sustained determines the maximum rate of usage under the constraints of the controlling factors.

Range science is the organized body of knowledge upon which the practice of range management is based.

E. The Ecosystem Manager

Since the range is an ecosystem, producing many goods and services, by whom should it be managed? On private lands, the answer is simple—

the rancher, combining experience with advice and counsel from many sources. However, on public lands there is disagreement. There are at least three alternatives:

(1) Decisions should be made by a land manager, a generalist, in many cases a social scientist, schooled in public relations and administration, supported by a staff of specialists; (2) decisions should be made by a team of specialists captained by a former specialist; or (3) decisions should be made by a range ecosystem manager supported by a staff of specialists. Discussion has been largely restricted to alternatives (1) and (2) for forest ecosystems. The comments of Duerr (1967) are particularly pertinent.

. . . I predict that in the future we shall have multiple use land management, but not multiple use land managers. I expect land managers to be single use people by training and by at least much of their experience. Multiple-use management will be performed largely as it is today: by a team of specialists captained by a former specialist. Surely the team members will have acquired, partly in school and partly on the job, an appreciation and a tolerance for the specialities other than their own. They will have learned about the biological and social systems of which the forest is a segment. Thus, they will be able to see the forest and its management in entirety and in context. They will have an attachment for multiple use. And multiple use will be maintained as it is today: primarily an ideal and an abstraction. . . .

The man trained from the outset as a generalist in resource management is apt to become simply a superficialist. The future holds more and more specialization, since it holds more and more knowledge. The land management team of the future, and the forest management team, will be notably larger than today.

Therefore, I predict that we shall achieve the land manager of many parts in the same way that we shall reach our other technological achievements in tomorrow's complex world—that is, collectively.

These points are well taken; however, alternative (3) is not considered. A range ecosystem manager, or a range management specialist as viewed in this paper, is not a generalist or a social scientist and certainly is not a superficialist, but rather a highly skilled, applied ecologist, thoroughly grounded in basic sciences and well-trained in the characteristics of range ecosystems: their potential uses, the impact of these uses on the ecosystem, the compatibility of these uses, and management for maximizing values and minimizing conflicts and costs; and schooled in the use of decision-making tools. Since range ecosystems are by definition either natural pastures or derived pastures, thorough grounding in grazing management and in ecosystem manipulation for the improvement of grazing values are basic and fundamental. A range ecosystem manager will need a group of advisors, including specialists in wildlife, watershed, recreation, and other disciplines. He may also need specialists in particular aspects of range management, such as plant control, range seeding, etc.

TABLE 3
RELATIONSHIP OF ECOSYSTEM MANAGER TO SOME
OTHER DISCIPLINES USING THE ECOSYSTEM

Kind of ecosystem	Decision maker	Staff specialist's or adviser's discipline
Forest	Forest manager	Silviculture Watershed Recreation Wildlife Livestock grazing
Range	Range or ranch manager	Livestock grazing Wildlife Recreation Watershed
Cultivated	Farm manager	Crop production Livestock grazing Watershed Wildlife Recreation
Urban-suburban	City planner	Landscape architect Recreation Watershed Wildlife

However, the range ecosystem manager is the decision maker, the man most knowledgeable about the resource under his supervision.

Because of the extremely diverse nature of terrestrial ecosystems it would probably be impossible to prepare a general ecosystem manager in the period usually allotted for formal education. However, suitable managers can be trained for forest, range, cultivated and urban-suburban ecosystems. The relationship of the ecosystem manager to staff specialists and advisers is shown in Table 3. Specialists in livestock grazing, wildlife, recreation, and watershed are similar in that their disciplines involve particular products or uses of forest, range, and cultivated ecosystems. All but livestock grazing may also be important in urban-suburban ecosystems. Specialists in these areas need a knowledge of each kind of ecosystem in which they work, but not in the depth required of the ecosystem manager. The ecosystem manager also requires a knowledge of each use made of the ecosystem, but not in the depth required of the specialist.

As knowledge and sophistication of management increase different managers may be required for different kinds of ecosystems, such as grassland, desert, tundra, or pasture ecosystems derived from forests. The rancher, in particular, will need to specialize in his particular kind of ecosystem because of the many faceted nature of his business.

F. Institutional Relationships

Academic and research institutions usually are commodity oriented rather than ecosystem oriented. Consequently, determination of the total research effort in range management is difficult (Agricultural Research Institute, 1966). Range management was not even listed as a subject-matter heading in the Library of Congress before 1968.

In our land-grant colleges and universities, departments are usually organized by products rather than by the ecosystem which produces them. Since range management is neither a plant science nor an animal science and since it involves the management of a total ecosystem and is concerned with more than one product, there is considerable confusion as to where it should be administered. Of the 24 schools affiliated with the Range Management Education Council that offered a Bachelor of Science degree in range management or equivalent in 1968 (Tueller, 1968), three were administered in a separate department, four were administered with forestry, four with animal science, four with agronomy, three with botany, three in a School of Natural Resources or equivalent, one was administered jointly by animal science and agronomy, one was a section within a plant science division, and one was administered with watershed management. Since range management has close affinities with product disciplines as well as basic sciences, departmental lines may frustrate cooperation in research and teaching. Cooperation between colleges on the same campus may be even more difficult, yet interdisciplinary research involving several disciplines—such as range management, animal science, botany, soil science, wildlife management, and economics—is often essential for problem solution. Consequently, the needs of range ecosystem management are best served by temporary task forces assembled for specific research, teaching, or extension functions. Thus, the same worker might be a part of several teams and his departmental location relatively unimportant.

G. Range Management in the Future

If range management develops as the author hopes it will, by 1985 all range and related ecosystems will be classified according to their potential values. Management for potential multiple- and coordinated-use values will be stressed by private landowners as well as public land administrators. More intensive management will be practiced using the knowledge and skills supplied by teams of specialists. The rancher will be a well-trained business man knowledgeable in all facets of his operation, using the services of many specialists either from public agencies or from private consultant firms. The public land administrator will be a specialist in the management of the particular kind of ecosystem under his administration with a staff of experts, including specialists in all natural resource disciplines and in different phases of range management, such as grazing management, plant control, and range seeding. Particularly difficult environmental problems will be delineated and diagnosed by analysts utilizing field investigations, extensive field data, computer hardware, and systems analysis. Analysts must be men with extensive field experience, well-trained in several natural resource disciplines, mathematics, systems ecology, and computer science. Complex research problems will be attacked by teams of competent specialists coordinated by a highly skilled multiple biologist. Instruments and procedures developed by the physical sciences, such as electronic devices, radioactive isotopes, and remote sensing, will be widely and commonly used both in research and application (Costello, 1964b). As our science develops, we will be deeply concerned with the formulation of principles, mathematical generalizations, and the prediction of events using measured inputs and mathematical models. Yet we will be continually involved in refining the descriptions of range ecosystems and in basic and problem-oriented research. The range profession will be articulate and politically effective in securing needed action in the problems of range ecosystems. Members of the range profession will be involved in research and education as well as in making management recommendations throughout the nations of the world.

III. THE ECOSYSTEM FRAMEWORK

A. Introduction

Although the concept of the ecosystem can be seen in some early writings (Möbius, 1877; Forbes, 1887), the English ecologist, Tansley (1935) introduced the term ecosystem which he defined as the system resulting from the integration of all the living and nonliving factors of the environment. The operation of such a system includes the ". . . circulation, transformation and accumulation of energy and matter through the medium of living things and their activities. Photosynthesis, decomposition, herbivory, predation, parasitism and other symbiotic activities . . ." are involved (Evans, 1956). The ecosystem must be studied as a whole in order to understand energy transformations, the hydrological cycle, or cycles of carbon, nitrogen, phosphorus, or other elements. The ecosystem, therefore, should be regarded as the fundamental unit of ecological study (Evans, 1956; E. P. Odum, 1959). Dyksterhuis (1958) stressed the importance of the ecosystem as a basic unit in range evaluation.

The ecosystem concept can be applied to any study area regardless of size. The concept may be applied to a large geographical area, an individual ranch, a range site on a particular ranch, or even to a small landscape unit such as the tessera discussed by Jenny (1958). Such ecosystems are not closed systems. Energy and matter are removed and replaced and one ecosystem is connected with another.

A three-dimensional graph or conceptual model made of styrofoam (Lewis, 1959) has been used for the past 10 years to introduce undergraduate students to some of the principles of range management in the framework of the ecosystem. A similar approach is used here. The graph is much simplified from reality and is applicable only to homogeneous, monogenetic, natural pasture ecosystems. However, principles are illustrated some of which are applicable to all range ecosystems.

B. Controlling and Dependent Factors

The concept of the soil-forming factors (Jenny, 1941) has been very useful in understanding ecological processes. Subsequent studies by Major (1951) and Crocker (1952) extended and sharpened the early formulation and showed that vegetation as well as soil is a function of the same factors. Jenny (1958), using imaginary phytotrons, carefully examined the dependencies of ecosystem properties, soil properties, and vegetation properties on the state factors climate, initial state of the system (usually the same as parent material), relief, the biotic factor, and time. Particular attention was given to distinguishing between the plant factor (which is independent of the other state factors, originates outside the system, and is approximated by the regional flora) and the vegetation (which is a dependent factor resulting from the interaction of the state factors).

Jenny (1961), using open-system analysis, derived the state factor equations from a consideration of the initial state of the system, the flux potentials of matter and energy, and time. The generalized state factor equation which he derived was

$$l, s, v, a = f(L_0, P_x, t)$$

Jenny discusses the equation as follows:

Ecosystem properties l, soil properties s, vegetation properties v, and animal properties a are related to, or are a function of the three state factors: initial state of system L_0 (namely, its assemblage of properties at time zero, when genesis starts); external flux potentials P_x; and age of the system t. This is the state factor equation in its most general form. The factors define the state of the ecosystem or soil system.

In accordance with customary descriptions of landscapes, climates, soils, and vegetation the two large groups of variables represented by L_0 and P_x shall be subdivided.

Subgroups of L_0—The initial *mineral and organic matrix* of the soil portion of the ecosystem, its mineralogical, chemical, and physical buildup, is known as parent material p. It is referred to an arbitrary standard state of pressure (e.g., vapor pressure) and temperature, and zero organisms. The *configuration* of the system, that is, its topographic features, specifically slope and exposure, is designated as r. Certain aspects of the water table below the boundary of the system, relating to water influx from below, are also conveniently included in r. When soil genesis begins, the initial slope may be changed and r then becomes a dependent variable, denoted by r'.

Subgroups of P_x—As indicated, the external potentials of the fluxes are environmental properties. One group represents the *climate cl* at the upper boundary, particularly precipitation and the temperature of the outside "heat reservoir." Defined in this manner *cl* is external climate and is akin to regional climate. The climates within the ecosystem, the soil climate and vegetation climate, given the symbol *cl'*, are entirely dependent upon the constellation of all state factors and upon the state of the system.

A second group of potentials is identified as *biotic factor o*. It comprises all species, active or dormant (eggs, seeds, spores) which may migrate or may be carried into the ecosystem. Also included are the species, if any, present initially in the ecosystem at time zero. As outlined previously vegetation actually growing inside the ecosystem is the result of system development and is conditioned by all the state factors; its array of species may mirror but part of the composition of the entire plant biotic factor. The letter o denotes potential vegetation.

Besides *cl* and o, which are universally operating, there are the numerous other P_x factors, such as dust storms, or floods, or annual additions of fertilizers. They are not assigned special symbols, except when needed in pertinent studies.

The extended state factor equation has the form

$$l, s, v, a = f(cl, o, r, p, t, \ldots)$$

The dots stand for unspecified components of the L_0-group and P_x-group. Note that *cl* and o may or may not be functions of t; but r and p are by definition never time dependent [symbols italicized by the author].

If one or more of the state factors change, a new cycle of development is initiated with a new t_0 and a new initial state (Crocker, 1952; Jenny, 1958). If one state factor is dominant over all others, six broad groups of functions may be written which result in six broad groups of ecosystem sequences. They may be written as follows:

$$l, s, v, a = f(\mathbf{cl}, o, r, p, t, \ldots) \text{ climofunction}$$
$$l, s, v, a = f(\mathbf{o}, cl, r, p, t, \ldots) \text{ biofunction}$$
$$l, s, v, a = f(\mathbf{r}, cl, o, p, t, \ldots) \text{ topofunction}$$
$$l, s, v, a = f(\mathbf{p}, cl, o, r, t, \ldots) \text{ lithofunction}$$
$$l, s, v, a = f(\mathbf{t}, cl, o, r, p, \ldots) \text{ chronofunction}$$
$$l, s, v, a = f(\ldots, cl, o, r, p, t) \text{ other functions}$$

In the sixth group the dots denote unspecified P_x factors ". . . such as dust storms giving rise to loess-functions, or phosphate additions producing yield functions."

In the present analysis time is considered as a dimension in which the other state factors interact, rather than an environmental factor (Billings, 1952). The state factors other than time are considered to be the controlling factors of the ecosystem and are grouped as climate, geological materials, and available organisms. Each of these controlling factors is a composite of many elements. Thus climate includes all of the climatic elements, including lightning-set fires. The geological materials include all of the properties of parent material, relief, and groundwater. The available organisms include all of the plants and animals, both macroscopic and microscopic, which are able to send disseminules into the area. The controlling factors are regarded as independent in some and relatively independent in other ecosystems (Sjors, 1955; Gorham, 1955).

The dependent factors of the ecosystem are those factors which can be expressed as a function of the state factors. These are the composite factors of soil, vegetation (the primary producers), consumer organisms (herbivores and carnivores), composer and transformer organisms (bacteria, fungi, etc.), and microclimate. Microclimate is considered as the climate where an organism lives and thus is different for different organisms. However, in range management, the unqualified term usually refers to the climate a few inches above the ground. The climate of the soil is referred to as soil climate under the broad category of microclimate. Each of the dependent factors is dynamically interdependent on the others. Each is a product of the controlling factors interacting through time. Consequently, the controlling factors and time determine the kind of natural pasture and the range site. The controlling factors, time, and the intervention of man determine the kind of derived pasture.

Man is both a controlling and a dependent factor of his ecosystem. He has the power and the skill to control, at least partially, his environment, yet he must live with the environmental changes which he produces. This may be the greatest challenge to the survival of mankind (Cole, 1968). Teilhard de Chardin (1956) discussed man's domination of his environment under the name "noosphere" (from the Greek *noos,* mind, and *sphaera,* sphere) which he considered coordinate with atmosphere, biosphere, lithosphere, etc. Cain (1966) discussed the noosphere as a special aspect of the biosphere worthy of consideration because of the all pervading influence of man through all categories of ecological classification. Man is not illustrated in the model, although the effects of his manipulations are shown.

The controlling factors, climate, geological materials, and available

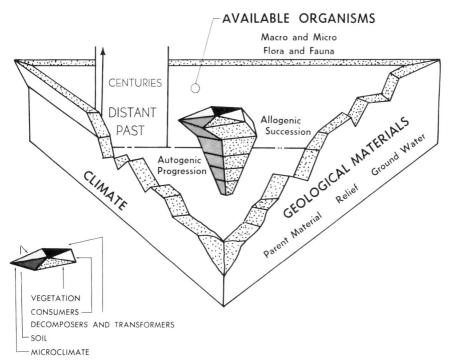

FIG. 1. Three-dimensional graph of a monogenetic natural pasture ecosystem, showing controlling factors, dependent factors, autogenic progression, and allogenic succession through time.

organisms are illustrated in the model (Fig. 1) as walls which contain or limit the dependent factors of the ecosystem. Time is the vertical dimension in the model. The walls have been cut away so that the graph of the dependent factors can be seen more clearly. The dependent factors are diagramed in the model as a horizontal pentagonal surface at varying locations in time. The favorability of a factor is diagramed as proportional to the distance from the center of the pentagon to the edge of the wedge representing that factor. Thus, the area of the pentagon is indicative of the level of energy utilization of the ecosystem.

C. Ecosystem Change

1. SPATIAL CHANGE

a. Continua and Discrete Boundaries. Each of the controlling factors is a composite of many separate elements, each of which is variable in time and space. Every change in a controlling agent produces a corresponding change in the dependent elements of the ecosystem, thus each forms a continuum with steep or gentle gradients. Competition between organisms may in some cases produce rather discrete communities with different dominants even when gradients of controlling factors are gentle and the ecological amplitude of the populations would permit both dominants to grow over the entire area. Usually there is a narrow sector, called an ecotone, where both dominants are present. Thus, evidence can be found both for and against the continuum concept. [Compare Daubenmire (1966), Vogl (1966), Cottam and McIntosh (1966), D. J. Anderson (1965), and McIntosh (1967).] Dyksterhuis (1958) stated that the second ecological principle to be recognized in range evaluation is that "Climates, plant communities and soils tend to cover the earth as a continuum with measurable horizontal gradients" and then showed how this concept served as a basis for the range site classification scheme used by the Soil Conservation Service in the Northern Great Plains.

b. Pattern. Soils are usually not homogeneous even in very uniform landscapes. Variations in the initial state of parent material and relief interacting with other controlling factors produce soil variability. Microrelief may be produced in interaction with biotic factors or by the volume change properties of the soil interacting with the other dependent and controlling factors. Soils on different micropositions may be greatly different and may support different kinds and amounts of vegetation (E. M. White, 1961a).

Pattern in vegetation is the nonrandom arrangement of plants. The most common type of pattern is the occurrence of individuals in clumps or groups. Such vegetation is said to be aggregated, overdispersed, or contagiously distributed (Greig-Smith, 1964). Plants may also occur distributed at random or in a more or less regular distribution such as trees in an orchard. This has been referred to as regular distribution or underdispersion. Patterns may occur on several scales. Large-scale patterns usually are seen readily especially when individuals of different life forms are involved, such as trees or shrubs in a grassland. Smaller patterns may require careful measurement to determine their existence. Causes of vegetation patterns were reviewed by Kershaw (1964), who summarized vegetation patterns as morphological (spacing of individuals along a rhizome, size and form of the individual plant, etc.), environmental (patterning of vegetation along macro-, meso- and microgradients), and sociological (product of the interaction of individuals and species with each other and with the environment). Sampling methods and statistical procedures to measure patterns and to minimize the effect of pattern on vegetation measurement were discussed by Grieg-Smith (1964). Usually vegetation is patterned more strongly during progression and more nearly approaches a random distribution at climax (Hanson and Churchill, 1961, p. 132).

Patterns in the distribution of consumer organisms are more complex because of mobility. Some mathematical aspects are discussed by C. B. Williams (1964). Most soil organisms appear to be contagiously distributed. Spatial patterns are an important attribute of range ecosystems that affect sampling and measurement and which might be used as indicators of successional stage or of range condition.

2. TEMPORAL CHANGE

Ecosystem change in time has been classified in many ways. Ecosystem change is classified herein primarily on the basis of vegetation change, which is an indicator of change in all of the dependent factors. The classification given in Table 4 is preferred to those with fewer categories because it allows greater precision of meaning in describing changes that occur in range ecosystems. The meaning of change is the same as the concept of succession expressed by Cooper (1926), but not by Clements (1916). Ecosystem change in time may be measured on different scales, for example, geological, recent, and contemporary. Most ecosystems have developed from previously existing ones during geological time

TABLE 4

ECOSYSTEM CHANGE IN TIME

Geological time
Recent and contemporary time
   Nondirectional
      Replacement
         Noncyclic
         Cyclic (intracommunity)
      Intercommunity cyclic
      Fluctuation
   Directional
      Progression
         Autogenic
            Primary
            Secondary
         Allogenic
         Induced
      Regression
         Autogenic
         Allogenic
         Induced

(Clements, 1936). Many characteristics of the grasslands of the Northern Great Plains today, for example, can be traced to its ancestry in geological time (Dix, 1964). Some apparently stable grassland ecosystems occurring in forest climates in the Northern Rocky Mountains may be residual from a xerothermic maximum period following glacial retreat. Adverse environmental conditions may be slowing the forest advance (Patten, 1963). In this article, however, we are mainly concerned with recent and contemporary change which can be classified as either directional or nondirectional (Churchill and Hanson, 1958).

a. Nondirectional Change. Nondirectional changes may be classified as replacement, intercommunity cyclic, and fluctuation change (Table 4). *Replacement change* may be cyclic or noncyclic (Hanson and Churchill, 1961, p. 142). Noncyclic replacement change is the usual replacement of an individual that dies by an individual of the same community. Cyclic intracommunity change occurs as a mosaic within the framework of a single community. In the cyclic system there is an upgrade series in which soil and vegetation develop to a peak followed by a downgrade series in which soil and vegetation are depleted leaving a bare area which then begins the upgrade series again (Kershaw, 1964, p. 63). Most reports are from European grass and shrub communities. However, Billings and Mooney (1959) reported the occurrence of a frost hummock cycle on alpine tundra in the Medicine Bow Mountains of southern Wyoming. Hanson and Churchill (1961) suggested that the patchy grassland vegetation on solodized solonetz complexes in western North Dakota was an example of cyclic replacement change. However, these do not necessarily begin with a saline soil as postulated (White, 1964a,b) and when they do, the time required for the cycle is probably too long to be considered as an example of cyclic replacement change. This type of change can occur within intercommunity cyclic, fluctuational and directional change.

"The intercommunity cycle is the kind of change whereby one type of community repeatedly changes to another type and then returns to the first one . . . ," such as cyclic change from grassland to forest and back again (Hanson and Churchill, 1961, p. !46). Such cycles may be the result of long-term fluctuation change in climate.

"Fluctuation change is a random fluctuation about a norm or average" (Hanson and Churchill, 1961, p. 146), and may be due to fluctuation in one or more of the controlling factors or to population dynamics. Fluctuation change is characteristic of natural equilibrium, but also occurs superimposed on directional change. For example, any range management plan must consider the reality of fluctuation change that may result in temporary improvement of a deteriorating range during wet years or temporary regression of an improving range during dry years.

b. Directional Change. "Directional change from less complex to more complex communities may be considered as progression. A directional change from more complex to less complex communities may be considered as regression" (Churchill and Hanson, 1958). If progression occurs when the controlling factors of the ecosystem are relatively stable, the change is due to the effect of the community on the habitat and is called autogenic progression. This is the same process referred to as succession by Clements (1916) or as autogenic succession by Tansley (1935). Although these workers were concerned primarily with vegetation and soil, corresponding changes in the other dependent factors were implied. Clements (1916, p. 4) formalized this process as nudation, migration, ecesis, competition, reaction, and stabilization. Clements and Shelford (1939) extended the concept to specifically include consumer organisms. The process may be very briefly described as follows. Pioneer plants grow; a microclimate is differentiated; organic matter is accumulated and soil formation begins. As the environment is altered, other organisms are better adapted than the pioneers and gradually replace them; microclimate becomes more favorable; populations of consumer, transformer, and decomposer organisms become larger and more diverse, often with complex food webs. Total biomass, total energy storage, and rate of recycling of nutrients are increased. Calcium and various other ions as well as fine-textured particles are removed from the upper portion of the soil and deposited at lower levels as horizon differentiation proceeds. Fertility elements become more available and soil productivity is increased. These changes are gradual and continuous. However, classes of communities can usually be recognized which follow one another on a given area until a relatively stable community is established which is in equilibrium with the controlling factors of the ecosystem. From an energy standpoint progression may be considered to occur when the anabolic processes of ecosystem metabolism exceed the catabolic ones (Schultz, 1967). This suggests that the steady state reached by progression is the one which maximizes the rate of energy flow (Watt, 1968, p. 39). For examples of soil chronosequences, see Bunting (1965, pp. 79–87) and Viereck (1966).

Primary autogenic progression is shown in the model (Fig. 1) as a steady increase in the favorableness of all five dependent factors and thus in the energy utilization of the total ecosystem. Each layer of the model represents a long period of time measured in hundreds or thousands of years. Because of the very long time periods involved, stages of primary autogenic progression are classified as different range sites.

Secondary autogenic progression occurs after denudation or destruction when the disturbing agent is removed (secondary succession: Clements, 1916) and usually proceeds rapidly since a soil mantle is already formed. This process is one of the tools which the range manager can use. Complete rest from grazing is one of the methods most commonly used for range improvement.

Autogenic regression occurs rarely, if ever, on water-deficient natural pastures. However, in water-surplus ecosystems continued leaching of soluble ions and fine-textured particles may result in regression.

During autogenic progression, the ecosystem controls may change and produce changes in the dependent factors. This process is allogenic succession and may be either progression or regression depending on whether the controlling factors change to a more favorable or a less favorable state. This definition is modified from Tansley (1929, 1935), who used allogenic succession to refer to successive changes brought about by factors external to the community. Dansereau (1957, p. 163) used allogenic succession to refer to changes in the vegetation produced by changes in the physical elements of the site, such as siltation or erosion. Allogenic succession as used here may be caused by such things as climatic change and attendant cycles of erosion, loess deposition, or the invasion or development of a new organism which was previously not available to enter the ecosystem. For example, the introduction of the rabbit and the cactus into Australia or *Bromus japonicus* and *Bromus tectorum* into North America are examples of changes in the controlling factor of available organisms which have produced allogenic succession. Introduction of the domestic horse into North America and its subsequent effect on the plains Indian and the buffalo is another instance. As Tansley (1935) suggested ". . . autogenic and allogenic factors are present in all successions. . . ." This is true because the controlling factors are not constant. They continuously show fluctuation change and frequently show directional change.

Allogenic succession is diagramed in the model as a change in the containing wall of climate (Fig. 1) which produces a corresponding change in each of the dependent factors. Although allogenic successions have probably occurred many times in any given ecosystem, for the sake of simplicity only one is diagramed.

The ecosystem is thus a product of its previous history. Changes in the controlling factors may be severe enough to be called destruction, or may only modify the ecosystem. Most ecosystems are polygenetic and are the result of several climatic changes, erosion cycles, and organism invasions (Crocker, 1952; Butler, 1950).

During progressive succession there is usually an increase in primary production, biomass, relative stability, and regularity of populations and diversity of species and life forms within the ecosystem. Finally, a relatively steady state is reached which is characterized by dynamic fluctuation rather than directional change in these characteristics (Fig. 2). This

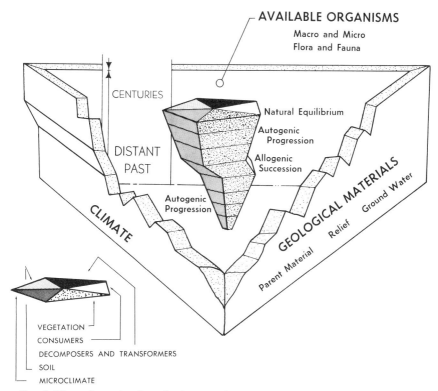

Fig. 2. Three-dimensional graph of a natural pasture ecosystem, showing autogenic progression and allogenic succession, culminating in natural equilibrium.

steady state, the end product of autogenic progression, is the climax (Clements, 1916, 1936) or biotic climax (Tansley, 1935). However, the term climax has been the subject of so much controversy and so many modifiers have been used to express different meanings (Whittaker, 1953; Sellick, 1960) that the term is seldom used in this paper. Instead, natural equilibrium is used to designate an ecosystem which is relatively stable and whose dependent factors are in balance with the controlling factors. The natural equilibrium is usually characterized by maximum diversity of species which permits maximum utilization of the resources of the environment on a sustained basis (Lindeman, 1942; Churchill and Hanson, 1958). Primary production is relatively high but not necessarily at a maximum (Costello, 1964a). However, the total biomass per unit of primary production is probably at a maximum (Margalef, 1963) and total energy flux is probably at a maximum (Watt, 1968, p. 39).

Ecosystem change produced by man and his activities is induced succession (Stoddart and Smith, 1955, p. 120) and may be either progression or regression. Range deterioration and improvement are classified here and will be discussed in a later section. Dansereau (1957, p. 165) considered natural biotic influences, such as the effect of the buffalo herds on the North American grassland, the effect of the elephant herds on the African savannas, the effect of a new organism, or the effect of man himself to be biotic succession. Here all of the dependent factors are considered to be the result of autogenic and allogenic succession with changes produced by human intervention classified as induced succession.

The end product of induced progression is the management equilibrium which is that state of an ecosystem which is in dynamic equilibrium with the controlling factors (including management) and which will produce the optimum sustained yield of the optimum combination of goods and services. This new balance, the goal of range management, will be discussed in a later section.

D. The Nature of Natural Equilibrium

The natural equilibrium is characterized by a diversity of different life forms, species, and biotypes which occupy every available functional niche (Cain, 1966). Organisms are genetically adapted to their environment, including each other. Biotypic, ecotypic, and clinal adaptation to environmental gradients are present (Heslop-Harrison, 1964; Ford, 1964) including climatic (McMillan, 1959, 1960) and edaphic gradients (Nixon and McMillan, 1964). Producer organisms show both stratification and periodicity (Dyksterhuis, 1958). Adjustments among and between organisms are very complex. Herbivores in great variety utilize different kinds of vegetation and occupy different areas. For example, in Northern Rhodesia, Darling (1960) reported 29 large grazing animals occupying various functional niches with a minimum of competition. Talbot and Swift (1966) reported that *Acacia* savanna in Kenya Masailand could support only 11,000 pounds of cattle, sheep, and goats per square mile yearlong compared with 70,000 to 100,000 pounds of wild ungulates. Food webs are extensive and interlocking (Fig. 3) with a variety of producers, herbivores, carnivores, parasites, decomposers, transformers, and organisms that function in more than one trophic level. Interactions among populations are extremely complex, but well adjusted (E. P. Odum, 1959; Klopfer, 1962). Since all niches are filled and all organisms are genetically adapted to one another, successful introduction or invasion without deterioration is rare (Elton, 1958). However, when a new organism is able to become a permanent part of the natural equilibrium by allogenic succession, it should be recognized as a constituent of natural equilibrium, rather than being considered as an invader.

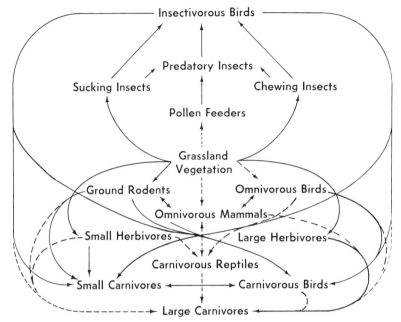

FIG. 3. Food web of the grassland biome (modified from Carpenter, 1940).

The natural equilibrium was doubtless continually shifting in dynamic equilibrium (see Murdoch, 1966; Ehrlich and Birch, 1967; Slobodkin *et al.*, 1967) set in motion by fluctuation change in the controlling factors and dynamic relationships between and within populations. These oscillations were self-limiting by means of feedback mechanisms, territoriality, and behavior patterns (Watt, 1968; Wynne Edwards, 1965; Klopfer, 1962), or by other mechanisms, possibly of a behavioral-physiological nature (Christian and Davis, 1964) or possibly of a genetic nature (Chitty, 1960). These changes involved populations of all living organisms in the ecosystem as well as the microclimate and soil. The more complex the relationships of the living organisms, the greater the variety of living things, generally the more stable the ecosystem can be expected to be (MacArthur, 1955; Elton, 1958; Schultz, 1967), although exceptions have been noted (Watt, 1968, pp. 39–50).

However, all ecosystems do show fluctuation change. A simplified cycle illustrating dynamic equilibrium is shown in the model as a series of lighter colored blocks each representing a period of time measured in years or decades (Fig. 4). A population of herbivores (bison, antelope, elk, rabbits, or others) increased in abundance (first light-colored block, Fig. 4) and overutilized the vegetation which resulted in a more severe

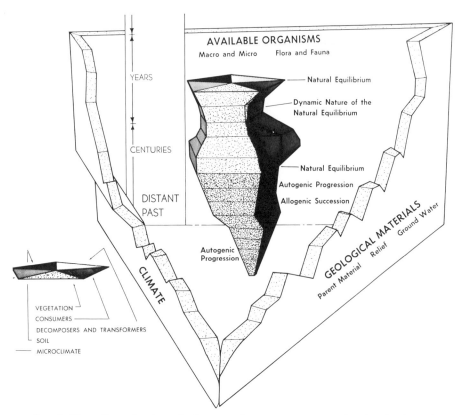

FIG. 4. Three-dimensional graph of a natural pasture ecosystem, showing the dynamic nature of natural equilibrium.

microclimate. The weakened vegetation in a less favorable microclimate produced less food. Less organic matter was available as food for decomposers, which in turn were less active in recycling nutrients in the soil. Soil erosion was accelerated by lack of protective cover of the vegetation but probably did not become severe in most ecosystems. The increased population of herbivores provided more food for the carnivores that preyed on them. The lowered food supply on the closely utilized range weakened the herbivores, reduced reproduction, and rendered them more susceptible to parasites and disease and more easily captured by predators, and the population was reduced below its previous numbers (second and third light-colored blocks, Fig. 4). The predators were consequently reduced. The vegetation was underutilized and improved in composition and vigor, resulting in a more favorable microclimate, larger populations of decomposers and transformers with improved recycling of nutrients and better soil protection (fourth light-colored block, Fig. 4). Under these conditions all of the dependent factors increased to a normal level (upper block, Fig. 4). This type of fluctuation change must have characterized the natural equilibrium and must have involved organisms in every level of organization.

At natural equilibrium, the rate of turnover and recycling of matter and nutrients are at a maximum, although irregularities resulting in extreme fluctuations often develop in very simple ecosystems such as the arctic tundra (Schultz, 1964). In more complex ecosystems, the biogeochemical cycles of water, carbon, oxygen, nitrogen, and other kinds of matter are extremely complex but remarkably stable, allowing relatively stable primary production. Energy flows through the ecosystem at a rapid rate; yet all of the energy fixed by the primary producers is dissipated in the maintenance cost of the life processes of the extensive and diverse herbivores, carnivores, top carnivores, parasites, scavengers, decomposers, reducers, etc. (Fig. 5). There is no net output from an ecosystem in natural equilibrium (H. T. Odum, 1957; H. T. Odum and Pinkerton, 1955).

Most ecosystems were at natural equilibrium when civilization entered or began. In the North American grassland, the natural equilibrium was destroyed before its marvelous complexities were ever studied. Today range scientists use the relict method, described by Clements (1934) to study the effect of livestock grazing and other disturbances on the range ecosystem. Generally, the criteria used to select a relict area for study

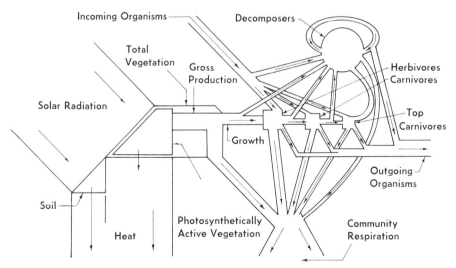

FIG. 5. Energy flow in a natural pasture ecosystem at natural equilibrium (adapted from H. T. Odum, 1957).

is that it should be (1) at equilibrium, or in a long-enduring stage of primary autogenic progression classified by range site; (2) relatively free from grazing by large grazing animals or by high concentrations of rodents and rabbits; and (3) undisturbed by recent fire. Such areas are in natural equilibrium, but they are not completely representative of the natural equilibrium that prevailed under grazing by buffalo, elk, antelope, etc. and that characterized most of the grassland. They should be thought of as relicts of the climax, considering the kind of soil developed locally. Such areas are valuable as a standard against which grazed areas can be measured to determine the departure from the climax which is one of the definitions given for range condition (Dyksterhuis, 1949). A large number of such relicts of climax have been studied. Notable among these are studies of relatively inaccessible areas, such as buttes, mesas, and areas isolated by lava flow (for example, see Larson and Whitman, 1942; Quinnild and Cosby, 1958; Passey and Hugie, 1963). The Nature Conservancy is active in helping to preserve natural areas. The National Park Service is attempting to remove the disturbing influence of man as much as possible and restore large areas of the parks and monuments in the western United States to pristine conditions (Reid, 1968). However, except in such withdrawn areas, the goal of range management is not and cannot be a relict of climax.

Man in the hunting cultures was a top carnivore, a food gatherer, and to a limited degree a manipulator. Fire was used by prehistoric man as a hunting method (Wedel, 1961, p. 76). However, fire use by primitive man as a dominant factor in the development of grasslands (Sauer, 1950; O. C. Stewart, 1956) has been overestimated, particularly in grassland climates (Borchert, 1950; Wedel, 1961, p. 55; Dix, 1964). Lightning-set fires are frequent in many range areas today and must have been important in determining the nature of natural equilibrium. Phillips (1965) reports naturally occurring fires in Africa due to lightning, sparks from falling rocks, and presumably, spontaneous combustion of moist, packed vegetation and mulch. Fire was certainly important in increasing the area of grassland and in sharpening the boundary between grassland and forest or grassland and desert shrubs (Humphrey, 1958, 1962). In drier climates burning reduces foliage yields (Daubenmire, 1968) and was regarded as harmful by the Indians of the Northern Great Plains who had regulations against burning (L. A. White, 1959, p. 151). Fire was also an important agent in erosion and contributed to the development of drainage patterns (E. M. White, 1961b). The Great Plains was probably fully stocked with buffalo (Roe, 1951) and other herbivores and range condition was maintained below that found in relicts of climax (Larson, 1940). Intermittent gully formation was probably typical (E. M. White and Lewis, 1967).

Under the impact of fire, recurrent drought, fluctuating animal populations and availability of water, range condition probably ranged from fair to excellent. Fair range condition probably occurred in areas near water, especially when high populations of herbivores concentrated on recently burned areas near water during and immediately following drought.

Because of isolating barriers some range ecosystems developed to natural equilibrium with few or sometimes no large herbivores. California grasslands were grazed only moderately by pronghorns, deer, and elk (Burcham, 1957, pp. 107–109). New Zealand had no grazing or browsing fauna other than birds (Howard, 1964). Such ecosystems were very easily damaged by domestic or exotic grazing animals.

E. Induced Regression

1. GENERAL CONCEPTS

Most ecosystems were in natural equilibrium when civilization entered or began. Primitive man functioned as a consumer organism, primarily as a top carnivore. Civilized man, likewise, is a consumer, a dependent factor of the ecosystem, but he has within his power the unique ability to modify the total ecosystem for better or for worse by modifying its components. Wise management demands a degree of understanding that civilizations have rarely possessed. The economy of civilized man demands that the ecosystem under his domination produce a removable product which can be exchanged with other groups. In order to achieve this goal under grazing, man has usually reduced the native herbivores, shortened food chains, and confined domestic livestock on the land in large numbers. The yield of livestock products was high at first, but then declined due to deterioration of vegetation, microclimate, decomposers, and finally the soil. This process is shown in the model (Fig. 6) as a series of blocks representing time periods measured in decades or centuries in which the wedge representing consumer organisms is disproportionately large in comparison with the other dependent factors. As a result the favorability of all of the dependent factors and the efficiency of energy utilization of the total ecosystem declined through time. However, this maladjustment was not self-correcting as in dynamic fluctuation of natural equilibrium (Fig. 3) because civilized man prevented migration and at least partially controlled predators, parasites, disease, and starvation. Regression did not retrace the earlier stages of autogenic progression but instead was deflected at an angle (Godwin, 1929). The resulting communities were called plagioseres or twisted seres (Tansley, 1935). These features are diagramed in the model by progressive displacement to the right of a

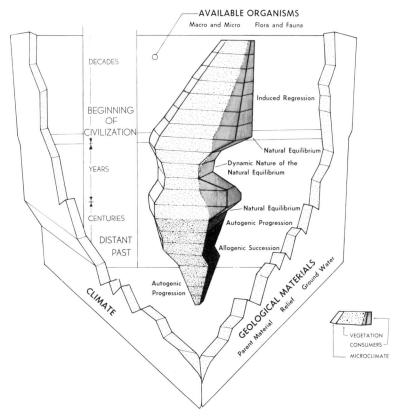

Fɪɢ. 6. Three-dimensional graph of a natural pasture ecosystem showing regression induced by poor grazing management.

series of progressively smaller blocks with a removable product of livestock that was initially high, but declined through time. Vegetation was drastically affected first, followed by microclimate, consumers, decomposers, and finally soil.

2. Regression Induced by Grazing

a. Effect of Induced Regression on the Range Resource. The same pattern of degradation (induced regression or exploitation) has occurred on every continent (Larin, 1956; Moore and Biddiscomb, 1964; R. E. Williams *et al.,* 1968) and is seen in operation over millions of acres of range in this country today. The vegetation is closely grazed. The plants most palatable to the particular kind of livestock at that season are grazed repeatedly in convenient areas first. Mulch cover is reduced, microclimate becomes drier and more severe, soil is trampled and puddled when wet

(Reynolds and Packer, 1963), water infiltration is reduced, runoff is increased and man-made drought is produced (Rauzi and Hanson, 1966; Rauzi *et al.*, 1968a). Rate of energy flow is reduced, stratification and periodicity of primary producers are disrupted, and the orderly operation of many biogeochemical cycles is altered, especially those of water, carbon, and nitrogen. Those plants which are very palatable, tall growing, whose apical meristems are elevated early, which have a large proportion of fruiting to vegetative stems (Booysen *et al.*, 1963; Vogel, 1966), or which are near their margin of tolerance for drought, are weakened by grazing use. Foliage and seedstalk production (Ellison, 1960; Jameson, 1963) and root and rhizome production (Troughton, 1957; Schuster, 1964) are decreased. Accumulated carbohydrates and other labile substances are reduced (May, 1960; Cook, 1966a; Milthorpe and Davidson, 1966). Plants thus weakened by grazing are replaced by other species better able to persist under close grazing. Often some plants of these decreasing species may persist for many years (Weaver and Hansen, 1941). However, these are probably either not readily accessible or are biotypes that are less palatable, shorter growing, or more resistant to grazing than the mean of the population (Stapledon, 1928; Peterson, 1962). The development and spread of such biotypes may take considerable time. At the Manitou Experimental Forest no grazing resistant biotypes of mountain muhly were found after 24 years of heavy grazing use (Quinn and Miller, 1967). The decreaser plants are replaced by those that escape grazing, are more resistant to grazing, or are more drought tolerant (increasers). However, as the relative palatability changes, some of the increasers receive more grazing pressure and may begin to decrease and be superseded by more grazing resistant increasers or by invaders (Dyksterhuis, 1946). The behavior of plants as decreasers or increasers is dependent upon range site, kind of stock, and season of use. Usually plants which are not grazable or which were not originally present on the range except in denuded areas invade and spread over the range (O. B. Williams, 1961). These are usually species that are capable of rapid migration, that have been adapted to a drier habitat than the original ecosystem, and that escape grazing because of low stature, short season of growth, low palatability, poisonous properties, or spines. In the final stages of degradation the microclimate is very dry and there may be extensive wind and water erosion. Once this stage has been reached and is in equilibrium at high grazing intensities, response to grazing manipulation may be very slow, especially in arid regions (Blydenstein *et al.*, 1957).

Characteristics of depleted ecosystems derived from natural equilibrium on the same range site will vary depending on the kinds of animals, season of grazing, intensity and duration of grazing; the sequence of grazing or the grazing system; and interactions with fluctuations in the controlling factors, especially periodic drought (Albertson *et al.*, 1957). Thus a multiplicity of apparently different range ecosystems may be developed from a single natural equilibrium.

Drought is a normal periodic feature of the climate of most natural pastures as shown by precipitation records, historical records, and dendroclimatic studies. For example, a 534-year tree ring record near Bismark, North Dakota (Wills, 1946), showed extensive clusters of wet and dry years ranging from as many as 35 wet years in sequence to as many as 14 dry years. Drought may cause regression even in the absence of grazing (Weaver and Albertson, 1956). Usually, however, grazing pressure is intensified during drought and both are involved in range deterioration. Defoliated grasses are less able to use water held at high soil moisture tensions (Jantii and Kramer, 1956). Plant and animal species vary in their resistance to drought and so are differentially affected, resulting in large population fluctuations (Coupland, 1959; Humphrey, 1962).

Where economic conditions are unfavorable, stocking rates tend to be maintained at wet year levels during normal or even dry years, and, in some cases, may be lowered only when regression has reduced primary production so that even at reduced stocking rates vegetation is overutilized. Thus range condition often declines sharply during drought, remains relatively constant during wet years, only to decline again with recurrent drought.

Protection from fire coupled with overgrazing, which reduces fuel and weakens palatable herbaceous plants, appears to have been largely responsible for extensive brush invasion on many predominantly grassland ranges (Humphrey, 1958; Box *et al.*, 1967) and for the formation of thickets in savannas (Dyksterhuis, 1957; Heady, 1960; Phillips, 1965).

Induced regression may proceed to the point of site destruction. Soil erosion that has removed the soil mantle, destroyed the A horizon, or resulted in gully formation which drained a water table are examples of this. The result is a change in the state of the ecosystem that necessitates a new range site designation.

b. Effect of Induced Regression on Grazing Animals. As grazing pressure is increased, competition between kinds of grazing animals, both livestock and wildlife, is intensified. Although the quantitative relationships have varied widely, most studies have shown that forage intake is decreased as available forage is reduced (Pieper *et al.*, 1959; Willoughby, 1959; Arnold, 1964). However, Wheeler *et al.* (1963) found no relationship between intake and available forage with sheep. As grazing pressure is increased nutritional value of the diet is decreased (Cook *et al.*, 1962, 1965), and grazing time is increased (Arnold, 1964). Dry lot and range studies have shown that as the plane of nutrition is lowered, animal condition is reduced; interval between birth and first estrus is increased; number of services per conception is increased; number of ova shed is reduced; reproductive rate is decreased; and birth weight, growth rate, weaning weight, weaning condition, and growth rate of weaned animals are reduced (Lewis *et al.*, 1956; Reed and Peterson, 1961; Houston and Woodward, 1966). When animals are placed on a higher plane of nutrition, compensatory growth will occur, but the effect of high grazing intensity will not be removed quickly and may be a permanent effect. When animal condition is reduced, parasite and disease problems, supplemental and emergency feed costs, and death losses are increased. When game populations are allowed to exceed the limits of their forage supply the same effects occur. This has been documented for deer in Utah (Julander *et al.*, 1961) and on the Kaibab Plateau in Arizona (Russo, 1964), for moose on Isle Royal in Lake Superior (Mech, 1966), for the Gallatin elk herd (Peek *et al.*, 1967) and for reindeer on St. Matthew Island in the Bering Sea (Klein, 1968). As a result of drought and heavy stocking by cattle, horses, and pronghorns in Trans-Pecos, Texas, pronghorns were forced to subsist mainly on three unpalatable browse species available on depleted range, and 57% of the herd died from malnutrition and tarbush poisoning (Hailey *et al.*, 1966).

The relationship of grazing pressure to individual animal gain and pasture gain. The relationship of gain per animal, gain per pasture, and stocking rate is extremely important in range management and has been studied in both range and pasture by several workers without definitive results. With growing animals, as stocking rate is increased, gain per animal is decreased. This relationship has been linear in some studies but not in others. Generally, it appears that at very light grazing pressures an increase in stocking rate produces no decrease or possibly even an increase in gain per animal. However, at very high grazing pressures, gain per animal decreases rapidly with slight increases in stocking rate. At intermediate grazing pressures the curve may approach linearity. Once the relationship of gain per animal to stocking rate is known, gain per acre can be calculated, since gain/acre = gain/animal × animals/acre. If gain per head decreases at an ever increasing rate with increasing stocking rate as postulated by Harlan (1958) then maximum gain per acre will be produced at extremely heavy stocking rates. Relatively small increases in stocking rate beyond this point will produce weight loss. If the gain per head curve is linear throughout, as suggested by Riewe *et al.* (1963), then the stocking rate may be increased substantially, as much as double, before weight losses occur. Practically all intensity of grazing studies on ranges have shown highest gains per acre at the heaviest stocking rate even though the range was deteriorating. One exception is the study on the Manitou Experimental Forest (D. R. Smith, 1967) in which moderately grazed pastures produced the greatest animal gain. However, these pastures appear to have had a higher potential than the heavily grazed pastures by virtue of a larger percentage of grassland and abandoned field and a lower percentage of timber.

As range deterioration proceeds under heavy use, the advantage of heavy grazing in gain per acre declines. However, gain per acre generally remains higher for heavy than for lighter grazing rates, except in dry years. The gain per head-gain per animal-stocking rate relationship is difficult to study because of the differences in the response of different animal species and classes (reproducing versus growing), the cumulative effects of intensity of grazing on the dependent factors interacting with nonuniform site factors, and the mathematical consideration that the factors being studied are ratios.

Too much emphasis has been placed on gain per acre rather than net income per acre and the change in the value of the range resource. Decrease in product value per head and an increase in animal connected costs will often result in net income from heavy grazing being reduced below that from moderate grazing even though gain per acre is higher. If the cost of range deterioration is considered, this is almost always true.

The effect of range condition on animal production. The effect of range condition on animal production has received very little attention. Higher production from ranges in higher range condition is usually anticipated. McCorkle and Heerwagen (1951) surveyed ranches in northeastern New Mexico and southeastern Colorado and found that the ranches with their ranges in good range condition were producing more animal products than those with ranges in fair or poor range condition. In an economic study of budgets of representative southwestern ranches stratified by vegetation type, size, kind of stock, and range condition, net ranch incomes from ranches in good range condition were about 5 times as great as those in poor in five comparisons. In one comparison net ranch income from ranches in good range condition was only $1\frac{1}{3}$ times as great as from those in poor range condition (Gray *et al.*, 1965). No ranches in excellent range condition were reported. Various management factors were doubtlessly confounded with range condition in this study. However, no experiments are known to the author in which the effect of range condition on animal production has been studied under a grazing management regime designed to secure maximum sustained animal production from that range condition category.

Range deterioration caused by one kind of herbivore may produce a better range for another kind of grazing animal. Usually, however, simplification of the flora and the decline in favorability of all of the dependent factors results in a poorer habitat for all animals and increases the degree of competition between different animal species. The insect and small animal populations are drastically changed by range deterioration. Often, pests are increased and compound the problem of range deterioration.

3. REGRESSION INDUCED BY OTHER DISTURBANCES

Induced regression may result from disturbing factors other than grazing by livestock or wildlife. Haying too frequently, burning too frequently, shifting cultivation, trampling by people, use of wheeled vehicles on unstable sites, and industrial pollution are some of the other important causes of range deterioration.

A large portion of the forests of the world, especially in the tropical and subtropical regions, has been converted to savanna or grassland by shifting cultivation, frequent burning, and overgrazing. For example, much of the central African equatorial forest, almost all of the island of Madagascar, most of the Philippine Islands, and a large part of Brazil that was originally forested is now in native derived pasture extensively managed, a kind of range (Bartlett, 1956). Practically all of India is either cropland or extensively managed derived pasture (Dabadghao, 1960). In general these lands have high potential, but are terribly depleted.

F. Induced Progression

1. GENERAL CONCEPTS

Induced regression due to grazing is continuing on many ranges. However, many millions of acres are improving through secondary autogenic progression or induced progression. Secondary autogenic progression occurs when the disturbing agent is removed. Resting the range utilizes this process to achieve improvement. Progression may be induced under managed grazing by manipulating one or more of the dependent and/or the controlling factors of the ecosystem (Table 5) in order to improve the efficiency of energy utilization (W. A. Williams, 1966). In general the manipulation of the dependent factors is relatively cheap and easy, the manipulation of the controlling factors is relatively expensive and difficult (Dyksterhuis, 1958). Induced progression is diagramed in the model (Fig. 7) as a series of increasingly larger blocks representing time periods measured in years or decades in which the wedge representing consumer organisms is small in comparison with vegetation and the other dependent factors. Thus, more energy is stored than is removed from the ecosystem by virtue of exclusion from grazing, reduced stocking rate, alteration of

TABLE 5

SOME MANIPULATIONS OF RANGE ECOSYSTEMS FOR INCREASED EFFICIENCY
OF ENERGY UTILIZATION

Factors	Manipulation
Controlling factors	
Climate	Weather modification, burning
Geological materials	Water spreading, land leveling, terracing, fertilization, groundwater recharge, drainage
Available organisms	Species elimination, introduction, and improvement
Dependent factors	
Consumers	
Native biota	Grazing management, wildlife management, insect and rodent control
Livestock	Grazing management, livestock management
Vegetation	Plant control, hay management, plant disease control, revegetation
Soil	Mechanical treatments, nitrogen fertilization
Decomposers and transformers	Antibiotics, growth stimulators(?)
Microclimate	Shades, shelters, mulch manipulation
Human factor	Objectives, management, use of goods and services, economics

season of use, or other manipulation to permit increased energy fixation. This diagram is oversimplified, but is consistent with Margalef (1963), who said ". . . succession is simply the exchange of an excess available energy in the present for a future increase of biomass." As energy storage proceeds the favorability of all of the dependent factors and the energy flux increases through time until a new equilibrium is reached. The series of blocks representing induced progression is shown progressively displaced to the left to show that range improvement does not simply reverse the pathway of regression.

2. RANGE EVALUATION

The first prerequisite for planning a range improvement program is a detailed knowledge of the kinds of range ecosystems which are involved and their present status with reference to their potential. In general, these requirements are met by classification into range sites and range condition classes.

a. Range Site Classification. Natural pasture ecosystems may differ at natural equilibrium because they developed under different climates, on different parent materials, with different initial configurations, in areas with different available organisms, for different periods of time. The dependent factors of the ecosystem are patterned along environmental gradients determined by the controlling factors. Once the relationship between the community and its habitat is understood, the nature of the climax community can be inferred from a knowledge of the habitat, even when the community is destroyed or changed by severe disturbance. This is the basis of the range site concept used by various federal agencies. "A range site is a distinctive kind of rangeland that differs from other kinds of rangeland in its potential to produce native plants" (Soil Conservation

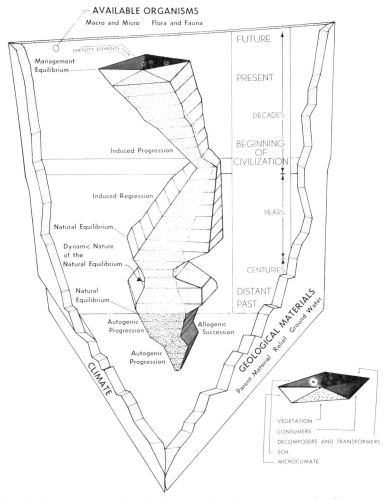

FIG. 7. Three-dimensional graph of a natural pasture ecosystem showing progression induced by range management culminating in the management equilibrium.

Service, 1967). Range sites are distinguished by differences in the total annual biomass accumulation measured at maturity of the most abundant species, and the kinds and proportions of plant species in the climax. Ecosystem behavior under various stresses should also be considered in distinguishing range sites. Range sites are characterized by measurements of pertinent climatic, edaphic, and topographic features. Habitats and communities exist as a continuum patterned along environmental gradients. Range sites, therefore, subtend arbitrary segments of these gradients and are defined on the basis of requirements for planning the use of range ecosystems. Two or more range sites may occur intermingled as a mosaic in the same mapping unit and thus be represented on planning maps as a complex (Aandahl and Heerwagen, 1964). For examples of the application of the range site concept, see K. L. Anderson and Fly (1955), Tomanek (1964), and Johnston *et al.* (1966). The range site concept has been developed for the grazing management of domestic livestock on natural pastures. In some cases modification will be needed when uses other than livestock grazing are emphasized.

   b. Range Condition Classification. Although a large number of range condition classification schemes have been devised, the two most widely used are those initially described by Dyksterhuis (1949) with the Soil Conservation Service and Parker (1951, 1954) with the U.S. Forest Service. Federal agencies have made various local and in-service modifications of these methods. "Range condition is the present state of vegetation of a range site in relation to the climax community for that site" (Soil Conservation Service, 1967). The U.S. Forest Service method relates range condition to vegetation types and subtypes instead of range sites. The ecological site method developed by Dyksterhuis uses estimates of botanical composition on a dry weight basis of different plant species which are classified on the basis of their response to heavy grazing as decreasers, increasers, or invaders. The Parker three-step method uses frequency of occurrence in a ¾-inch loop of different plant species classified as desirable, intermediate, or undesirable and supplemented with other measurements of vegetational and soil surface characteristics. The objective is more the measurement of long-term trends using permanent transects than determination of current range condition. In the ecological site method, relics of climax by kinds of range sites are used and departures of current vegetation are calculated. Composition of climax vegetation must sometimes be interpolated or extrapolated from known relics along environmental gradients. Both approaches use relics of climax for inference of range condition classes but neither uses climax vegetation as the goal of management. Management equilibrium would be a more useful standard and departures measured by distance measures such as the D^2 statistic (Rao, 1952), discriminant functions, or other statistical procedures should be more satisfactory for range condition evaluation in the future, especially where uses other than livestock grazing are important.

c. Other Measurements. Measurements of degree of use, forage residue, grazing pressure, net primary productivity, short-term trend in range condition, and vegetation change as well as many other measurements of both controlling and dependent factors may be useful for planning.

3. MANIPULATION OF DEPENDENT FACTORS

The dependent factors which can be manipulated to induce progression in range ecosystems are the various consumer organisms, vegetation, soil, microclimate, and decomposer-transformer organisms.

a. Manipulation of Consumer Organisms. Herbivorous organisms are natural components of most range ecosystems and under proper management are beneficial for ecosystem function by stimulating growth of some plant species, increasing the rate of nutrient cycling, reducing mulch accumulation on wetter ranges, disseminating and planting seed, as well as by providing a food base for man and other predators (Heady, 1967; compare Ellison, 1960). A herbivore may also condition the environment for other herbivores and so increase the complexity of the niche structure, which in turn increases the biomass and the energy flux. In the range ecosystem livestock and game animals should be managed for optimum production while inducing maximum production of the kinds of vegetation which are most valuable for the management objectives. These objectives are met by grazing management, livestock and game management, and the management of other wildlife.

i. Grazing management to induce progression. Grazing management is the central portion, the heart and core of range management, and is one means by which most ecosystem factors can be manipulated. It determines the kinds of plants that are grazed and the degree, time, frequency, and uniformity of grazing, which in turn require a knowledge of the value of range plants for different uses, the autecology of the desirable and undesirable plant species, and the characteristics, requirements, and behavior patterns of the grazing animals. Grazing management is accomplished by selecting the number, the kinds, and the proportions of grazing animals and controlling their distribution over the range in space and time. Grazing management is basic to the success of other range management practices and must be carefully integrated with them.

(*a*) Kinds and proportions of grazing animals to induce progression. The proper kinds and proportions of grazing animals are determined by the characteristics of the plant community, food and area preferences of the different animals, season of grazing by each, economic and social values, and, on private lands, the size of the operation and the managerial ability and interests of the operator. Many studies have reported qualitative differences in grazing habits and preferences for cattle, sheep, goats, horses, camels, swine, poultry, and different species of wildlife, including jackrabbits and grasshoppers. In range areas where there is much heterogeneity of habitat and forage, common use grazing is highly desirable because different species of grazing animals have different orders of forage preference and habitat preference, resulting in reduced competition and higher stocking rates without overutilization. However, careful attention must be paid to degree of use and trend in range condition to avoid damage. All herbivores at all seasons must be included in range management planning. Overutilization forces more direct competition between grazing animals.

In the early days of the western range competition between sheep and cattle caused many range wars, yet studies have shown that sheep and cattle have somewhat different forage preferences even on very homogeneous areas (Van Dyne and Heady, 1965; Bedell, 1968). On Utah summer range well-suited to cattle use, Cook (1954) used utilization data from single-use pastures to calculate that with cattle and sheep together twice as many animal units could be run as with sheep alone or 1.1 times as many as with cattle alone without overuse. Today cattle and sheep are run together on many ranches, especially in the Northern Great Plains, and cattle, sheep, and goats are grazed together on most Edwards Plateau ranches (Gray, 1968, pp. 36–57).

Most conflicts between kinds of large grazing animals have arisen between livestock and big game, especially where both populations are migratory and seasonal ranges limit the size of the population. For example, in the Intermountain Basin where the foothills are winter range for deer and spring-fall range for cattle, severe conflicts have developed in spite of the fact that deer and cattle have widely different forage and habitat preferences. Julander (1955) lists three principles to determine optimum proportions of cattle and deer: ". . . 1) proper use of perennial grasses limits cattle stocking, 2) proper use of forbs and browse palatable to deer limits maximum deer stocking, and 3) proper use of forbs and browse palatable to both kinds of animals limits maximum combined deer-cattle stocking."

However, optimum proportions of game and livestock are also determined by economic and social pressures. On privately owned land, even though more animal units can be carried with common use grazing, if game values are not an income source, the rancher cannot be expected to manage for game. Hunting leases are an important source of income in some areas, such as the Edwards Plateau where Ramsey (1965), using data from the Kerr Wildlife Refuge, calculated average net income for 6 years of $37.42 per animal unit from deer and $10.08 per animal unit from livestock. However, deer numbers were limited by severe competition with cattle, sheep, and goats in areas in poor range condition even under light grazing by livestock (McMahan and Ramsey, 1965). In the same area with higher range condition using a four-pasture rotational deferment system, deer numbers are increasing in pastures grazed by cattle, sheep, and goats (Straub, 1962). In areas where private and public lands are intermingled, hunting leases are difficult to sell. In these areas public compensation for game use would result in better range management for game and would ease sportsman-rancher tensions.

The situation with elk is similar to that with deer except that elk are more adaptable in their species and area preferences and because of their large size may cause more damage to haystacks and fences (Julander and Jeffrey, 1964, Skovlin *et al.*, 1968). The pronghorn is a plains and prairie animal that consumes mainly forbs and browse with grass being chosen to any extent only during periods of rapid grass growth (Severson *et al.*, 1968). Yet, most of the antelope in the United States today are produced in the grasslands. Buechner (1950) reported 19% competition between cattle and pronghorns and 33% between sheep and pronghorns under proper use in Trans-Pecos, Texas, but almost 100% competition under heavy grazing by sheep, with the antelope giving way to the sheep. In the Red Desert of Wyoming, Severson *et al.* (1968) found that 1.8 times as many animal units of sheep and antelope could be stocked with common use as with sheep alone or 1.4 times as many as with antelope alone.

The outstanding example of common use, of course, is East Africa where the savanna supports a wild animal population composed mainly of 9 herbivores and 1 carnivore with a biomass of 70,000–100,000 lb per square mile yearlong with no evidence of overgrazing. This is 6 times the biomass of cattle, sheep, and goats supported in the same area under native herding with moderate to severe overgrazing and 3 times the biomass supported under European-type cattle ranching with slight to moderate overgrazing (Talbot and Talbot, 1963; Maloiy, 1965). Petrides and Swank (1965) state that this far exceeds the best livestock ranges and may be the highest on record for a terrestrial ecosystem.

Methods for calculating optimum proportions of kinds of grazing animals for maximum sustained animal production and maximum net returns are given by Cook (1954), Hopkin (1954), A. D. Smith (1965), and Gray (1968, pp. 144–151). However, manipulating kinds of grazing animals may have special values for other purposes such as control of regrowth of oak brush and juniper in central Texas (Magee, 1957) or fringed sagewort in the Northern Great Plains.

(*b*) Control of grazing distribution to induce progression. Distribution of grazing by cattle has been shown to be affected by the factors of steepness of slope, length of slope, distance from water, distance from salt, palatability of forage, and thickness of brush—alone and in combination (Mueggler, 1965; Cook, 1966b). Methods for securing good cattle distribution may be classified as (1) methods to make more areas convenient to the grazing animal through water development, salting stations, trail construction, and opening of brush barriers; (2) methods to entice grazing animals into lightly used areas, such as feeding of supplements, shade placement, water-spreading, burning, fertilizing, and the interseeding of highly palatable forage species; and (3) methods to force the distribution of grazing animals, such as fencing and herding. The more commonly used distribution tools such as salting, water development, fencing, herding, and trail construction were used by the pioneers (Jardine and Anderson, 1919). Modern applications are discussed by R. E. Williams (1954) and Skovlin (1965). A large part of the benefit of rest-rotation grazing systems has come from improved grazing distribution (for example, see Johnson, 1965). Good distribution may be impossible on some ranges with single use. For example, in the Northern Great Plains sheep in fenced pastures tend to graze regrowth wherever possible, make intensive use of ridges near the bedground and areas near water, and tend to avoid ungrazed areas with lush grass. In such cases common use by more than one species is indicated to improve grazing distribution.

Workman and Hooper (1968) compared the economics of different kinds of water developments, trail construction, fencing, salting, and herding to secure cattle distribution on a mountain range in Utah. Salting was the most profitable method, increasing animal unit months (AUM's) of grazing by 18%. However, in some areas salt is ineffective as a distribution tool because of excess salt in the forage or the water. The increased AUM's resulting from riding just barely exceeded the increase required to break even because of the high cost of labor. Because of the high initial cost of construction, cross-fencing to secure distribution was not profitable. Fencing may be profitable when other cheaper methods will not secure good grazing distribution, or where uses other than distribution are considered, such as separation of animals by species, age, sex, or production; separation of different range sites which are difficult to manage together; separation of pastures with special values or with special problems; or subdivision in order to initiate a grazing system. Fences should be placed with careful consideration of pasture size and shape, range site and range condition compatibility, prevailing wind direction, topography, and water location. However, fences are expensive from the standpoint of initial investment, depreciation, and repair. Consequently, they should be used only where needed rather than being used to secure grazing distribution when other, cheaper methods are effective. Needed water developments on Utah ranches, either springs or wells, returned about 15% on the investment, assuming a 20-year life. Optimum spacing of water developments is a function of animal requirements, costs of water development, value of the increased forage utilization gained, and the effectiveness of cheaper methods to secure good grazing distribution. Consequently, where water development is expensive, optimum spacing is usually inversely proportional to forage production.

(c) Selecting the season of grazing to induce progression. Plants are easily damaged by grazing too soon after the initiation of spring growth. This was recognized by the pioneers (Sampson, 1913; Jardine and Anderson, 1919) and the physiological basis for such damage has been explained. The recommended stage of development of the plant community prior to grazing has been related to the flowering of showy plants (range readiness indicators) in many areas, especially in the mountains. Some plant species are easily damaged by grazing at the early reproductive stage, or when the apical meristem has just been elevated (Booysen *et al.*, 1963; Vogel, 1966). Vogel and Bjugstad (1968) found that the forage yields and spring-initiated tillers of the following year were increased by clipping little bluestem, big bluestem, and yellow Indian grass after dormancy.

Since vegetation is most nutritious for grazing animals when it is actively growing, ranges which are in the desired range condition class should be used when the vegetation is most suitable for grazing. Where diverse vegetation types with different seasonal values occur on the same ranch or administrative unit, the grazing of each at the optimum time will permit higher stocking rates with better nourished animals. This method of grazing is widely practiced and has been termed the "seasonal suitability grazing system" by Valentine (1967). However, fully used seasonal ranges may need a periodic change of season of use in order to maintain the vigor of the key management species. When ranges are depleted, deferment of grazing during the major period of growth of the key management species will induce progression. For example, on Northern Great Plains ranges where western wheatgrass is a key management species, deferment of grazing from green-up until mid-July is nearly as effective as complete rest for securing range improvement. In occasional years with good fall precipitation, fall deferment is also needed.

Season of use is especially important on ranges with a long growing season where different plant species grow at different times of the year. Important vegetation changes can be induced on such ranges even with the same kind of grazing animal, the same season of use, and the same stocking rate by varying the grazing pressure during the season. Drawe

TABLE 6
CLASSIFICATION OF GRAZING SYSTEMS[a]

I. All subunits grazed one or not more than two periods per year	
A. Continuous occupation by grazing animals during the same grazing season each year	Continuous grazing
B. Grazing deferred on one or more subunits	
1. Deferment not rotated	Deferred grazing
2. Deferment rotated	
a. Grazed subunits continuously occupied	Rotational deferment
b. Grazing rotated in subunits	Deferred rotation
II. One or more subunits rested	
A. Rest not rotated	Rest
B. Rest rotated	Rest rotation
III. All subunits grazed more than two periods per year	Rotational grazing

[a] Variations within these categories are discussed in the text.

and Box (1968) reported widely different forage ratings (% utilization × % frequency of use × % cover) of the same species for deer and also for cattle at different sampling dates. Opportunities to induce progression by manipulation of season of use and grazing pressure with different kinds of grazing animals in such range ecosystems are tremendous.

(*d*) Grazing systems to induce progression. Grazing systems were among the first range improvement practices studied (J. G. Smith, 1899). The kinds of grazing systems which have been proposed since that time are legion and the terminology is diverse. In the hope of simplification, the classification shown in Table 6 is suggested. Continuous grazing is continuous occupation whether the same or different animals are present and thus includes continuous stocking, fixed stocking, and set stocking (Spedding, 1965a). Variations included in continuous grazing are year-long and seasonlong grazing as well as seasonal changes in stocking density, and kind or class of grazing animal. The widely used seasonal suitability system of Valentine (1967) is continuous grazing on pastures fenced by range site and/or range condition class, so that they can be grazed when they are most suitable. Deferment of grazing is delayed grazing, usually until after seed maturity of the key management species. In rotational deferment, deferment is rotated from pasture to pasture in different years while the remainder of the area is continuously grazed. In deferred-rotation grazing, deferment is also rotated from pasture to pasture in different years, but the season of use is rotated on the remaining units. Rest is complete nonuse for a year or more. Rotation grazing involves alternating periods of grazing and nonuse, with each subdivision being grazed more than two times per season. Rotation grazing includes strip grazing, zero grazing, rational grazing, and some forms of alternate grazing. In general these definitions and the classification in Table 6 are consistent with the Range Term Glossary (American Society of Range Management, 1964), Spedding (1965a), and E. W. Anderson (1967a,b).

Specialized grazing systems (systems other than continuous grazing) usually require additional water developments and fencing into pastures with approximately equal potential stocking rates. However, some rotational deferment and rest rotation schemes do not require equal size units, although equal size units simplify planning. Important variations in specialized grazing systems include concentration of grazing in few pastures at a time or dispersal of grazing over many pastures at a time; variation in length of periods of grazing and nongrazing; criteria for determining times of use and nonuse, such as calendar date, degree of use, amount of herbage residue, morphological and/or physiological state of the key management species; provision for more than one herd; flexible stocking rates; creep grazing; forage conservation; use of fertilization, irrigation, or other practices in addition to the grazing system; or the inclusion of leased grazing at definite seasons, tame pastures, crops, and dry-lot feeding. The use of burning in connection with specialized grazing systems has been a very important variation in the African savannas (West, 1955; Heady, 1960) and in the open pine forests of the southern United States, where grazing is controlled in a rotational deferment grazing system by burning subunits in late winter, spring, and midsummer with rotation of the time of burning. Cross-fencing is not required since animals concentrate on the new burns. Relatively high quality forage is maintained during a large part of the year (Duvall, 1969). In most studies grazing by wildlife has not been controlled, but has been measured in some experiments.

Most experiments have shown no advantage in livestock production for a specialized grazing system over continuous grazing, while many studies have shown the definite superiority of continuous grazing (Heady, 1961; McIlvain and Shoop, 1969). Seasonlong or yearlong grazing at a reduced stocking rate has also been shown to be an economical range improvement practice on ranges where the most important species at the desired range condition class and the most important range forage species after deterioration are the same. Examples are blue grama-dominated ranges on medium-textured uplands in the 10–14-inch precipitation zone of eastern Colorado (Klipple and Bement, 1961) and western wheatgrass–green needlegrass-dominated ranges on heavy clay soils with poor surface structure in western South Dakota (E. M. White and Lewis, 1969; Nichols, 1969).

The results of a number of experiments and a large body of rancher experience have shown faster range improvement with specialized grazing systems than with continuous grazing. However, in many reports the improvement may have been induced by other types of management associated with, but not a part of the grazing system, such as better grazing distribution, proper stocking rates, plant control, etc. Confining animals on smaller units also facilitates management, especially at breeding and at calving or lambing. As range improvement occurs and forage production is increased, increased animal production per acre can be expected. However, very few research studies have shown that a grazing system induced progression and also improved individual animal performance. One striking example is a four pasture rotational deferment system at the Ranch Experiment Station, Sonora, Texas, in the western Edwards Plateau in which each pasture is grazed by cattle, sheep, and goats at 43 animal units per section for 12 months and then rested for 4 months with each rest occurring at a different 4-month period (Straub, 1962; U.S. Forest Service, 1967). Under the rotational deferment system range improvement occurred and stocking rates were increased under full use during drought. Also, both animal gain per head and per acre were superior to continuously grazed pastures stocked at 32 or 48 aninal units per section per year. Furthermore, 10 to 12 animal units of deer used the rotational deferment pastures while none were present in the continuously grazed pastures stocked at 48 animal units per section. Considering moderate continuous grazing as 100%, average net returns per acre for 7 years were 61 and 56% for light and heavy continuous grazing, respectively, and 142% for rotational deferment. Corresponding stocking rates were 32, 16, 48, and 43 animal units per section (Merril, 1969). Comparison of a two-pasture switchback system and a three-pasture rotational deferment with continuous grazing at the Texas Experimental Range near Throckmorton showed that the two rotational deferment schemes induced progression and gave better animal gains both per head and per acre than continuous yearlong grazing. These three systems were all tested in areas with long growing seasons and involved continuous grazing for periods of 3 months to a year with each period of nonuse lasting at least 3 months.

The primary purpose of grazing systems involving deferment and rest is to minimize the damage resulting from selective grazing by providing an opportunity for the forage species which are most palatable and productive at each season to become healthy, vigorous, and competitive (Hormay and Talbot, 1961). The purpose of these grazing systems has often been thwarted by resting or by deferring too large a portion of the range which results in sometimes severe overutilization of the grazed portion. Severe overuse may also result from a lack of flexibility in the criteria for moving grazing animals from one pasture to another, especially in areas with erratic herbage growth due to highly variable growing season precipitation. Overuse may cause more damage than subsequent rest periods can overcome. It is especially damaging when it coincides with critical periods in the life cycle of the desirable forage species, such as early spring growth, growth following drought dormancy, shooting of seedstalks, or elevation of the shoot apex on vegetative shoots. Livestock concentration may aggravate parasite and disease problems. Furthermore, forage intake by grazing animals is reduced when the amount of available forage falls below some asymptotic value (Arnold, 1964) or when only unpalatable forage is available (Raymond, 1966). Likewise with severe utilization there is less opportunity for selection, nutritional value of the diet is reduced, and animal production is lowered. This effect is likely to be greater with young, weaned animals than with nursing animals (Spedding, 1965b); is likely to be greater with sheep than with cattle; and will be least evident with mature, low-producing animals such as dry females. When overutilization occurs wildlife habitat is endangered and watershed values are lost.

The purpose of deferment may also be thwarted by too short a period of nonuse. In general nonuse of at least 3 consecutive months geared to the physiology of the key management species is desirable. However, some grazing systems used in South Africa provide for periodic nonuse from the flowering of a key management species until the seed is ripe (Scott, 1955). An inherent undesirable feature of deferred rotation and rest rotation systems where grazing is rotated within a season is the loss in forage nutritional value that occurs with increasing maturity. Furthermore, if the system involves frequent moving of animals, disturbance may reduce production. Careful planning and management are required to minimize these undesirable features; ". . . it is pertinent to emphasize: 1) know the resource; 2) tailor the grazing system to fit the resources, not to a standardized format; 3) allow for flexibility so that needed adjustments can be made in time to comply with unpredicted changes in weather, markets, and so on" (E. W. Anderson, 1967b).

Rotation grazing systems are designed to stimulate production of nutritious forage, as well as to maintain a desirable balance of forage species. Special attention is given to the promotion of regrowth and tillering which require favorable conditions of light, temperature, soil moisture, and soil nutrients. Definite intervals of nonuse are essential for high production (Ivins, 1966). However, the length of nonuse periods should be variable depending upon rate of regrowth and, ideally, periods of grazing should be short enough to avoid the grazing of new regrowth during the same grazing period (Voisin, 1959). Grazing periods should also be planned so that shoot apices are removed when lateral buds are sufficiently mature, so that tillering is induced without a period of dormancy (Teel, 1956, as reported by D. Smith, 1962; Bridgens Bridgens, 1968). Rotation grazing systems have been used successfully on tame pastures and derived pastures in humid and subhumid climates (Voisin, 1959, 1960; McMeekan, 1960) and should be applicable to natural pastures in these regions as well as to those in range sites with favorable moisture relationships such as wetlands, subirrigated and overflow. On upland ranges in semiarid regions, where there is a pronounced rainy season, there may be a place for rotation grazing coupled with systems involving deferment and/or rest.

Larin (1960, pp. 94–165) described grazing systems for natural pastures in the forest, forest-steppe, steppe, semidesert, and desert zones of the USSR. Except in the desert, Larin recommended a combination of rotation and deferred rotation grazing. One plan for the steppe region called for five fields with six enclosures per field with one fifth of the area grazed three times during the summer and the remainder grazed only once. Each field was grazed three times per summer only one year in five. For the other four years it was grazed once, but at different seasons. On depleted pastures, a year of rest was recommended. On desert ranges deferred-rotation grazing was recommended.

On the same ranch or administrative unit year round grazing management may involve integrated grazing of forest openings, tame pastures, and forage crops, as well as natural pastures on various range sites and in various range condition classes. Each of these kinds of grazing lands may require different grazing systems for best utilization. The planned combination of natural pastures and tame pastures has been referred to as complementary grazing systems by Lodge (1963).

As the morphological and physiological characteristics of the key management species become better known, hopefully, grazing systems can be designed which will harmonize the animal and forage requirements and so induce rapid progression and increase energy flux through both primary producers and primary consumers.

(*e*) Control of stocking rate to induce progression. The number of grazing animals on a given area for a given time period is probably the most important single factor influencing energy flow in range ecosystems, yet a simple statement of stocking rate has little meaning until the kind and proportion of grazing animals, grazing distribution, season of use, and the grazing system are specified. The proper stocking rate depends upon the range site (including climatic, topographic, and edaphic factors), range condition, phase of the climatic cycle, current weather, unmanaged herbivores sharing the forage, and the objectives of management (compare Stoddart, 1960). The objectives of management may include rapid progression (resulting in a light stocking rate or even nonuse coupled with other practices), slow progression, or full use at the present range condition class without further improvement. The objectives of management might include plant control through grazing management, resulting in a high stocking density of a particular kind of animal at a specific time when the undesirable plant species is most susceptible to grazing damage. Generally, however, the management objectives are concerned with maximum energy flow.

An average initial recommended stocking rate can be estimated by direct comparison with a known area properly stocked or by indirect comparison using (1) range surveys (Brown, 1954), or (2) a regional stocking rate table with values interpolated from grazing experiments, conditioned by rancher and range technician experience. For those using the ecological site method of rating range condition, average initial recommended stocking rates are tabulated by range site and range condition, while values are tabulated by vegetation type and range condition within a region for those using the Parker three-step method. However, the average initial recommended stocking rate is only a guide and must be adjusted for any particular situation guided by degree of use and trend in range condition with full consideration for other uses of the range ecosystem. The degree of use which is optimum for a given environment, a certain management system, and a specific set of management objectives is known in a general way (Hedrick, 1958). However, the low or unknown accuracy of presently available methods of measuring utilization (Brown, 1954; National Academy of Sciences, 1962), as well as the variable nature of range ecosystems make precise recommendations based on degree of use impossible. Residue remaining at the end of the grazing season on a specific kind of range has been used as a guide to the proper stocking rate (Bement, 1969). Guidelines for determining short term trend in range condition are given by Ellison et al. (1951) and a widely used procedure for measuring long-term changes in range condition is described by Parker (1951, 1954). Other methods for measuring vegetation change are also available (Brown, 1954; National Academy of Sciences, 1962; Strickler and Stearns, 1962).

The proper stocking rate is not a constant but changes as the factors affecting it change. The most important variable affecting proper stocking rate is precipitation. Recurrent wet and dry periods which require adjustments in management are characteristic of most range ecosystems, but as yet cannot be successfully forecasted. Equations have been developed for predicting forage production several months in advance (for examples, see Sneva and Hyder, 1962; Dahl, 1963; Currie and Peterson, 1966). These equations can be useful in range planning to minimize drought effects (Gray, 1968, pp. 401–407). If wet and dry years occur intermingled, conservative stocking rates, stored or purchased feeds, and cash or credit reserves are adequate to minimize drought effects. However, in longer droughts and in areas characterized by the bunching of several years of above or below normal precipitation, such as the Northern Great Plains, these practices are inadequate and a premium is placed on flexibility in stocking rate and in ranch organization. For example, in wet cycles a cow-calf-yearling steer-deer operation may be used, whereas in dry cycles yearlings may be eliminated, the cow herd closely culled, and deer populations reduced to keep animal numbers in balance with the forage supply. In severe drought it may be more economical for the rancher to disperse the herd and borrow money to live on rather than to keep the herd intact, borrow money to buy feed, and induce severe range regression. Drylot feeding of reproducing animals on hay, silage, or even on all-concentrate rations for extended periods of time is possible (Thomas and Durham, 1964; Schuster and Albin, 1966) and may be preferable to either herd dispersal or overuse of the range. Movement of livestock to nondrought areas may be economical also. Use of emergency forages such as Russian thistle, ground mesquite, burned pricklypear, etc., have helped avoid disaster in some situations. Decision theory may be a useful planning tool for adjusting ranch operations to a fluctuating forage supply (Rogers and Peacock, 1968). Regardless of the care exercised in management, occasional overuse will occur and in some cases may be preferable to a reduction in livestock numbers with temporarily low livestock prices. However, moderate overuse for one year is usually not serious, especially on pastures in high range condition when followed by compensatory management such as deferment, rest, or a reduced stocking rate.

ii. Livestock and game management to induce progression. Livestock management is the domain of the animal scientist and game management that of the wildlife scientist. Both are concerned with the characteristics and requirements of different kinds of animals and how to study them. They are concerned with reproductive efficiency, growth and survival of offspring, disease and parasite control, nutrition, and management for more efficient animal production.

Since animal scientists have more control over the animals they manage, they are intimately concerned with animal breeding, involving mating systems and the selection of genetically better animals, thus increasing the frequency of superior genes in the population. They are also concerned with animal nutrition, involving the requirements of different kinds of animals for various nutrients at different stages of the life cycle for different kinds of production; the characteristics of animal feeds and how to combine them to meet these requirements economically; manipulation of the endocrine system; and the use of chemotherapeutic agents so that production is maximized. Animal scientists are involved in manipulating animal physiology to increase reproductive efficiency and control animal behavior. Study of the characteristics of meat, milk, wool, hides, and various other products, how they are affected by various manipulations, and how they can be marketed most efficiently are all very important aspects of livestock management. In the future game managers will be more concerned with these matters. The strong kinship between livestock and game management should be apparent to all.

The range ecosystem manager is especially interested in the nutritional characteristics of the basal diet selected by the grazing animals; how these characteristics are affected by kind of range, range site, range condition, season of the year, seasonal weather, and various range management practices; how the basal diet can be supplemented economically for optimum production; how plant poisoning can be minimized; and how emergency feeds can be used most effectively when range forage is not available because of snow cover, hail, fire, drought, management decisions, etc., so that energy flow through the most desirable primary consumers can be maximized.

The development of the esophageal fistula technique (Van Dyne and Torell, 1964) for obtaining samples of diets actually selected by grazing animals has opened the door for a more thorough investigation of animal diets (Cook, 1964). The technique has been used in Australia (see McManus *et al.*, 1968) and East Africa (Bredon *et al.*, 1967) as well as in many parts of the United States. Collection of diet samples by rumen evacuation is preferred by some workers. In this method the contents of the rumen and reticulum are removed thoroughly. The animal is allowed to graze, the sample is removed and then the original contents are replaced (Harris *et al.*, 1967). Botanical composition of the samples has been determined successfully using plant frequency with a low-power binocular microscope on fresh material (Heady and Van Dyne, 1965) and on dried, ground material using frequency of occurrence of epidermal tissue in 20 microscope fields on each of five slides with 125-power magnification. Density was calculated from frequency and relative density was highly correlated with percent by weight ($r^2 = 0.97$ to 0.98: Sparks and Malechek, 1968). A similar method has been used on feces to qualitatively study the food preferences of wild herbivores in East Africa (D. R. M. Stewart, 1967) and to study grasshopper diets (Brusven and Mulkern, 1960). Development and refinement of *in vitro* digestion procedures (Tilley and Terry, 1963; Dent *et al.*, 1967) permit rapid evaluation of the nutritional value of material actually selected by the grazing animal. *In vivo* microdigestion, using nylon bags suspended in the rumen, is simpler in operation but results have been less satisfactory thus far (Harris *et al.*, 1967). Van Dyne and Meyer (1964) developed a method for determining the dry matter intake of grazing animals using (1) *in vitro* cellulose microdigestion of esophageal fistula samples adjusted by regression to a live animal basis, (2) fecal cellulose percentage, and (3) weight of feces excreted. Dry matter intake was calculated by dividing the weight of cellulose excreted by the percent cellulose not digested. If animals with both esophageal and rumen fistulas are used and the inoculum for digesting the esophageal sample is taken from the same animal that ate the forage, then forage intake can be calculated on an individual animal basis. This procedure is difficult to use in range environments because it requires the use of fecal collection bags. The technique is especially difficult with female animals, and is probably impossible with many wild animals in their natural environment. Further work is needed to obtain greater precision in the use of external and internal indicators to measure fecal output and digestibility of grazing animals without fecal collection bags (Harris *et al.*, 1967).

The techniques described above can be used to sample the basal diet as a guide to supplementation; to compare the amount and the nutritional value of forage grazed on different range sites in different range condition classes at different seasons; to determine the effect of different range manipulations on energy transfer from primary producers to primary consumers; and to study the degree of competition between different herbivores.

iii. Management of other native fauna. The interrelationships of the other native fauna, including invertebrate animals (grasshoppers and other insects, spiders, etc.), small mammals (especially rabbits and rodents), and carnivores, with other factors of the range ecosystem are very complex (Ellison, 1960), involving range manipulations, prey–predator relationships, disease, weather fluctuations, and other factors with complex interactions and lag effects (Watt, 1968, pp. 133–168). The diets of several small herbivores have been studied, including prairie dogs (Koford, 1958; R. E. Smith, 1958), jackrabbits (Currie and Goodwin, 1966; Bear and Hansen, 1966; Sparks, 1968), ground squirrels (Howard *et al.*, 1959), pocket gophers (Hansen and Ward, 1966), and grasshoppers (Mulkern, 1967). These small herbivores often compete rather directly for forage with livestock and game and may become extremely abundant, especially when their predators are reduced and eliminated. They may also increase if range deterioration improves the habitat for them as it often does. However, extreme care must be used in manipulating these components to keep from upsetting the natural balance, or introducing toxic residues into the ecosystem. Care must be taken to avoid the eradication of endangered species, such as the black-footed ferret in the Northern Great Plains (Henderson *et al.,* 1969).

Pest control through direct manipulation is often uneconomical, especially on low-producing range sites. Biological control offers great promise for some pest problems, but must be handled with caution (Watt, 1968, pp. 59–69). Insects play an important role in the decomposition of carrion (Payne, 1965) and probably in most other types of organic matter decomposition. The effects of insecticide stress on plant–arthropod–mammal relationships have been studied in a simplified grain crop ecosystem by Barrett (1968). Similar studies are urgently needed on range ecosystems. Burrowing rodents may play a useful role in loosening the soil. While no data are available, it may be necessary to provide periodic mechanical soil treatments to replace the rodent effect which was present in natural equilibrium, in order to keep certain range sites in high production.

Manipulation of consumer organisms lies at the heart of range ecosystem management and must receive first consideration if the manipulation of other ecosystem components is to succeed. Management involving the manipulation of other ecosystem factors is considered in less detail in the following pages.

b. Manipulation of Vegetation. Direct manipulation of range vegetation to induce progression is accomplished mainly through plant control, range seeding, and disease and parasite control.

i. Plant control to induce progression. A very high degree of plant control in suppressing some plant species and encouraging others can be achieved through grazing management by the careful selection of kind of grazing animals, season of use, and grazing pressure. Additional plant control may be economical and may even be required (1) when long-lived highly competitive plant species (such as sagebrush, mesquite, juniper, or oak) occupy the site, slow the rate of progression induced by grazing management, and prevent the management equilibrium from being reached in an economically short period of time; (2) to replace the original fire effect in the transition zone between grass and either shrub or forest ecosystems so that the area of grassland, savanna, or open forest is maximized; (3) on pastures derived from forest ecosystems to prevent succession to forest and so maintain the ecosystem as a derived pasture; (4) to reduce the abundance of poisonous or injurious plants; and (5) to shift the balance of plant populations to achieve a vegetation which is most desirable for the objectives of management. Chemical methods of control are most widely used followed by mechanical treatments, prescribed burning, and rarely biological control.

The control of introduced prickly pear in Australia by the moth borer, *Cactoblastis cactorum,* and the control of St. Johnswort in California by the leaf-feeding beetles, *Chrysolina hyperici* and *C. gemellata,* and the root borer, *Agrilus hyperici,* are classic examples of the biological control of introduced weeds (Crafts and Robbins, 1962). The difficulties in finding the right organisms for biological control are enormous (Huffaker, 1959), yet this is probably the most feasible means of control for many exotic problem plants.

Mechanical methods have been used mainly on trees and shrubs and include chaining, railing, rootplowing, bulldozing, and the use of brushland plows, choppers, beaters, and shredders. Cost and effectiveness vary widely depending on the species being controlled, the equipment, soil, terrain, and time of application. Reseeding after treatment is often required. Prescribed burning has been used effectively and economically for the control of nonsprouting woody plants, especially when coupled with grazing with the proper kind of grazing animal at the proper season (West, 1965; Box *et al.,* 1967).

The literature on the use of selective herbicides for plant control has been reviewed by House *et al.* (1967), who list 54 pages of references on the use of herbicides with most of the reports from temperate North America; a section on rangeland is included. The use of selective herbicides on rangeland is a very active research area. For example, 12 of the 92 articles published in Vol. 21 of the *Journal of Range Management* were concerned with various aspects of chemical plant control. Almost every state agricultural experiment station has a weed manual (which usually includes many desirable range plants) with recommendations for control or eradication. Selective herbicides are very powerful and useful tools for inducing range progression and improving grazing values for livestock and game, wildlife habitat, watershed, and recreation. By the use of selective herbicides the proportions of different perennial grasses can be changed and the relative abundance of forbs and shrubs in relation to grass can be increased or decreased. However, good forage plants may be eliminated by the same sprays used to control undesirable vegetation. For example, control of big sagebrush with 2,4-D and 2,4,5-T may also kill palatable browse and associated forbs, thus losing valuable forage (Wilson, 1969) and perhaps resulting in the destruction of food supplies for other animals such as the sage grouse (Klebenow and Gray, 1968). Changes in populations of primary producers can be expected to cause changes in populations of various kinds of herbivores.

Sometimes great ecological wisdom is needed to determine the kind of vegetation which is best for the goals of management and to select the time and the methods of control to achieve the balance of plant populations which are desired. Critical evaluation of the existing literature from an ecological standpoint with interpretations presented through mathematical models is urgently needed. Thorough long-term studies of the effect of various herbicides on the function of different kinds of ecosystems should be initiated. Research should be intensified on developing herbicides that are more selective and that can be used with greater precision in the "fine-tuning" of range ecosystems.

ii. Range seeding to induce progression. Range seeding may be required (1) where range regression has proceeded so far that desirable range plant species or desirable biotypes of species are so rare that the management equilibrium cannot be reached in an economically short period of time by manipulating the other ecosystem factors; (2) when the area has been denuded by cultivation, flooding, etc.; (3) when a special use native pasture is needed; and (4) to speed the development of a derived pasture from chaparral, savanna, or woodland. Range seedings may be classified as broadcast, full seeding, or interseeding.

Broadcasting seeds of grasses and legumes on unplowable land has been successful in California and Israel by prescribed burning, then seeding into the ash at a time of high rainfall frequency, followed by chemical control of competing brush (Naveh and Ron, 1966; J. R. Bentley *et al.,* 1966). In humid climates oversowing with grasses and legumes into existing sod may be successful if competition is reduced by close grazing or chemical suppression (Cullen, 1966; Winch *et al.,* 1966), or with close grazing and hoof cultivation (Kikuchi *et al.,* 1966). However, if possible grasses should be drilled even in humid climates. In semiarid and arid regions chances of success with broadcast seedings are very low.

Success with full seeding in range environments is still not high. Unfavorable climatic and microclimatic factors, problems of seed dormancy, low seed viability, and poor seedling vigor in many native plants, unfavorable soils, seed removal or seedling damage by range fauna, and disease all combine to make range seeding a hazardous venture except on favorable soils in years with favorable precipitation. On most ranges natural seedling establishment occurs only in a series of favorable years (O.B. Williams, 1961). Close attention to the following factors will increase the probability of seeding success, especially in dry climates: adapted species and ecotypes, seed quality, treatment of seed with a fungicide, proper seedbed preparation, time of seeding, seed placement, seeding rate, row spacing, reduction of competition, grazing management, and protection from other fauna (compare Hull and Holmgren, 1964). The best time of seeding is immediately before the time of highest precipitation probability. Planting on summer fallow or into a thin stand of a cover crop with a drill equipped with depth bands, seed agitator, and provision for packing the soil will increase chances of success (Bement *et al.,* 1965). Moisture relationships in dry regions can be improved with an asphalt emulsion mulch (Bement *et al.,* 1961). Vernalization (Troughton, 1960) and pre-planting moisture treatment of the seed (Keller and Bleak, 1969) may improve establishment. Nixon and McMillan (1964) observed edaphic ecotypes in range grasses in Texas and McMillan (1967) found that population samples of grasses that originated in central Texas showed selective superiority during extended drought to population samples collected further away in Texas and New Mexico. These studies lend credence to the recommendation that for range seeding seed should not be used more than 300 miles north or more than 200 miles south of its origin and that it should be used in precipitation zones and on soils relatively similar to those from which it came (Atkins and Smith, 1967). Most range seedings have been made with only one or two native grasses and should be considered as derived pastures rather than native pastures because of the simplified vegetation. Over long time periods productivity and stability can be expected to be less than in natural pastures in excellent range condition. In some cases a large portion of the original vegetation has been incorporated in the new seeding by mulching with late-cut hay from the same range site in excellent range condition. The importance of including browse for game is emphasized by Plummer *et al.* (1968). Seeding with introduced or exotic species is discussed in a later section.

Interseeding (in which the seed is drilled in a tilled strip in otherwise undisturbed vegetation) is used when (1) the environment is severe and there is danger of severe erosion if a full seedbed is prepared, or (2) the objective is to increase the amount of a particular plant species in the stand without destroying the rest of the community (Becker *et al.,* 1957; Schumacher, 1964). To ensure success, management must be based on the interseeded species until they are fully established. The investment in interseeding is lower than in full seeding, but improvement is slower and failures are more frequent.

A range seeding should be permanent and it should not be necessary to include cost of depreciation. Nevertheless, seeding is an expensive practice. In fact, when all expenses are considered (Gray, 1968, pp. 422–427), seeding costs will sometimes exceed the value of the land. Federal cost-share programs often make seedings possible that otherwise would not be economical. A seeding program should be begun only after careful consideration and planning and should begin on good sites where the probability of success is reasonably high.

iii. Parasite and disease control to induce progression. Range plants are subject to a wide range of diseases and parasites, many of which are continually present. Many biotypes and selected strains show genetic resistance to pathogens (Braverman, 1967). However, research efforts have been meager and little is known about control measures which are feasible for range ecosystems. In some cases, control would probably result in large increases in productivity. However, care must be taken not to disrupt the ecological balance of the decomposers. Research in range plant disease and parasite control is urgently needed with special reference to low-cost methods suitable for large areas under extensive management.

c. Manipulation of Soil. Direct manipulation of soil as a dependent factor to induce progression includes various mechanical treatments and nitrogen fertilization.

i. Mechanical soil treatments to induce progression. In many cases mechanical treatments, such as pitting (Barnes *et al.,* 1958; Rauzi *et al.,* 1962), contour furrowing (Branson *et al.,* 1966), ripping or subsoiling (Hickey and Dortignac, 1958), and discing (Thatcher, 1966), can be used along with grazing management to hasten range improvement. Mechanical treatments are of little value on coarse-textured soils, or on very fine clays where the soil surface becomes loose and powdery because of freezing and thawing or wetting and drying (Nichols, 1969). Discing is most applicable on gentle slopes, pitting on slopes up to about 10 or 12% while contour furrows are effective on slopes up to 20% or more if properly designed. To induce range progression effectively contour furrows should be established on the contour and blocked at intervals to prevent gullying. They should be at least 4 inches wide by 4 inches deep and not more than 5 feet apart. Pits should be at least 4 inches deep and should be about 16–32 inches apart. Range improvement is almost directly proportional to the degree of surface disturbance. Deferment from grazing during the growing period of the key management species is essential for best results (Soil Conservation Service, 1962). Pits and contour furrows hold water and if properly designed and constructed should substantially reduce rainfall runoff. In addition, these practices as well as the others listed loosen the soil, increase the rate of water infiltration and movement, improve soil-plant-water relationships, and initially reduce the plant cover, change the botanical composition, and in most cases increase primary productivity.

On depleted ranges in the Mixed Prairie, on medium and fine-textured soils where a highly competitive shortgrass sod has developed, mechanical treatments coupled with cool season deferment or complete rest from grazing usually result in rapid range improvement, due to the increase of western wheatgrass which is a cool season rhizomatous midgrass. However, if western wheatgrass plants are very sparse (less than about one stem per square yard), or if heavy grazing pressure is continued, improvement is very slow and the bare areas are colonized by buffalograss, a warm season stoloniferous shortgrass that increases with range deterioration. In these circumstances, seeding is required for economical range improvement.

In many situations where regression has not proceeded too far, mechan-

ical treatments coupled with wise grazing management will improve
botanical composition and increase energy flow economically without
other manipulations. Where populations of burrowing animals are small,
periodic but infrequent retreatment may be necessary to maintain good
tilth and a high rate of energy flow.

   ii. Nitrogen fertilization to induce progression. Soil nitrogen is a dependent factor of the range ecosystem, and thus increases during autogenic progression, changes with allogenic succession, and would be expected to stabilize at natural equilibrium at a level which would maximize
energy flow under the constraints of the available plant species, the occurrence of periodic drought, and other controlling factors. It is probable
that in pristine conditions soil nitrogen stabilized at a level which was
suboptimum in wet years but did not hinder survival in dry years.

   Nitrogen enters the ecosystem through accession of ammonia and
nitrate in precipitation, animal immigration, soil and plant deposition by
wind or water, and fixation by microorganisms. Symbiotic fixation by
microorganisms in the root nodules of legumes is the most important
source of nitrogen in most ecosystems. While there are about 200 legume
species that are used in crops and tame pastures there are some 12,000
species distributed through range and forest ecosystems. Of those which
have been examined, 89% are nodule-bearing and probably contribute to
soil fertility (W. D. P. Stewart, 1967). There is good evidence that symbiotic nitrogen fixation also occurs in the root nodules of 13 genera of angio-
sperms which are not legumes, most of which are woody pioneers that
grow in areas with low soil nitrogen where they have a competitive advantage. Symbiotic nitrogen fixation also occurs in some plant species
that bear leaf nodules. Furthermore, blue-green algae, free-living nitrogen-fixing bacteria, and possibly some fungi, actinomycetes, and yeasts
are also involved in nitrogen fixation (W. D. P. Stewart, 1967). Any or all
of these nitrogen inputs may be effective in any given range ecosystem.

   Nitrogen leaves the ecosystem by animal emigration, wind- or water-
borne organic matter, soil erosion, leaching, and volatilization. Leaching
is relatively unimportant in arid and semiarid regions, but may be an important source of loss in pastures derived from forest ecosystems in
humid climates or in natural pastures with periodically high water tables
or with run-in water. Volatilization losses may occur due to ammonia
release, chemical or bacterial denitrification of organic nitrogen compounds to free nitrogen or nitrous oxide, burning of herbage and mulch,
and gaseous losses from urine and manure (Broadbent and Clark, 1965).

   Total soil nitrogen is closely correlated with soil organic matter, increases with increasing clay content, and is higher in grassland than in
forest soils. Total soil nitrogen increases with increasing precipitation,

with decreasing soil temperature and increases with any other factor that retards bacterial decomposition. Most of the soil nitrogen is present in organic compounds, many of them complex polymers, and is released very slowly by bacterial decomposition, especially when soil temperatures are low. When soils are warm and moist, mineralization proceeds more rapidly. Flushes of mineralization also occur with wetting following desiccation. Usually, as soon as ammonium ions are released by mineralization, they are absorbed by plant roots or oxidized by nitrifying bacteria to NO_3^-. Nitrate oxidation is usually the fastest process, mineralization the slowest, and oxidation of ammonia intermediate (Alexander, 1965). Nitrifying organisms are consequently limited in number by the available energy supply, especially if the pH is acid (Robinson, 1963). The amount of inorganic nitrogen in natural grassland soils is always small, even though the total amount of soil nitrogen may be quite large. Some workers have attributed the low levels of nitrifying bacteria to inhibitors produced by grass roots (Meiklejohn, 1968). Rice (1964) attributed long enduring stages of secondary autogenic progression on abandoned fields to the production of inhibitors by plants with low nitrogen requirements that maintain the site in a low nitrogen status by inhibiting nitrogen-fixing and nitrifying bacteria and retarding the nodulation of inoculated legumes. Extracts were prepared from 20 plant species and 13 showed inhibition of test bacteria. The inhibitors extracted from *Ambrosia elatior, Euphorbia corollata, Helianthus annuus,* and *Bromus japonicus* have been partially identified (Rice, 1965; Rice and Parenti, 1967).

During autogenic progression, native legumes and other plants have a competitive advantage and consequently are more abundant in early than in later stages of progression when soil organic matter and nitrogen have accumulated. In the Northern Great Plains on weakly developed soils that are low in nitrogen, warm season grasses, such as little bluestem and sideoats grama, are able to compete better than cool season grasses, such as western wheatgrass and green needlegrass, because they grow during the period of maximum mineralization (E. M. White, 1961a). As soil development proceeds the proportion of cool season grass increases while warm season grasses and legumes decrease. Legumes usually occur, but are not abundant, in natural grasslands with fully developed soils. In relics of climax nitrogen losses are small and accessions from precipitation may be sufficient to maintain nitrogen balance (Stevenson, 1965). Legumes probably occur in these relics because they are able to compete by virtue of the type of root system, the season of growth, or the production of inhibitors, rather than by virtue of nitrogen fixation. Since combined nitrogen has been observed to depress nodulation and nitrogen fixation in cultivated legumes (Nutman, 1965), it is possible that native legumes on normal soils may fix nitrogen only in years following disturbance by fire, prolonged drought, or severe grazing so that nitrogen levels drop below some threshold value.

In natural equilibrium net removal of nitrogen was probably not appreciable in large ecosystems, although doubtless there was uneven cycling and irregular distribution of excreta by the large animal populations. Gaseous losses of nitrogen by prairie fires and from feces, urine, and the bodies of dead animals may have been large. During drought cycles or following prairie fires, wind and water erosion could have resulted in losses of nitrogen which were of sufficient magnitude to give some competitive advantage to nitrogen-fixing legumes, permitting more rapid nitrogen accumulation during the recovery period.

On depleted ranges the same losses listed above occur and in addition the nitrogen content of the animal products which are exported is lost. Also with fewer kinds of grazing animals distribution of excreta is probably more irregular resulting in greater losses (Hilder, 1966). Many native legumes are decreasers or high-level increasers that are either absent or too sparse to be an important source of nitrogen on depleted ranges. In water-deficient ecosystems, water is usually the first limiting factor followed by nitrogen. Addition of water increases plant production, but yield is increased more if nitrogen is added. However, when nitrogen is added, yield is increased appreciably only if adequate soil moisture is available.

In general nitrogen fertilizer studies on rangelands have shown the following trends.

(a) Increases in dry matter yield due to nitrogen fertilization are strongly influenced by the amount and seasonal distribution of available soil moisture. Fertilization of natural pastures has generally been profitable from a dry matter yield or a stocking rate standpoint in the humid, subhumid, and in the wetter portions of the semiarid region (see Rogler and Lorenz, 1966). In the drier portions of the semiarid and in the arid region, responses usually have not been economical (see Rauzi *et al.*, 1968b). Many negative results were never published. However, some large responses have been noted in wet years, or in years when soil moisture was favorable, soil temperatures were cool and mineralization was slow. Responses have been essentially nil in drought years, especially when soils were dry during the cool spring period (see Cosper and Thomas, 1962). In warm climates where total amounts of soil nitrogen are small, responses are proportional to soil moisture and time of application is very critical (Stroehlein *et al.*, 1968). In these areas responses are affected very little by mineralization.

(b) With nitrogen fertilization of individual plant species leaf area, top:root ratio, seedstalk production, crude protein content, and palatability to grazing animals are increased; time of maximum growth rate and maximum plant heights are reached sooner (Van Dyne and Thorsteinsson, 1966); soil moisture is used more efficiently (Black, 1968), but is exhausted quicker and drought is thus prolonged and made more severe. The effect of drought has been documented in unpublished data collected by Burzlaff in experiments at Scottsbluff, Nebraska.

(c) When natural pastures are fertilized with nitrogen in the fall or spring, cool-season plants are stimulated more than warm-season ones. Cool-season annual grasses, such as *Bromus japonicus,* make large responses and usually require control which is difficult and expensive. Many weedy annual and perennial forbs, such as *Artemisia frigida,* are stimulated so that control is needed, while others, especially legumes, are decreased. In areas where the key management species (decreasers and high-level increasers) are cool season, such as in the Northern Mixed Prairie, fertilization coupled with rest or cool-season deferment will induce rapid progression. However, when large amounts of nitrogen are used and treatment is prolonged, the warm-season grasses will probably be eliminated by competition for water, light, and nutrients, resulting in an essentially pure stand of cool-season species (Rogler and Lorenz, 1966). This may be desirable if a cool-season pasture meets the management objectives. However, such a pasture would be a derived pasture, not a natural pasture, and under intensive management with periodic fertilization and plant control would be classified as a cultivated ecosystem. Usually if this degree of environmental control is economically feasible, better results would be obtained with an intensively managed cool-season tame pasture. However, in some studies significant changes in botanical composition such as these have not occurred (Black, 1968). In areas such as the True Prairie where the key management species are warm season and some of the less desirable species are cool season, nitrogen fertilization will almost surely result in regression when applied on pastures in high range condition. When applied on pastures in low range condition, yield of cool-season grasses and weeds probably will be increased, but the desirable climax dominants will be suppressed. During drought, death loss of the more mesic plant species can be expected to be higher on fertilized than on unfertilized ranges.

(d) In arid and semiarid ecosystems where leaching losses are minimal, fertilizer nitrogen which is not used during the year of application may accumulate and be used later when soil moisture is available. Recent work at Mandan, North Dakota (Power, 1967, 1968), indicates that a pool of soluble mineral nitrogen can be maintained in soils under smooth bromegrass, thus eliminating nitrogen as a growth factor. This type of fertilizer management, if used in range ecosystems, may cause extensive changes in botanical composition as indicated above. High nitrogen fertilization rates may also contribute to the eutrophication of streams and reservoirs when runoff occurs following snow melt or heavy rains, especially in ponds that are subject to high evaporation losses and infrequent flushing.

(e) Research data are not adequate to determine the effect of fertilizing with nitrogen at different seasons on dry matter yield and botanical composition. However, a pool of mineral nitrogen will probably be developed with prolonged nitrogen fertilization of semiarid or arid ecosystems regardless of the time at which the fertilizer is applied. Once the pool is formed the cool-season grasses would be stimulated and the warm-season grasses depressed unless the competitive ability of the cool-season grasses is retarded by close spring grazing or other means. Surface applications and foliar applications of nitrogen fertilizers in varying amounts carefully timed with soil moisture, plant phenology, and grazing management deserve research attention in range ecosystems.

Legumes may supply needed nitrogen on a very timely basis and avoid some of the problems involved in fertilization. Additional research is needed on the effectiveness of native legumes in the nitrogen cycle, as well as on the ecological problems associated with the use of various introduced legumes, especially grazing-type alfalfas (Rumbaugh *et al.*, 1965). Thorough studies of the effect of nitrogen fertilization or legume culture on energy flow, on nutrient cycling, and on the grazing and detritus food webs are notably lacking in range ecosystems.

d. Manipulation of Microclimate. Manipulation of each of the dependent factors and each of the controlling factors of the ecosystem will affect the microclimate. Grazing management, especially, has been shown to have a profound effect. However, direct microclimatic manipulation is probably not feasible, except for such items as the construction of windbreaks and shades, the planting of shelterbelts, and seeding of tall grasses in rows to reduce wind velocities and hold snow. Mulch has been directly manipulated in some studies, but the treatments were really designed to simulate the effect of grazing.

e. Manipulation of Decomposer and Transformer Organisms. The importance of the detritus food web in energy flow in grassland ecosystems is emphasized by Golley (1960), who estimated that the decomposers in a bluegrass pasture used about 70% of the net primary production. Macfadyen (1963) calculated that 66% of the respiration of soil organisms in an English meadow and 73% of that in a limestone grassland resulted from the activity of decomposers. The soil fauna (including nematodes, annelids, snails, isopods, millipedes, mites, spiders, collembolans, and certain insects) and the soil microflora (including bacteria, actinomycetes, fungi, algae, and lichens) are associated together in the surface and below-ground portions of range ecosystems. The detritus food web is far more complex and difficult to study than the grazing food web. This highly complex, closely integrated system is highly responsive to environmental changes and yet remarkably stable. Direct manipulations to induce progression are probably not feasible. However, care must be taken not to interfere with the orderly functioning of the detritus web by treatments applied to the above ground portions of the ecosystem. The possible effect of pesticides, herbicides, fertilizers, and other chemicals should be considered carefully before they are used.

4. Manipulation of the Controlling Factors

The controlling factors which can be manipulated to induce progression are the various elements of climate, geological materials, and available organisms. With few exceptions these factors are slow and difficult to change. Some of those which can be modified more easily are discussed below.

a. Manipulation of Climatic Elements to Induce Progression. Either protection from fire or burning may be considered as the manipulation of a climatic factor. Burning is one of the earliest environmental manipulations practiced by man, often resulting in regression. In recent years protection from fire has been practiced, in some cases resulting in improvement, and in others resulting in the encroachment of undesirable woody species, especially when the herbaceous vegetation is weakened by overgrazing. The scientific use of fire in the management of forest and range ecosystems is a recent development. Prescribed burning, coupled with wise grazing management, can be used to reduce undesirable brush (West, 1965; Box *et al.,* 1967) in grasslands and savannas, to modify botanical composition and increase productivity in humid and subhumid grasslands (Kucera *et al.,* 1967); to speed up nutrient cycling and energy flow in chaparral ecosystems; to increase the productivity of grass, forb, and shrub components of forests; and to maintain pastures derived from forest ecosystems. "Fire is a bad master, but a good servant. . . . Let us learn without delay how better to thwart the master but much more effectively to employ the servant" (Phillips, 1965)!

Weather modification is a very active area of research. Although results have been disappointing, a small but significant increase in the precipitation falling on some range ecosystems may be a reality in the near future. Even small increases, if sustained, would affect all of the dependent factors of the ecosystem and would have far reaching implications for management. Effects would be most noticeable in ecotones between kinds of ecosystems. Forests would expand into grasslands and the change from range to cultivated ecosystems would be accelerated, especially on the better soils. Increased precipitation would permit increased stocking rates and more intensive management practices and would require changes in ranch organization. If the variability of precipitation were decreased, the change from range to cultivated ecosystems would be wise in many cases. However, if the variability of precipitation were increased by weather modification, the difficulty of determining the best kind of land use would be accentuated, the need for long-range forecasts would be more critical, and a greater premium would be placed on flexibility in ranch and farm operations.

Far reaching changes in range ecosystems would also be produced if the temperature regime or any other weather element were significantly altered.

b. Manipulation of Geological Materials to Induce Progression. Water spreading may be regarded as manipulation of the controlling factor of relief to produce an area with run-in water. Water-spreading systems are highly individualized and vary from crude water diversion and wild flooding systems to sophisticated irrigation systems with detention storage and elaborate delivery systems. Where water and soils are favorable, intensively managed crops or tame pastures will usually be more economical uses of the spreading area than range. However, if the water supply is inadequate, undependable, or of poor quality, if the soil is unsuitable because of salinity, low permeability, or topography, or if extensive management is preferred, use as a range ecosystem will be more satisfactory. Water-spreading systems should be fenced separately for grazing management, otherwise animals will concentrate on the lush vegetation. Nitrogen fertilization is often feasible. Extensive changes in botanical composition and plant growth are induced by these treatments (Houston, 1960) until a new equilibrium is reached. Stocking rate increases or increases in hay production are extremely variable. The practice is usually economical if the collecting area is large enough to supply sufficient water, the spreading area is large enough to manage separately, and the costs of installation are not excessive.

Land leveling and terracing may be regarded as manipulations of the controlling factor of relief, but they are usually too expensive for use in range ecosystems. Level bench terraces have been constructed for crop production in dry areas using runoff water from the steeper slopes which are kept as range. In this case the principal objectives of range ecosystem management are water yield and water quality.

The controlling factor of a water table may be modified by drainage, thus converting a wetland or a subirrigated range site into some other kind of a range site. While drainage may be desirable if the management objectives include the development of a cultivated ecosystem, the practice usually results in regression in range ecosystems. Partial drainage to change an open water marsh into a wetland or subirrigated range site may enlarge the area of range. Unplanned drainage caused by gully erosion often occurs in wet mountain meadows as a result of regression induced by grazing. Whenever drainage occurs changes are produced in all of the dependent factors of the ecosystem until a new equilibrium is reached. In some topographical situations ground water can be recharged through the management of surface runoff, thus reconstituting a water table. However, such opportunities are rare.

Fertilization with phosphorus, trace elements, or other elements which were limiting in the parent material may produce striking increases in productivity and forage nutritional value and result in extensive changes in all of the dependent factors of the ecosystem. Vast areas of phosphorus-deficient range in dry regions respond to phosphorus fertilization, but the practice is not economical because rapid chemical combination of the fertilizer with soil chemicals renders the phosphorus unavailable or because increases in productivity are limited by low precipitation. Phosphorus supplementation of grazing animals is usually mandatory in these areas. Striking responses to fertilization and to animal supplementation with trace elements, such as copper, cobalt, molybdenum, and others, have been reported from various parts of the world. Low-producing range sites which do not respond to other range improvement practices should be screened for such limiting factors. However, proper formulations and the proper relative proportions of different elements must be used to secure increases in productivity and avoid causing new deficiencies as a result of ion antagonism.

Range ecosystems with solodized solonetz soils are less productive and support plant and animal communities with different characteristics than those with ordinary upland soils. Solodized solonetz soils differ among themselves in the amount of sodium on the ion exchange complex, the thickness of the A horizon, and the depth and compactness of the B horizon. The productivity of such soils may be improved by ripping to shatter the columnar structure, to improve soil-plant-water relationships, and subsequently to increase soil organic matter. If the sodium content is low, the columnar structure may not re-form or will redevelop slowly, resulting in relatively permanent improvement. However, if the colloids are dispersed by high sodium levels, the improvement will be very temporary. In some solodized solonetz soils where columnar development is not deep and is underlaid by material containing nonsodium salts, deep furrowing (Branson et al., 1966) will break up the structure, mix the subsoil with the high sodium B horizon, and result in long-lasting range improvement. Solodized solonetz soils with different characteristics occur in small- or large-scale mosaics (E. M. White, 1964a,b); consequently, careful evaluation is required to predict the response to treatment. In more humid climates where the potential productivity of range ecosystems with normal soils is great enough to warrant the expense, solodized solonetz soils with good drainage may be improved by large applications of gypsum (Richards, 1954). Large quantities of organic matter or manure may also result in permanent improvement. Permanent or semipermanent improvement of these soils will require a change in their range site classification.

c. Manipulation of Available Organisms to Induce Progression. Actually man himself was not native in most range ecosystems, but established himself as a hunter, pastoralist, or agriculturalist. One of the first modifications made by the herding cultures was the elimination of many native herbivores, as well as predators and animals regarded as pests, and the introduction of domesticated animals. The impact of such changes was severe in New Zealand (Clark, 1956), in Australia (O. B. Williams, 1961), and in the central valley of California (Burcham, 1957). In Africa the wisdom of such action is open to serious question (Dasmann, 1964; Watt, 1968, pp. 69–73). In North America under good management the changes have resulted in significant increases in animal production without severe repercussions in the various components of the ecosystem. Genetic modification of domestic animals through selection and mating systems has resulted in permanent improvement in ecosystem productivity as long as careful management is continued. Such practices may also be applied to game animals. Indeed, genetic change induced by the activities of man, especially hunting, is undoubtedly occurring in game animals at the present time (Allen, 1966).

Range improvement through the introduction of exotic or domesticated forage plants, either in pure stands or interseeded, was one of the first range improvement methods tried in the western United States. However, plant communities in range ecosystems in high range condition have developed over thousands of years. Each species is genetically adapted to the vagaries of its environment, showing clinal changes along environmental gradients. The various species are closely integrated showing periodicity, stratification, and numerous interactions with other plant species, grazing animals, parasites, disease, and organisms of the detritus food web. Successful invasion of such communities by exotics is rare unless disturbance destroys the integration of the community and its environment, thus opening ecological niches for colonization. Introduction of exotic plants into range ecosystems usually requires suppression of the native community. Furthermore, proper management for the exotic is usually not proper management for the native community. Introduction of clovers into some range ecosystems with humid climates or with high water tables has been accomplished easily and appears to be beneficial. "Creeping" or pasture-type alfalfas have been introduced into depleted natural pastures in the subhumid areas of South Dakota. These introductions appear to be successful, especially in pastures dominated by Kentucky bluegrass, which is probably another exotic. Such ecosystems are really derived pastures and, when intensively managed, are classified as cultivated ecosystems. The possible advantage of introducing alfalfa into native range in high range condition has not been demonstrated. Pasture-type alfalfas can be interseeded into depleted ranges in years with favorable precipitation in the semiarid regions of the Northern Great Plains (Rumbaugh and Thorn, 1965). However, further research is needed on their effectiveness in range improvement compared with other methods, their integration with the various factors of the range ecosystem, and the modifications in grazing management required by their presence.

Pure stands of introduced forage species intensively managed with agronomic techniques and reestablished at intervals are classified as cultivated ecosystems. They may be very valuable seasonal additions to native range (Lodge, 1963; Smoliak, 1968) when grown on arable soils in climates to which they are adapted. In the Northern Great Plains crested wheatgrass is especially useful for spring (and also fall) grazing and Russian wild rye is very desirable for fall grazing. In some cases where all of the land is suitable for cultivation, a combination of sown pastures to provide choice forage throughout the growing season and harvested crops for winter feed may produce greater net income than range pastures alone (see Lang and Landers, 1960; Jeffries et al., 1967). However, these are cultivated ecosystems managed according to agronomic principles and should be used on arable lands. When costs and returns from cultivated ecosystems are compared with those from range ecosystems, comparisons should be made on the same soils and at comparable levels of management. If the cultivated ecosystem is given a high level of management, the range ecosystem should be at the optimum range condition class and given the best grazing management. Costs should be carefully tallied and should encompass all the direct and indirect costs of seeding, including such frequently overlooked items as risk of failure, risk of drought kill, depreciation of the practice, cost of inputs required during the utilization period, and additional taxes.

Pure stands of well-adapted introduced forage species, managed as permanent vegetation using ecological principles, are regarded as extensively managed derived pastures—a kind of range. Various species have been used including weeping lovegrass, smooth bromegrass, Russian wild rye, and various kinds of wheatgrasses. It has been estimated that more than 10 million acres of crested wheatgrass have been seeded in the western United States (Newell, 1955). Most of these seedings were intended to be permanent and have been managed like native range. Some stands have persisted and remained productive for more than 50 years (Hull and Klomp, 1966). Others have been invaded by big sagebrush, rabbitbrush, and other species (Frischknecht and Harris, 1968). Grazing management appears to be very important in maintaining productive stands which are resistant to invasion (Frischknecht, 1968). Although interpretation is complicated by the overriding effects of variable precipitation, crested wheatgrass stands appear to be more productive during the early years following establishment and to decline in productivity after about 2–5 years (Hull and Klomp, 1966). Some workers have used the term sod-bound to refer to such stands which is similar to the term years of depression used by Voisin (1960), Klapp (1964) and other European workers to describe the decline in productivity of sown pastures. They attribute this decline to the destruction of soil organisms and the alteration of the orderly recycling of nutrients through the detritus food web. These declines in production, in crested wheatgrass stands at least, can be prevented by nitrogen fertilization, accompanied by plant control if needed (Lorenz and Rogler, 1962; McGinnies, 1968). On untreated stands crested wheatgrass may occupy the site for many years before being invaded by native species. Although these old stands without fertilization may still be more productive than depleted natural pastures, they are usually much less productive than natural pastures in high range condition. McWilliams and Van Cleave (1960) found that after 17 years abandoned cropland seeded to a mixture of range grasses yielded $1\frac{1}{2}$ times as much dry matter per acre as adjacent areas in the same pastures which were seeded to crested wheatgrass. Native species were beginning to invade the crested wheatgrass planting. Stands of introduced grasses can probably be kept productive longer without fertilization or reestablishment if alfalfa can be maintained with them. This has not been possible with hay-type alfalfa. However, grazing-type alfalfas should remain longer. Monocultures or even bicultures with low ground cover and the same season of growth are certainly less desirable for multiple use than more diverse communities. In range ecosystems where permanent vegetation is desired for grazing, wildlife habitat, recreation, watershed, scenic beauty, or right-of-way protection, seeding to establish native communities is preferable to planting single species or simple mixtures of introduced species. However, once long-lived, well-adapted exotics are established they can be managed extensively on an ecological basis as a derived pasture—a kind of range.

Wherever man has gone he has brought domestic animals with him which are usually not native to that ecosystem. These animals usually thrive only under management by man. However, in North America the horse, the pig, the donkey, and to a lesser degree, the goat, and the cow have escaped and established themselves as feral animals. ". . . more than 15 species of exotic animals are now maintaining themselves in the wild in the United States. Thirteen foreign big game species can be found in the State of Texas alone . . ." (Box, 1968), where they are used in commercial hunting programs on private ranches (Schreiner, 1968).

In order for any introduced species to survive a functional niche must be available to it. If the niche is already occupied by a native animal the introduced species must be able to displace it in order to survive (Dasmann, 1968). However, if the new species is successful it will induce changes in the vegetation and create new niches. Grazing by domestic livestock in North America has changed many grasslands and savannas into brush lands which are not adequately utilized by livestock or native game animals. Some workers have advocated introducing exotics that will utilize these brushlands more efficiently. However, ". . . new animals can be added only to the extent that they utilize vegetation not taken by resident animals" (Box, 1968). Some forage species may be preferred by several kinds of animals and so must be sacrificed or the vegetation underutilized. ". . . North American ungulates such as white-tailed deer switch their food preferences throughout the year in order to keep an annual diet of high quality. Competition from an exotic for only six weeks during a critical period of conception or birth could cause severe damage to the native population. Therefore, it is not only necessary to balance numbers of all animals combined with the carrying capacity of the range, but seasonal food preferences must be balanced against plant phenology" (Box, 1968). Thus, the introduction of exotic wild animals to increase net income and energy flow through range ecosystems will require very careful testing and screening before release and very careful management thereafter.

5. PLANNING TO INDUCE PROGRESSION

The planning process is essentially the same on both public and private lands and involves (1) clear-cut management objectives; (2) thorough inventory of the available resources; (3) determination of the optimum proportion of crop, tame pasture, natural pasture, extensively and intensively managed derived pastures, and forest; (4) selection of the optimum combination and proportion of uses; (5) a clear statement of the present and anticipated problems; (6) a comparison of those alternatives that will solve the problems and meet the management objectives; (7) selection of those alternatives that will fit together to optimize the management objectives; (8) implementation of the plan; (9) revision of the plan; and (10) bearing the consequences. Budgeting, and linear, quadratic, and dynamic programing are useful tools at various points in the planning process. However, the newer techniques described by Watt (1968) as systems analysis, description, simulation, and optimization will permit more realistic planning.

The rate of progression is determined by the kind of ecosystem, the extent of depletion, climatic fluctuations, the improvement plan, and the efficiency of management. Usually, the less favorable the environment, the slower the improvement. Furthermore, the rate of improvement is usually directly proportional to the size of the investment, other factors being equal. Thus, economic considerations are of major importance in determining the planned rate of progression. The success of the plan depends on its soundness and the efficiency with which it is carried out. Soundness of planning is more critical when environmental conditions are less favorable. As Leopold (1966, p. 204) commented, "You can get away with murder in the glaciated northern United States and go back 50 years later and revise the whole system of land use and the country will absorb it. But you make one mistake in the arid west and it is semi-permanent. You make one bad mistake in the Far North and it is permanent for all practical purposes."

On privately owned lands advisory technicians can assist with planning, but the owner-operator must be the decision maker since he must implement the plan and bear the consequences. On public lands the range ecosystem manager, assisted by a staff of specialists, is the man best qualified to direct the planning process. However, society must bear the consequences. The restoration of depleted land is the concern of both individuals and society. The land is the heritage of generations yet unborn—the wealth of all mankind, the breadbasket of the world. Lowdermilk (1962) said, ". . . the condition of land and its natural resources, is a measure of the stability, of the success, and of the promise of a people. . . . it is no more possible to build a safe and prosperous social structure on eroding or wasting lands, than to build a house on sinking sands." Truly, society has an interest in range improvement. Nevertheless, the individual landowner has the primary responsibility for the improvement of the resource. The stewardship of the land is vested in him by reason of the title which he holds. Society has the obligation to educate this citizen as well as all others in the wise use and conservation of our land resources.

Induced progression requires an investment which will provide a return both monetary and esthetic. Usually range improvement and wise resource use mean better income to the land owner. However, when the rate of return is not sufficiently high to attract private capital or when time discounts reduce the expected net return so that alternative investments are favored, range improvement must be justified on esthetic values or by the long range values of society. When investment in range improvement does not yield a monetary or esthetic return sufficient to secure the conserving action of educated landowners, then it is necessary for society to bear part of the financial burden in order to secure the improvement and conservation of the land for future generations. As an expression of this concern of society for our privately owned land resources, the United States Congress has authorized cost share payments for range improvement practices through the Agricultural Stabilization and Conservation Service and through the Soil Conservation Service. In addition adult education and technical assistance is provided through the Agricultural Extension Service and the Soil Conservation Service. Likewise, the administration of the public lands is devoted to the improvement and wise use of the resource.

Just as a nation is impoverished by depleted range resources, so the entire world is poorer for every deteriorating acre of range. Since nearly half of the earth's land surface is range, the perspective of the range management profession must be global.

G. Management Equilibrium

Progression in range ecosystems cannot be induced by man indefinitely. A ceiling is placed on potential productivity by the limitations of those controlling factors which man either cannot manipulate or should not because of economic constraints. This end product, the culmination of induced progression, is the management equilibrium (Fig. 7), which is that state of an ecosystem which is in dynamic equilibrium with the controlling factors (including management) and which will produce the optimum sustained yield of the optimum combination of goods and services. This new balance differs from the old natural equilibrium in that native species are represented in different proportions. Some plant and animal species have been eliminated or replaced by domesticated or newly established species. Plant and animal genetics may have been improved by selection and mating systems. Although plant communities have been simplified, sufficient complexity has been retained to provide a high degree of stability. Food webs have been simplified and shortened, metabolic waste has been reduced and an optimum set of removable products has been obtained, consistent with the limitations of the controlling factors and the management objectives. Energy flow is rapid and nutrient cycling is orderly. Since products are being continually removed, soil nutrients derived from the parent material may eventually become limiting and require an addition of essential nutrients. On derived pastures where the natural equilibrium was a forest, periodic woody plant suppression will be required to maintain the management equilibrium and prevent the development of a forest. Although, there will be exceptions, generally management to maintain the management equilibrium will be optimum management for livestock and wildlife as well as watershed, recreation, and other values. Range management can be defined as the manipulation of the dependent and/or the controlling factors of one or more range ecosystems in order to reach and maintain the management equilibrium.

The characteristics of this new balance will differ with different kinds of ranges, with different management objectives, and must be determined by experimentation and observation. As man's ecological wisdom, scientific understanding, and technology increase, the management equilibrium will be changed to be more efficient in meeting human needs over long time periods. The outcome of various management alternatives will be predicted by range ecosystem managers using mathematical models and high-speed computers. Optimum solutions will be sought that will maximize human values, minimize input costs, and maintain orderly cycling of matter and rapid energy flow on a permanent basis. Land classification will distinguish between different kinds of ecosystems and careful attention will be paid to optimum methods of coupling them. While arable lands will be managed more intensively for the production of crops and harvested forages, the vast range ecosystems will support great herds of livestock and game, provide a sustained water supply with a minimum of siltation, provide opportunities for wholesome recreation, and yet maintain itself as an efficient, smoothly functioning ecosystem for future generations. However, if we fail to develop scientific understanding of range ecosystems and ecological wisdom to plan their wise use, man's well-being will be jeopardized. We cannot plunder the resources of nearly half of the land of this planet with impunity. There are feedback mechanisms that in time will bring retribution. Induced progression must be our aim and the management equilibrium our goal.

ACKNOWLEDGMENTS

Appreciation is expressed to Dr. E. J. Dyksterhuis, Range Science Department, Texas A&M University, Dr. D. F. Burzlaff, Agronomy Department, University of Nebraska, Dr. K. E. Severson, Department of Wildlife and Fisheries Sciences, South Dakota State University, and to Dr. G. M. Van Dyne, Natural Resource Ecology Laboratory, Colorado State University for critical reading of the manuscript. This article was approved by the Director of the South Dakota Agricultural Experiment Station as Journal Series No. 887.

REFERENCES

Aandahl, A. R., and A. Heerwagen. 1964. Parallelism in the development of soil survey and range site concepts. *Am. Soc. Agron., Spec. Publ.* **5**, 137–146.

Agricultural Research Institute. 1966. "A National Program of Research for Agriculture." Agr. Res. Inst., Washington, D.C. 271 pp.

Albertson, F. W., G. W. Tomanek, and A. Riegel. 1957. Ecology of drought cycles and grazing intensity on grasslands of Central Great Plains. *Ecol. Monographs* **27**, 27–44.

Alexander, M. 1965. Nitrification. *In* "Soil Nitrogen" (W. V. Bartholomew and F. E. Clark, eds.), Monograph No. 10, pp. 309–346. Am. Soc. Agron., Madison, Wisconsin.

Allen, D. L. 1966. The preservation of endangered habitats and vertebrates of North America. *In* "Future Environments of North America" (F. F. Darling and J. P. Milton, eds.), pp. 22–37. The Natural History Press, Garden City, New York.

American Society of Range Management. 1964. "A Glossary of Terms Used in Range Management." Portland, Oregon. 32 pp.

Anderson, D. J. 1965. Classification and ordination in vegetation science: Controversy over a nonexistent problem? *J. Ecol.* **53**, 521–526.

Anderson, E. W. 1967a. Rotation of deferred grazing. *J. Range Management* **20**, 5–7.

Anderson, E. W. 1967b. Grazing systems as methods of managing the range resource. *J. Range Management* **20**, 383–388.

Anderson, K. L., and C. L. Fly. 1955. Vegetation-soil relationships in Flint Hills bluestem pasture. *J. Range Management* **8**, 163–169.

Arnold, G. W. 1964. Factors within plant associations affecting the behavior and performance of grazing animals. *In* "Grazing in Terrestrial and Marine Environments" (D. C. Crisp, ed.), pp. 133–154. Blackwell, Oxford.

Atkins, M. D., and J. E. Smith, Jr. 1967. Grass seed production and harvest in the Great Plains. *U.S. Dept. Agr., Farmers' Bull.* **2226**, 1–30.

Barnes, O. K., D. Anderson, and A. Heerwagen. 1958. Pitting for range improvement in the Great Plains and the Southwest desert regions. *U.S. Dept. Agr., Prod. Res. Rept.* **23**, 1–17.

Barrett, G. W. 1968. The effects of an acute insecticide stress on a semi-enclosed grassland ecosystem. *Ecology* **49**, 1019–1035.

Bartlett, H. H. 1956. Fire, primitive agriculture, and grazing in the Tropics. *In* "Man's Role in Changing the Face of the Earth" (W. L. Thomas, ed.), pp. 692–720. Univ. of Chicago Press, Chicago, Illinois.

Bear, G. D., and R. M. Hansen. 1966. Food habits, growth and reproduction of white-tailed jackrabbits in southern Colorado. *Colo., Agr. Expt. Sta., Tech. Bull.* **90**, 1–59.

Becker, C. F., R. L. Lang, and F. Rauzi. 1957. New methods to improve shortgrass range. *Wyoming, Univ., Agr. Expt. Sta., Bull.* **353**, 1–12.

Bedell, T. E. 1968. Seasonal forage preferences of grazing cattle and sheep. *J. Range Management* **21**, 291–296.

Bement, R. E. 1969. A stocking-rate guide for beef production on blue grama range. *J. Range Management* **22**, 83–86.

Bement, R. E., D. F. Hervey, A. C. Everson, and L. O. Hylton, Jr. 1961. Use of asphalt-emulsion mulches to hasten grass seedling establishment. *J. Range Management* **14**, 102–109.

Bement, R. E., R. D. Barmington, A. C. Everson, L. O. Hylton, Jr., and E. E. Remenga. 1965. Seeding of abandoned croplands in the Central Great Plains. *J. Range Management* **18**, 53–59.

Bentley, H. L. 1902. Experiments in range improvement in central Texas. *U.S. Bur. Plant Ind. Bull.* **13**, 1–72.

Bentley, J. R., L. R. Green, and A. B. Evanko. 1966. Principles and techniques in converting native chaparral to stable grassland in California. *Proc. 10th Intern. Grassland Congr., Helsinki, 1966* pp. 867–871. Finnish Grassland Assoc., Helsinki, Finland.

Billings, W. D. 1952. The environmental complex in relation to plant growth and distribution. *Quart. Rev. Biol.* **27**, 251–264.

Billings, W. D., and H. A. Mooney. 1959. An apparent frost hummock-sorted polygon cycle in the alpine tundra of Wyoming. *Ecology* **40**, 16–20.

Black, A. L. 1968. Nitrogen and phosphorus fertilization for production of crested wheatgrass and native grass in northeastern Montana. *Agron. J.* **60**, 213–216.

Blydenstein, J., C. R. Hungerford, G. I. Day, and R. R. Humphrey. 1957. Effect of domestic livestock exclusion on vegetation in the Sonoran desert. *Ecology* **38**, 522–526.

Booysen, P. de V., N. M. Tainton, and J. D. Scott. 1963. Shoot apex development in grasses and its importance in grassland management. *Herbage Abstr.* **33**, 209–213.

Borchert, J. R. 1950. The climate of the central North American Grassland. *Ann. Assoc. Am. Geographers* **40**, 1–39.

Box, T. W. 1967. Range management's share of agricultural research. *J. Range Management* **20**, 205–206.

Box, T. W. 1968. Introduced animals and their implications in range vegetation management. *In* "Introduction of Exotic Animals" (J. G. Teer, ed.), Caesar Kleberg Res. Program Wildlife Ecol., pp. 17–20. Texas A. and M. Univ., College Station, Texas.

Box, T. W., J. Powell, and D. L. Drawe. 1967. Influence of fire on south Texas chaparral communities. *Ecology* **48**, 955–961.

Branson, F. A., R. F. Miller, and I. S. McQueen. 1966. Contour furrowing, pitting, and ripping on rangelands of the western United States. *J. Range Management* **19**, 182–190.

Braverman, S. W. 1967. Disease resistance in cool-season forage, range, and turf grasses. *Botan. Rev.* **33**, 329–378.

Bredon, R. M., D. T. Torell, and B. Marshall. 1967. Measurement of selective grazing of tropical pastures using esophageal fistulated steers. *J. Range Management* **20**, 317–320.

Bridgens Bridgens, A. 1968. Aspects of shoot apex morphogenesis, development and behaviour in grasses with reference to the utilization and management of natural grassland. *Rept. S. Africa Dept. Agr. Tech. Serv., Tech. Commun.* **67**, 1–44.

Broadbent, F. E., and F. Clark. 1965. Denitrification. *In* "Soil Nitrogen" (W. V. Bartholomew and F. E. Clark, eds.), Monograph No. 10, pp. 347–362. Am. Soc. Agron., Madison, Wisconsin.

Brown, D. 1954. Methods of surveying and measuring vegetation. *Commonwealth Bur. Pasture Field Crops, Bull.* **42**, 1–223.

Brusven, M. A., and G. B. Mulkern. 1960. The use of epidermal characteristics for the identification of plants recovered in fragmentary condition from the crop of grasshoppers. *N. Dakota Agr. Expt. Sta., Res. Rept.* **3**, 1–11.

Buechner, H. K. 1950. Life history, ecology, and range use of the pronghorn antelope in Trans-Pecos Texas. *Am. Midland Naturalist* **43**, 257–354.

Bunting, B. T. 1965. "The Geography of Soil." Aldine Publ. Co., Chicago, Illinois. 213 pp.

Burcham, L. T. 1957. California range land. An historico-ecological study of the range resources of California. *Calif., Dept. Nat. Resources, Sacramento, California* 1–261.

Butler, B. E. 1950. A theory of prior streams as a causal factor of soil occurrence in the Riverine Plain of Southeastern Australia. *Australian J. Agr. Res.* **1**, 231–252.

Cain, S. A. 1966. Biotope and habitat. *In* "Future Environments of North America" (F. F. Darling and J. P. Milton, eds.), pp. 38–54. The Natural History Press, Garden City, New York.

Campbell, R. S., R. Price, and G. Stewart. 1944. The history of western range research. *Agr. Hist.* **18**, 127–143.

Carpenter, J. R. 1940. The grassland biome. *Ecol. Monographs* **10**, 617–684.

Chapline, W. R. 1951. Range management history and philosophy. *J. Forestry* **49**, 634–638.

Chapline, W. R., M. Drosdoff, M. L. Cox, A. Johnston, C. M. McKell, A. A. Adegbola, G. W. Tomanek, and C. K. Pearse. 1966. Range management worldwide. *J. Range Management* **19**, 321–340.

Chitty, D. 1960. Population processes in the vole and their relevance to general theory. *Can. J. Zool.* **38**, 99–113.

Christian, J. J., and D. E. Davis. 1964. Endocrines, behavior, and population. *Science* **146**, 1550–1560.

Churchill, E. D., and H. C. Hanson. 1958. The concept of climax in arctic and alpine vegetation. *Botan. Rev.* **24**, 127–191.

Civil Service Commission. 1967. Range conservationist examination. *U.S. Civil Serv. Comm. Ann.* **SP-7-55.**

Clark, A. H. 1956. The impact of exotic invasion on the remaining new world mid-latitude grasslands. *In* "Man's Role in Changing the Face of the Earth" (W. L. Thomas, ed.), pp. 737–777. Univ. of Chicago Press, Chicago, Illinois.

Clawson, M., R. B. Held, and C. H. Stoddard. 1960. "Land for the Future." Johns Hopkins Press, Baltimore, Maryland. 570 pp.

Clements, F. E. 1916. Plant succession: An analysis of the development of vegetation. *Carnegie Inst. Wash. Publ.* **242**, 1–512.

Clements, F. E. 1920. Plant indicators: The relation of plant communities to process and practice. *Carnegie Inst. Wash. Publ.* **290**, 1–388.

Clements, F. E. 1934. The relict method in dynamic ecology. *J. Ecol.* **22**, 39–68; reprinted in "Dynamics of Vegetation" (B. W. Allred and E. S. Clements, eds.), pp. 161–200. H. W. Wilson Co., New York, 1949.

Clements, F. E. 1936. Nature and structure of the climax. *J. Ecol.* **24**, 252–284; reprinted in "Dynamics of Vegetation" (B. W. Allred and E. S. Clements, eds.), pp. 119–160. H. W. Wilson Co., New York, 1949.

Clements, F. E., and V. E. Shelford. 1939. "Bio-ecology." Wiley, New York. 425 pp.

Cole, L. C. 1968. Man and the air. *Population Bull.* **24**, 103–113.

Cook, C. W. 1954. Common use of summer range by sheep and cattle. *J. Range Management* **7**, 10–13.

Cook, C. W. 1964. Symposium on nutrition of forages and pastures: Collecting forage samples representative of ingested material of grazing animals for nutritional studies. *J. Animal Sci.* **23**, 265–270.

Cook, C. W. 1966a. Carbohydrate reserves in plants. *Utah State Univ., Agr. Expt. Sta., Res. Ser.* **31**, 1–47.

Cook, C. W. 1966b. Factors affecting utilization of mountain slopes by cattle. *J. Range Management* **19**, 200–204.

Cook, C. W., K. Taylor, and L. E. Harris. 1962. The effect of range condition and intensity of grazing upon daily intake and nutritive value of the diet on desert ranges. *J. Range Management* **15**, 1–6.

Cook, C. W., M. Kothman, and L. E. Harris. 1965. Effect of range condition and utilization on nutritive intake of sheep on summer ranges. *J. Range Management* **18**, 69–73.

Cooper, W. S. 1926. The fundamentals of vegetation change. *Ecology* **7**, 391–413.

Cosper, H. R., and J. R. Thomas. 1962. Influence of supplemental water and fertilizer on production and chemical composition of native forage. *J. Range Management* **15**, 292–297.

Costello, D. F. 1957. Application of ecology to range management. *Ecology* **38**, 49–53.

Costello, D. F. 1964a. Range dynamics control—an ecological urgency. *In* "Grazing in

Terrestrial and Marine Environments" (D. J. Crisp, ed.), pp. 91–107. Blackwell, Oxford.

Costello, D. F. 1964b. Current trends in range management research. *Herbage Abstr.* **34,** 75–78.

Cottam, G., and R. P. McIntosh. 1966. Vegetational continuum. *Science* **152,** 546–547.

Coupland, R. T. 1959. Effect of changes in weather condition upon grasslands in the Northern Great Plains. *In* "Grasslands," Publ. No. 53, pp. 291–306. Am. Assoc. Advance. Sci., Washington, D.C.

Crafts, A. S., and W. W. Robbins. 1962. "Weed Control: A Textbook and Manual," 3rd ed. McGraw-Hill, New York. 660 pp.

Crocker, R. L. 1952. Soil genesis and the pedogenic factors. *Quart Rev. Biol.* **27,** 139–168.

Cullen, N. A. 1966. Pasture establishment on unploughable hill country in New Zealand: Factors influencing establishment of oversown grasses and clovers. *Proc. 10th Intern. Grassland Congr., Helsinki, 1966* pp. 851–855. Finnish Grassland Assoc., Helsinki, Finland.

Currie, P. O., and D. L. Goodwin. 1966. Consumption of forage by black-tailed jackrabbits on salt-desert ranges of Utah. *J. Wildlife Management* **30,** 304–311.

Currie, P. O., and G. Peterson. 1966. Using growing season precipitation to predict crested wheatgrass yields. *J. Range Management* **19,** 284–288.

Dabadghao, P. M. 1960. Types of grass covers of India and their management. *Proc. 8th Intern. Grassland Congr., Reading, Engl., 1960* pp. 226–230. Grassland Res. Inst., Hurley, England.

Dahl, B. E. 1963. Soil moisture as a predictive index to forage yield for the sandhills range type. *J. Range Management* **16,** 128–132.

Dansereau, P. 1957. "Biogeography: An Ecological Perspective." Ronald Press, New York. 394 pp.

Darling, F. F. 1960. "Wildlife in an African Territory—A Study Made for the Game and Tsetse Control Department of Northern Rhodesia." Oxford Univ. Press, London and New York. 160 pp.

Dasmann, R. F. 1964. "African Game Ranching." Macmillan, New York. 75 pp.

Dasmann, R. F. 1968. Game ranching potentials in North America. *In* "Introduction of Exotic Animals" (J. G. Teer, ed.), Caesar Kleberg Res. Program Wildlife Ecol., pp. 11–12. Texas A. and M. Univ., College Station, Texas.

Daubenmire, R. A. 1966. Vegetation: Identification of typal communities. *Science* **151,** 291–298.

Daubenmire, R. A. 1968. Ecology of fire in grasslands. *Advan. Ecol. Res.* **5,** 209–266.

Davis, R. B. 1961. Wildlife and range biology—a single problem. *J. Range Management* **14,** 177–179.

Dent, J. W., D. T. A. Aldrich, and U. Silvey. 1967. Systematic testing of quality in grass varieties. I. An assessment of the degree of precision obtainable in comparing *in vitro* digestibility figures. *J. Brit. Grassland Soc.* **22,** 270–276.

Dickinson, R. E. 1966. The process of urbanization. *In* "Future Environments of North America" (F. F. Darling and J. P. Milton, eds.), pp. 463 and 606. The Natural History Press, Garden City, New York.

Dix, R. L. 1964. A history of biotic and climatic changes within the North American grassland. *In* "Grazing in Terrestrial and Marine Environments" (D. J. Crisp, ed.), pp. 71–90. Blackwell, Oxford.

Drawe, D. L., and T. W. Box. 1968. Forage ratings for deer and cattle on the Welder Wildlife Refuge. *J. Range Management* **21,** 225–228.

Duerr, W. A. 1967. The changing shape of forest resource management. *J. Forestry* **65,** 526–529.

Duvall, V. L. 1969. Grazing systems for a pine forest range in the South. *Am. Soc. Range Management Abstr.* **22**, 23.

Dyksterhuis, E. J. 1946. The vegetation of the Fort Worth Prairie. *Ecol. Monographs* **16**, 1–29.

Dyksterhuis, E. J. 1949. Condition and management of range land based on quantitative ecology. *J. Range Management* **2**, 104–115.

Dyksterhuis, E. J. 1955. What is range management? *J. Range Management* **8**, 193–196.

Dyksterhuis, E. J. 1957. The savannah concept and its use. *Ecology* **38**, 435–441.

Dyksterhuis, E. J. 1958. Ecological principles in range evaluation. *Botan. Rev.* **24**, 253–272.

Ehrlich, P. R., and L. C. Birch. 1967. The "balance of nature" and "population control." *Am. Naturalist* **101**, 97.

Ellison, L. 1960. Influence of grazing on plant succession of rangelands. *Botan. Rev.* **26**, 1–78.

Ellison, L., A. R. Craft, and R. W. Bailey. 1951. Indicators of condition and trend on high range-watersheds of the Intermountain Region. *U.S. Dept. Agr., Agr. Handbook* **19**, 1–66.

Elton, C. 1958. "The Ecology of Invasions by Animals and Plants." Methuen, London. 181 pp.

Evans, F. C. 1956. Ecosystem as the basic unit in ecology. *Science* **123**, 1127–1128.

Eyre, S. R. 1963. "Vegetation and Soils—A World Picture." Aldine Publ. Co., Chicago, Illinois. 324 pp.

Forbes, S. A. 1887. The lake as a microcosm. *Bull. Peoria Sci. Assoc.* pp. 77–78; reprinted in Kormondy, E. J. 1965. "Readings in Ecology." pp. 168–170. Prentice-Hall, Englewood Cliffs, New Jersey.

Ford, E. B. 1964. "Ecological Genetics." Wiley, New York, 335 pp.

Frischknecht, N. C. 1968. Factors affecting halogeton invasion of crested wheatgrass range. *J. Range Management* **21**, 8–12.

Frischknecht, N. C., and L. E. Harris. 1968. Grazing intensities and systems on crested wheatgrass in central Utah: Response of vegetation and cattle. *U.S. Dept. Agr., Tech. Bull.* **1388**, 1–47.

Godwin, H. 1929. The sub-climax and deflected succession. *J. Ecol.* **17**, 144–147.

Golley, F. B. 1960. Energy dynamics of a food chain of an old field community. *Ecol. Monographs* **30**, 187–206.

Gorham, E. 1955. Vegetation and the alignment of environmental forces. *Ecology* **36**, 514–515.

Gray, J. R. 1968. "Ranch Economics." Iowa State Univ. Press, Ames, Iowa. 534 pp.

Gray, J. R., T. M. Stubblefield, and N. K. Roberts. 1965. Economic aspects of range improvements in the southwest. *New Mexico, Agr. Expt. Sta., Bull.* **498**, 1–47.

Greig-Smith, P. 1964. "Quantitative Plant Ecology," 2nd ed. Butterworth, London and Washington, D.C. 256 pp.

Hailey, T. L., J. W. Thomas, and R. M. Robinson. 1966. Pronghorn die-off in Trans-Pecos Texas. *J. Wildlife Management* **30**, 488–496.

Hansen, R. M., and A. L. Ward. 1966. Some relations of pocket gophers to rangelands on Grand Mesa, Colorado. *Colo., Agr. Expt. Sta., Tech. Bull.* **88**, 1–20.

Hanson, H. C. 1962. "Dictionary of Ecology." Philosophical Library, New York. 382 pp.

Hanson, H. C., and E. D. Churchill. 1961. "The Plant Community." Reinhold, New York. 218 pp.

Harlan, J. R. 1958. Generalized curves for gain per head and gain per acre in rate of grazing studies. *J. Range Management* **11**, 140–147.

Harris, L. E., G. P. Lofgren, C. J. Kercher, R. J. Raleigh, and V. R. Bohman. 1967. Techniques of research in range livestock nutrition. *Utah State Univ., Agr. Expt. Sta., Bull.* **471**, 1–86.

Heady, H. F. 1960. "Range Management in East Africa." Government Printer, Nairobi, Kenya, 125 pp.

Heady, H. F. 1961. Continuous vs. specialized grazing systems: A review and application to the California annual type. *J. Range Management* **14**, 182–193.

Heady, H. F. 1967. "Practices in Range Forage Production." Univ. of Queensland Press, Brisbane, Australia. 81 pp.

Heady, H. F., and G. M. Van Dyne. 1965. Prediction of weight composition from point samples on clipped herbage. *J. Range Management* **18**, 144–148.

Hedrick, D. W. 1958. Proper utilization—a problem in evaluating the physiological response of plants to grazing use: A review. *J. Range Management* **11**, 34–43.

Hedrick, D. W. 1966. What is range management? *J. Range Management* **19**, 111.

Henderson, F. R., P. F. Springer, and R. Adrian. 1969. "The Black-Footed Ferret in South Dakota." South Dakota Dept. Game, Fish, Parks, Pierre. 37 pp.

Heslop-Harrison, J. 1964. Forty years of genecology. *Advan. Ecol. Res.* **2**, 159–247.

Hickey, W. C., Jr., and E. J. Dortignac. 1958. "An Evaluation of Soil Ripping and Soil Pitting on Runoff and Erosion in the Semiarid Southwest," Extract Publ. 65, pp. 22–33. Intern. Assoc. Sci. Hydrology. Land Erosion, Precipitations, Hydrometry, Soil Moisture Section.

Hilder, E. J. 1966. Rate of turnover of elements in soils: The effects of stocking rate. *Wool Technol. Sheep Breeding* **13**, 11–16.

Hopkin, J. A. 1954. Economic criteria for determining optimum use of summer range by sheep and cattle. *J. Range Management* **7**, 170–175.

Hormay, A. L., and M. W. Talbot. 1961. Rest-rotation grazing—a new management system for perennial bunchgrass ranges. *U.S. Dept. Agr., Prod. Res. Rept.* **51**, 1–43.

House, W. B., L. H. Goodson, H. M. Gadberry, and K. W. Dockker. 1967. "Assessment of Ecological Effects of Extensive or Repeated Use of Herbicides." Midwest Res. Inst., Kansas City, Missouri. 369 pp.

Houston, W. R. 1960. Effects of water spreading on range vegetation in eastern Montana. *J. Range Management* **13**, 289–293.

Houston, W. R., and R. R. Woodward. 1966. Effects of stocking rates on range vegetation and beef cattle production in the Northern Great Plains. *U.S. Dept. Agr., Tech. Bull.* **1357**, 1–58.

Howard, W. E. 1964. Introduced browsing mammals and habitat stability in New Zealand. *J. Wildlife Management* **28**, 421–429.

Howard, W. E., K. A. Wagnon, and J. R. Bentley. 1959. Competition between ground squirrels and cattle for range forage. *J. Range Management* **12**, 110–115.

Huffaker, C. B. 1959. Biological control of weeds with insects. *Ann. Rev. Entomol.* **4**, 251–276.

Hull, A. C., Jr., and R. C. Holmgren. 1964. Seeding southern Idaho rangelands. *U.S. Dept. Agr., Forest Serv., Res. Paper, INT* **INT-10**, 1–31.

Hull, A. C., Jr., and G. J. Klomp. 1966. Longevity of crested wheatgrass in the sagebrush-grass type in southern Idaho. *J. Range Management* **19**, 5–11.

Humphrey, R. R. 1958. The desert grassland: A history of vegetational change and an analysis of causes. *Botan. Rev.* **24**, 193–252.

Humphrey, R. R. 1962. "Range Ecology." Ronald Press, New York. 234 pp.

Ivins, J. D. 1966. Systems of management of grasses for improved yield and quality. *In* "The Growth of Cereals and Grasses" (F. L. Milthorpe and J. D. Ivins, eds.), pp. 335–345. Butterworth, London and Washington, D.C. 359 pp.

Jameson, D. A. 1963. Responses of individual plants to harvesting. *Botan. Rev.* **29**, 532–594.

Jantii, A., and P. J. Kramer. 1956. Regrowth of pastures in relation to soil moisture and defoliation. *Proc. 7th Intern. Grassland Congr., Palmerston, New Zealand, 1956*, pp. 33–44. Dept. Agr., Wellington, New Zealand.

Jardine, J. T., and M. Anderson. 1919. Range management on the national forests. *U.S. Dept. Agr. Bull.* **790**, 1–98.

Jeffries, N. W., R. L. Lang, M. May, and L. Landers. 1967. Cow-calf production on seeded and native range. *Wyoming, Univ., Agr. Expt. Sta., Bull.* **472**, 1–6.

Jenny, H. 1941. "Factors of Soil Formation." McGraw-Hill, New York. 281 pp.

Jenny, H. 1958. Role of the plant factor in the pedogenic functions. *Ecology* **39**, 5–16.

Jenny, H. 1961. Derivation of state factor equations of soils and ecosystems. *Soil Sci. Soc. Am. Proc.* **25**, 385–388.

Johnson, W. M. 1965. Rotation, rest-rotation, and season-long grazing on a mountain range in Wyoming. *U.S. Dept. Agr., Forest Serv., Res. Paper, RM* **RM-14**, 1–16.

Johnston, A. 1966. Strengthening range management technical assistance—the advisor. *J. Range Management* **19**, 327–330.

Johnston, A., S. Smoliak, L. M. Forbes, and J. A. Campbell. 1966. "Alberta Guide to Range Condition and Recommended Stocking Rates." Alberta Dept. Agr., Edmonton, Alberta. 17 pp.

Julander, O. 1955. Deer and cattle range relations in Utah. *Forest Sci.* **1**, 130–139.

Julander, O. 1962. Range management in relation to mule deer habitat and herd productivity. *J. Range Management* **15**, 278–281.

Julander, O., and D. E. Jeffrey. 1964. Deer, elk and cattle range relations on summer range in Utah. *North Am. Wildlife Natl. Res. Conf., Proc.* **29**, 404–414.

Julander, O., W. L. Robinette, and D. A. Jones. 1961. Relationship of summer range condition to mule deer herd management. *J. Wildlife Management* **25**, 54–60.

Keller, W., and A. T. Bleak. 1969. Root and shoot growth following preplanting treatment of grass seed. *J. Range Management* **27**, 43–46.

Kershaw, K. A. 1964. "Quantitative and Dynamic Ecology." Arnold, London. 183 pp.

Kikuchi, H., T. Kagomi, and K. Hori. 1966. Research on pasture reclamation by hoof cultivation. *Proc. 9th Intern. Grassland Congr., São Paulo, 1965* pp. 231–236. Secretary of Agriculture, State of São Paulo, São Paulo, Brazil.

Klapp, E. 1964. Features of a grassland theory. *J. Range Management* **17**, 309–322.

Klebenow, D. A., and G. M. Gray. 1968. Food habits of juvenile sage grouse. *J. Range Management* **21**, 80–93.

Klein, D. R. 1968. The introduction, increase, and crash of reindeer on St. Matthew Island. *J. Wildlife Management* **32**, 350–367.

Klipple, G. E., and R. E. Bement. 1961. Light grazing—is it economically feasible as a range improvement practice? *J. Range Management* **14**, 57–62.

Klopfer, P. H. 1962. "Behavioral Aspects of Ecology." Prentice-Hall, Englewood Cliffs, New Jersey. 173 pp.

Knapp, R. 1965. "Die vegetation von Nord-und Mittelamerika und der Hawaii-Inseln." Fischer, Stuttgart. 373 pp.

Koford, C. B. 1958. "Prairie Dogs, Whitefaces, and Blue Grama," *Wildlife Monographs* **3**, 1–78.

Kucera, C. L., R. C. Dahlman, and M. R. Koelling. 1967. Total net productivity and turnover on an energy basis for tallgrass prairie. *Ecology* **48**, 536–541.

Küchler, A. W. 1964. Potential natural vegetation of the conterminous United States. *Am. Geograph. Soc., Spec. Publ.* **36**, 1–116 (with map).

Lang, R. L., and L. Landers. 1960. Beef production and grazing capacity from a combination of seeded pastures versus native range. *Wyoming, Univ., Agr. Expt. Sta., Bull.* **370**, 1–12.

Larin, I. V. 1956. "Pasture Economy and Meadow Cultivation" (transl. from Russian by A. Lapid and H. Aylat). Israel Program Sci. Transl., Jerusalem, 1962. 641 pp.

Larin, I. V. 1960. "Pasture Rotation—System for the Care and Utilization of Pastures" (transl. from Russian by M. Raymist and J. Gale). Israel Program Sci. Transl., Jerusalem, 1962. 204 pp.

Larson, F. 1940. The role of the bison in maintaining the short-grass plains. *Ecology* **21**, 113–121.

Larson, F., and W. W. Whitman. 1942. A comparison of used and unused grassland mesas in the Badlands of South Dakota. *Ecology* **23**, 438–445.

Leinweber, C. L. 1967. Range education needs for the next 20 years. *Am. Soc. Range Management Abstr.* **20**, 8.

Leopold, A. S. 1966. General discussion following Session II. Regions: Their developmental history and future. *In* "Future Environments of North America" (F. F. Darling and J. P. Milton, eds.), p. 204. The Natural History Press, Garden City, New York.

Lewis, J. K. 1959. The ecosystem concept in range management. *Am. Soc. Range Management Abstr.* **12**, 23–25.

Lewis, J. K., G. M. Van Dyne, L. R. Albee, and F. W. Whetzal. 1956. Intensity of grazing: Its effect on livestock and forage production. *S. Dakota State Coll., Agr. Expt. Sta., Bull.* **459**, 1–44.

Lindeman, R. L. 1942. The trophic-dynamic aspect of ecology. *Ecology* **23**, 399–418.

Lodge, R. W. 1963. Complementary grazing systems for sandhills of the Northern Great Plains. *J. Range Management* **16**, 240–244.

Lorenz, R. J., and G. A. Rogler. 1962. A comparison of methods of renovating old stands of crested wheatgrass. *J. Range Management* **15**, 215–219.

Love, R. M. 1961. The range—natural plant communities or modified ecosystems? *J. Brit. Grassland Soc.* **16**, 89–99.

Lowdermilk, W. C. 1962. History of civilization without soil and water management planning. *Intern. Seminar Soil Water Util., S. Dakota State Coll., Brookings, 1962* pp. 5–10.

MacArthur, R. 1955. Fluctuations of animal populations, and a measure of community stability. *Ecology* **36**, 533–536.

McCorkle, J. S., and A. Heerwagen. 1951. Effect of range condition on livestock production. *J. Range Management* **4**, 242–248.

Macfadyen, A. 1963. The contribution of the fauna to the total soil metabolism. *In* "Soil Organisms" (J. Doeksen and J. Van der Driff, eds.), pp. 3–17. North-Holland Publ., Amsterdam.

McGinnies, W. J. 1968. Effects of nitrogen fertilizer on an old stand of crested wheatgrass. *Agron. J.* **60**, 560–562.

McIlvain, E. H., and M. C. Shoop. 1969. Grazing systems in the Southern Great Plains. *Am. Soc. Range Management Abstr.* **22**, 21.

McIntosh, R. P. 1967. The continuum concept of vegetation. *Botan. Rev.* **33**, 131–187.

McMahan, C. A., and C. W. Ramsey. 1965. Response of deer and livestock to controlled grazing in central Texas. *J. Range Management* **18**, 1–7.

McManus, W. R., G. W. Arnold, and J. Bull. 1968. The effect of physiological status on diet selection by grazing ewes. *J. Brit. Grassland Soc.* **23**, 223–227.

McMeekan, C. P. 1960. Grazing management. *Proc. 8th Intern. Grassland Congr., Reading, Engl., 1960* pp. 21–26. Grassland Res. Inst., Hurley, England.

McMillan, C. 1959. The role of ecotypic variation in the distribution of the central grassland of North America. *Ecol. Monographs* **29**, 285–307.

McMillan, C. 1960. Ecotypes and community function. *Am. Naturalist* **94**, 245–255.

McMillan, C. 1967. Phenological variation within seven transplanted grassland community fractions from Texas and New Mexico. *Ecology* **48**, 807–812.

McWilliams, J. L., and P. E. Van Cleave. 1960. A comparison of crested wheatgrass and native grass mixtures seeded on rangeland in eastern Montana. *J. Range Management* **13**, 91–94.

Magee, A. C. 1957. Goats pay for clearing Grand Prairie rangelands. *Texas Agr. Expt. Sta., Misc. Publ.* **MP-206**, 1–8.

Major, J. 1951. A functional factorial approach to plant ecology. *Ecology* **32**, 392–412.

Maloiy, G. M. O. 1965. African game animals as a source of protein. *Nutr. Abstr. Rev.* **35**, 903–908.

Margalef, R. 1963. On certain unifying principles in ecology. *Am. Naturalist* **97**, 357–374.

May, L. H. 1960. The utilization of carbohydrate reserves in pasture plants after defoliation. *Herbage Abstr.* **30**, 239–245.

Mech, L. D. 1966. "The Wolves of Isle Royale," U.S. Fauna Ser. 7. Fauna Natl. Parks. U.S. Govt. Printing Office, Washington, D.C. 210 pp.

Meiklejohn, J. 1968. Numbers of nitrifying bacteria in some Rhodesian soils under natural grass and improved pastures. *J. Appl. Ecol.* **5**, 291–300.

Merril, L. B. 1969. Grazing systems in the Edwards Plateau of Texas. *Am. Soc. Range Management Abstr.* **22**, 22–23.

Milthorpe, F. L., and J. L. Davidson. 1966. Physiological aspects of regrowth following defoliation. *In* "The Growth of Cereals and Grasses" (F. L. Milthorpe and J. D. Ivins, eds.), pp. 241–255. Butterworth, London and Washington, D.C.

Möbius, K. 1877. "An Oyster-Bank is a Biocönose, or a Social Community," translated from "Die Auster und die Austerwirthschaft." Wiegundt, Hempel & Parey, Berlin. H. J. Rice. 1880. Report U.S. Commission of Fisheries, pp. 683–751. Excerpt reprinted in Kormondy, E. C., ed. 1965. "Readings in Ecology." pp. 121–124. Prentice-Hall, Englewood Cliffs, New Jersey.

Moore, R. M., and E. F. Biddiscombe. 1964. The effects of grazing on grasslands. *In* "Grasses and Grasslands" (C. Barnard, ed.), pp. 221–235. Macmillan, New York.

Mueggler, W. F. 1965. Cattle distribution on steep slopes. *J. Range Management* **18**, 255–257.

Mulkern, G. B. 1967. Food selection by grasshoppers. *Ann. Rev. Entomol.* **12**, 59–78.

Murdoch, W. W. 1966. "Community structure, population control and competition"—a critique. *Am. Naturalist* **100**, 219–226.

National Academy of Sciences. 1962. Basic problems and techniques in range research. *Natl. Acad. Sci.—Natl. Res. Council, Publ.* **890**, 1–341.

National Park Service. 1968. "Compilation of the Administrative Policies for the National Parks and National Monuments of Scientific Significance (Natural Area Category)." U.S. Dept. Interior, Natl. Park Serv. Washington, D.C. 138 pp.

Naveh, Z., and B. Ron. 1966. Agro-ecological management of Mediterranean ecosystems. *Proc. 10th Intern. Grassland Congr., Helsinki, 1966* pp. 871–874. Finnish Grassland Assoc., Helsinki, Finland.

Newell, L. C. 1955. Wheatgrasses in the West. *Crops Soils* **8**, 7–9.

Nichols, J. T. 1969. Range improvement practices on deteriorated dense clay wheatgrass range in western South Dakota. *S. Dakota State Coll., Agr. Expt. Sta., Bull.* **552**, 1–23.

Nicholson, E. M. 1966. Discussion following Session I. The organic world and its environments. *In* "Future Environments of North America" (F. F. Darling and J. P. Milton, eds.), pp. 110–112. The Natural History Press, Garden City, New York.

Nixon, E. S., and C. McMillan. 1964. The role of soil in the distribution of four grass species in Texas. *Am. Midland Naturalist* **71**, 114–140.

Nutman, P. S. 1965. Symbiotic nitrogen fixation. *In* "Soil Nitrogen" (W. V. Bartholomew and F. E. Clark, eds.), Monograph No. 10, pp. 360–383. Am. Soc. Agron., Madison, Wisconsin.

Odum, E. P. 1959. "Fundamentals of Ecology," 2nd ed. Saunders, Philadelphia, Pennsylvania. 546 pp.

Odum, H. T. 1957. Trophic structure and productivity of Silver Springs, Florida. *Ecol. Monographs* **27,** 55–112.

Odum, H. T., and R. C. Pinkerton. 1955. Times speed regulator, the optimum efficiency for maximum output in physical and biological systems. *Am. Sci.* **43,** 331–343.

Parker, K. W. 1951. A method for measuring trend in range condition on national forest ranges. Processed. U.S. Dept. Agr., Forest Serv. Washington, D.C. 22 pp.

Parker, K. W. 1954. Application of ecology in the determination of range condition and trend. *J. Range Management* **7,** 14–23.

Passey, H. B., and V. K. Hugie. 1963. Some plant-soil relationships on an ungrazed range area of Southwestern Idaho. *J. Range Management* **16,** 113–118.

Patten, D. T. 1963. Vegetational pattern in relation to environments in the Madison Range, Montana. *Ecol. Monographs* **33,** 375–406.

Payne, J. A. 1965. A summer carrion study of the baby pig *Sus scrofa* Linnaeus. *Ecology* **46,** 592–602.

Pearse, C. K. 1966. Expanding horizons in worldwide range management. *J. Range Management* **19,** 336–340.

Pechanec, J. F. 1967. The range society is at the crossroads. *J. Range Management* **20,** 125–128.

Peek, J. M., A. L. Lovaas, and R. A. Rouse. 1967. Population changes within the Gallatin elk herd, 1932–65. *J. Wildlife Management* **31,** 304–316.

Penfound, W. T. 1967. A physiognomonic classification of vegetation in the conterminous United States. *Botan. Rev.* **33,** 289–326.

Peterson, R. A. 1962. Factors affecting resistance to heavy grazing in needle-and-thread grass. *J. Range Management* **15,** 183–189.

Petrides, G. A., and W. G. Swank. 1965. Population densities and the range-carrying capacity for larger mammals in Queen Elizabeth National Park, Uganda. *Zool. Africana* **1,** 209–225.

Phillips, J. 1965. Fire—as master and servant: Its influence in bioclimatic regions of trans-Saharan Africa. *Proc. 4th Tall Timbers Fire Ecol. Conf., Tallahassee, Florida.* pp. 7–109. Tall Timbers Research Station, Tallahassee, Florida.

Pieper, R. C., C. W. Cook, and L. E. Harris. 1959. Effect of intensity of grazing upon nutritive content of the diet. *J. Animal Sci.* **18,** 1031–1037.

Plummer, A. P., D. R. Christensen, and S. B. Monsen. 1968. Restoring big-game range in Utah. *Utah Div. Fish Game Publ.* **68-3,** 1–183.

Poulton, C. E. 1967. What are we going to do about it? *J. Range Management* **20,** 63.

Power, J. F. 1967. The effect of moisture on fertilizer nitrogen immobilization in grasslands. *Soil Sci. Soc. Am. Proc.* **31,** 223–226.

Power, J. F. 1968. Mineralization of nitrogen in grass roots. *Soil Sci. Soc. Am. Proc.* **32,** 673–674.

Pratt, D. J., P. J. Greenway, and M. D. Gwynne. 1966. A classification of African rangeland, with an appendix on terminology. *J. Appl. Ecol.* **3,** 369–382.

Quinn, J. A., and R. V. Miller. 1967. A biotic selection study utilizing *Muhlenbergia montana. Bull. Torrey Botan. Club* **94,** 423–432.

Quinnild, C. L., and H. E. Cosby. 1958. Relics of climax vegetation on two mesas in western North Dakota. *Ecology* **39,** 29–32.

Ramsey, C. W. 1965. Potential economic returns from deer as compared with livestock in the Edwards Plateau region of Texas. *J. Range Management* **18**, 247–250.

Rao, C. R. 1952. "Advanced Statistical Methods in Biometric Research." Wiley, New York. 390 pp.

Rauzi, F., and C. L. Hanson. 1966. Water intake and runoff as affected by intensity of grazing. *J. Range Management* **19**, 351–356.

Rauzi, F., R. L. Lang, and C. F. Becker. 1962. Mechanical treatments on shortgrass rangeland. *Wyoming, Univ., Agr. Expt. Sta., Bull.* **396**, 1–16.

Rauzi, F., C. L. Fly, and E. J. Dyksterhuis. 1968a. Water intake on midcontinental rangelands as influenced by soil and plant cover. *U.S. Dept. Agr., Tech. Bull.* **1390**, 1–58.

Rauzi, F., R. L. Lang, and L. I. Painter. 1968b. Effects of nitrogen fertilization on native rangeland. *J. Range Management* **21**, 287–290.

Raymond, W. F. 1966. Biochemical aspects of quality in grasses. *In* "The Growth of Cereals and Grasses" (F. L. Milthorpe and J. D. Ivins, eds.), pp. 259–271. Butterworth, London and Washington, D.C.

Reed, M. J., and R. A. Peterson. 1961. Vegetation, soil and cattle responses to grazing on Northern Great Plains range. *U.S. Dept. Agr., Tech. Bull.* **1252**, 1–79.

Reid, N. J. 1968. Ecosystem management in national parks. *North Am. Wildlife Natl. Res. Conf., Proc.* **33**, 160–168.

Renner, F. G., E. C. Crafts, T. C. Hartman, and L. Ellison. 1938. A selected bibliography on management of western ranges, livestock and wildlife. *U.S. Dept. Agr., Misc. Publ.* **281**, 1–468.

Reynolds, H. G., and P. E. Packer. 1963. Effects of trampling on soil and vegetation. *U.S. Dept. Agr., Misc. Publ.* **940**, 117–122.

Rice, E. L. 1964. Inhibition of nitrogen-fixing and nitrifying bacteria by seed plants. *Ecology* **45**, 824–837.

Rice, E. L. 1965. Inhibition of nitrogen-fixing and nitrifying bacteria by seed plants. II. Characterization and identification of inhibitors. *Physiol. Plantarum* **18**, 255–268.

Rice, E. L., and R. L. Parenti. 1967. Inhibition of nitrogen-fixing and nitrifying bacteria by seed plants. V. Inhibitors produced by *Bromus japonicus* Thunb. *Southwestern Naturalist* **12**, 97–103.

Richards, L. A., ed. 1954. Diagnosis and improvement of saline and alkali soils. *U.S. Dept. Agr., Agr. Handbook* **60**, 1–160.

Riewe, M. E., J. C. Smith, J. H. Jones, and E. C. Holt. 1963. Grazing production curves. 1. Comparison of steer gains on gulf ryegrass and tall fescue. *Agron. J.* **55**, 367–369.

Robinson, J. B. 1963. Nitrification in a New Zealand grassland soil. *Plant Soil* **19**, 173–183.

Roe, F. G. 1951. "The North America Buffalo: A Critical Study of the Species in its Wild State." Univ. of Toronto Press, Toronto. 957 pp.

Rogers, L. F., and W. S. Peacock, III. 1968. Adjusting cattle numbers to fluctuating forage production with statistical decision theory. *J. Range Management* **21**, 255–258.

Rogler, G. A., and R. J. Lorenz. 1966. Nitrogen fertilization of natural grasslands in the Northern Great Plains of the United States. *Proc. 9th Intern. Grassland Congr., São Paulo, 1965* pp. 1327–1330. Secretary of Agriculture, State of São Paulo, São Paulo, Brazil.

Rumbaugh, M. D., and T. Thorn. 1965. Initial stands of interseeded alfalfa. *J. Range Management* **18**, 258–261.

Rumbaugh, M. D., G. Semeniuk, R. A. Moore, and J. D. Colburn. 1965. Travois—an alfalfa for grazing. *S. Dakota State Coll., Agr. Expt. Sta., Bull.* **525**, 1–8.

Russo, J. P. 1964. The Kaibab north deer herd—its history, problems, and management. *Ariz., Game Fish Dept., Wildlife Bull.* **7**, 1–195.

Sampson, A. W. 1913. Range improvement by deferred and rotation grazing. *U.S. Dept. Agr., Bull.* **34**, 1–16.

Sampson, A. W. 1914. Natural revegetation based upon growth requirements and life history of the vegetation. *J. Agr. Res.* **3**, 93–147.

Sampson, A. W. 1923. "Range and Pasture Management." Wiley, New York. 421 pp.

Sampson, A. W. 1924. "Native American Forage Plants." Wiley, New York. 435 pp.

Sampson, A. W. 1928. "Livestock Husbandry on Range and Pasture." Wiley, New York. 411 pp.

Sampson, A. W. 1952. "Range Management—Principles and Practices." Wiley, New York. 570 pp.

Sampson, A. W. 1954. The education of range managers. *J. Range Management* **7**, 207–212.

Sampson, A. W. 1955. Where have we been and where are we going in range management? *J. Range Management* **8**, 241–246.

Sauer, C. O. 1950. Grassland, climax, fire, and man. *J. Range Management* **3**, 16–21.

Schreiner, C., III. 1968. Use of exotic animals in a commercial hunting program. *In* "Introduction of Exotic Animals" (J. G. Teer, ed.), Caesar Kleberg Res. Program Wildlife Ecol., pp. 13–16. Texas A. and M. Univ., College Station, Texas.

Schultz, A. M. 1964. The nutrient-recovery hypothesis for arctic microtine cycles. II. Ecosystem variables in relation to arctic microtine cycles. *In* "Grazing in Terrestrial and Marine Environments" (D. J. Crisp, ed.), pp. 57–68. Blackwell, Oxford.

Schultz, A. M. 1967. The ecosystem as a conceptual tool in the management of natural resources. *In* "Natural Resources: Quality and Quantity" (S. V. Ciriacy-Wantrup and J. J. Parsons, eds.), pp. 139–161. Univ. of California Press, Berkeley, California.

Schumacher, C. M. 1964. Range interseeding in Nebraska. *J. Range Management* **17**, 132–137.

Schuster, J. L. 1964. Root development of native plants under three grazing intensities. *Ecology* **45**, 63–70.

Schuster, J. L., and R. C. Albin. 1966. Drylot wintering of range cows—adaptation to the ranching operation. *J. Range Management* **19**, 263–268.

Scott, J. D. 1955. Principles of pasture management. *In* "The Grasses and Pastures of South Africa" (D. Meredith, ed.), pp. 601–623. Central News Agency, Cape Times Ltd., Parow C. P., Union of South Africa.

Sellick, G. W. 1960. The climax concept. *Botan. Rev.* **26**, 534–545.

Severson, K., M. May, and W. Hepworth. 1968. Food preferences, carrying capacities, and forage competition between antelope and domestic sheep in Wyoming's red desert. *Wyoming, Univ., Sci. Monograph* **10**, 1–51.

Sjörs, H. 1955. Remarks on ecosystems. *Svensk Botan. Tidskr.* **49**, 155–169.

Skovlin, J. M. 1965. Improving cattle distribution on western mountain rangelands. *U.S. Dept. Agr., Farmers' Bull.* **2212**, 1–14.

Skovlin, J. M., P. J. Edgerton, and R. W. Harris. 1968. The influence of cattle management on deer and elk. *North Am. Wildlife Natl. Res. Conf., Proc.* **33**, 169–181.

Slobodkin, L. B., F. E. Smith, and N. G. Hairston. 1967. Regulation in terrestrial ecosystems and the implied balance of nature. *Am. Naturalist* **101**, 109–124.

Smith, A. D. 1965. Determining common use grazing capacities by application of the key species concept. *J. Range Management* **18**, 196–201.

Smith, D. 1962. Physiological considerations in forage management. *In* "Forages: The Science of Grassland Agriculture" (H. D. Hughes, M. E. Heath, and D. S. Metcalfe, eds.), 2nd ed., pp. 401–409. Iowa State Univ. Press, Ames, Iowa.

Smith, D. R. 1967. Effects of cattle grazing on a ponderosa pine-bunchgrass range in Colorado. *U.S. Dept. Agr., Tech. Bull.* **1371**, 1–60.

Smith, J. G. 1899. Grazing problems in the Southwest and how to meet them. *U.S. Dept. Agr., Div. Agrost. Bull.* **16**, 1–47.

Smith, R. E. 1958. Natural history of the prairie dog in Kansas. *Univ. Kansas, Museum Nat. Hist., State Biol. Surv., Misc. Publ.* **16**, 1–36.

Smoliak, S. 1968. Grazing studies on native range, crested wheatgrass, and Russian wildrye pastures. *J. Range Management* **21**, 47–50.

Sneva, F. A., and D. N. Hyder. 1962. Estimating herbage production on semiarid ranges in the Intermountain Region. *J. Range Management* **15**, 88–93.

Soil Conservation Service. 1962. "Technical Guide for South Dakota," mimeo. Soil Conserv. Serv., Huron, South Dakota.

Soil Conservation Service. 1967. National handbook for range and related grazing lands. *U.S. Dept. Agr., Soil Conserv. Ser., Range Memo.* **7-67**, 1–85.

Sparks, D. R. 1968. Diet of black-tailed jackrabbits. *J. Range Management* **21**, 203–208.

Sparks, D. R., and J. C. Malechek. 1968. Estimating percentage dry weight in diets using a microscopic technique. *J. Range Management* **21**, 264–265.

Spedding, C. R. W. 1965a. Grazing management for sheep. *Herbage Abstr.* **35**, 77–84.

Spedding, C. R. W. 1965b. The physiological basis of grazing management. *J. Brit. Grassland Soc.* **20**, 7–14.

Stapledon, R. G. 1928. Cocksfoot grass (*Dactylis glomerata*): Ecotypes in relation to the biotic factor. *J. Ecol.* **16**, 71–104.

Stevenson, F. J. 1965. Origin and distribution of nitrogen in soil. *In* "Soil Nitrogen" (W. V. Bartholomew and F. E. Clark, eds.), Monograph No. 10, pp. 1–42. Am. Soc. Agron., Madison, Wisconsin.

Stewart, D. R. M. 1967. Analysis of plant epidermis in faeces: A technique for studying the food preferences of grazing herbivores. *J. Appl. Ecol.* **4**, 83–112.

Stewart, O. C. 1956. Fire as the first great force employed by man. *In* "Man's Role in Changing the Face of the Earth" (W. L. Thomas, ed.), pp. 115–133. Univ. of Chicago Press, Chicago, Illinois.

Stewart, W. D. P. 1967. Nitrogen-fixing plants. *Science* **158**, 1426–1432.

Stoddart, L. A. 1960. Determining correct stocking rate on range land. *J. Range Management* **13**, 251–255.

Stoddart, L. A. 1967. What is range management? *J. Range Management* **20**, 304–307.

Stoddart, L. A., and A. D. Smith. 1943. "Range Management." McGraw-Hill, New York, 547 pp.

Stoddart, L. A., and A. D. Smith. 1955. "Range Management," 2nd ed. McGraw-Hill, New York. 433 pp.

Straub, C. 1962. Rotating the range. *Farm Quart.* **17**, 102, 103, and 124–126.

Strickler, G. S., and F. W. Stearns. 1962. The determination of plant density. *U.S. Dept. Agr., Misc. Publ.* **940**, 30–40.

Stroehlein, J. L., P. R. Ogden, and B. Billy, 1968. Time of fertilizer application on desert grasslands. *J. Range Management* **21**, 86–89.

Talbot, L. M., and L. W. Swift. 1966. Production of wildlife in support of human populations in Africa. *Proc. 9th Intern. Grassland Congr., São Paulo, 1965* pp. 1355–1359. Secretary of Agriculture, State of São Paulo, São Paulo, Brazil.

Talbot, L. M., and M. H. Talbot. 1963. The high biomass of wild ungulates on East African Savanna. *North Am. Wildlife Natl. Res. Conf., Proc.* **28**, 465–476.

Tansley, A. G. 1929. Succession: The concept and its values. *Proc. 1st Intern. Congr. Plant Sci., Ithaca, 1926* pp. 677–686. Geo. Banta Publ. Co., Menasha, Wisconsin.

Tansley, A. G. 1935. The use and abuse of vegetational concepts and terms. *Ecology* **16**, 284–307.

Teilhard de Chardin, P. 1956. The antiquity and world expansion of human culture. *In*

"Man's Role in Changing the Face of the Earth" (W. L. Thomas, ed.), pp. 103–112. Univ. of Chicago Press, Chicago, Illinois.

Thatcher, A. P. 1966. Range production improved by renovation and protection. *J. Range Management* **19**, 382–383.

Thomas, G. W., and R. M. Durham. 1964. Drylot all-concentrate feeding—an approach to flexible ranching. *J. Range Management* **17**, 179–185.

Tilley, J. M. A., and R. A. Terry. 1963. A two stage technique for the *in vitro* digestion of forage crops. *J. Brit. Grassland Soc.* **18**, 104–111.

Tomanek, G. W. 1964. Some soil-vegetation relationships in western Kansas. *Am. Soc. Agron., Spec. Publ.* **5**, 158–164.

Troughton, A. 1957. The underground organs of herbage grasses. *Commonwealth Bur. Pasture Field Crops, Bull.* **44**, 1–163.

Troughton, A. 1960. Further studies on the relationship between shoot and root systems of grasses. *J. Brit. Grassland Soc.* **15**, 41–47.

Tueller, P. T. 1968. Student enrollment in range management. *J. Range Management* **21**, 346–347.

U.S. Department of Agriculture. 1962. Basic statistics of the national inventory of soil and water conservation needs. *U.S. Dept. Agr., Statist. Bull.* **317**, 1–164.

U.S. Forest Service. 1967. "Rotation Grazing: Examples and Results." U.S. Dept. Agr. Forest Serv., Southwestern Region, Albuquerque, New Mexico. 23 pp.

Valentine, K. A. 1967. Seasonal suitability, a grazing system for ranges of diverse vegetation types and condition classes. *J. Range Management* **20**, 395–397.

Van Dyne, G. M. 1966. Ecosystems, systems ecology, and systems ecologists. *U.S. At. Energy Comm., Oak Ridge Natl. Lab.* **ORNL-3957**, 1–40.

Van Dyne, G. M. 1968. Some mathematical models of grassland ecosystems. *1st IBP Grassland Inform. Syn. Workshop, Centennial, Wyoming, 1968* pp. 1–41.

Van Dyne, G. M., and H. F. Heady. 1965. Botanical composition of sheep and cattle diets on a mature annual range. *Hilgardia* **36**, 465–492.

Van Dyne, G. M., and J. H. Meyer. 1964. A method for measurement of forage intake of grazing livestock using microdigestion techniques. *J. Range Management* **17**, 204–208.

Van Dyne, G. M., and I. Thorsteinsson. 1966. Intra-seasonal changes in the height growth of plants in response to nitrogen and phosphorus. *Proc. 10th Intern. Grassland Congr., Helsinki, 1966* pp. 924–929. Finnish Grassland Assoc., Helsinki, Finland.

Van Dyne, G. M., and D. T. Torell. 1964. Development and use of the esophageal fistula: A review. *J. Range Management* **17**, 7–19.

Viereck, L. A. 1966. Plant succession and soil development on gravel outwash of the Muldrow Glacier, Alaska. *Ecol. Monographs* **36**, 181–199.

Vogel, W. G. 1966. Stem growth and apical meristem elevation related to grazing resistance of three prairie grasses. *Proc. 9th Intern. Grassland Congr., São Paulo, 1965* pp. 345–348. Secretary of Agriculture, State of São Paulo, São Paulo, Brazil.

Vogel, W. G., and A. J. Bjugstad. 1968. Effects of clipping on yield and tillering of little bluestem, big bluestem, and indian grass. *J. Range Management* **21**, 136–140.

Vogl, R. J. 1966. Vegetational continuum. *Science* **152**, 546.

Voisin, A. 1959. "Grass Productivity." Philosophical Library, New York. 353 pp.

Voisin, A. 1960. "Better Grassland Sward: Ecology-Botany-Management." Crosby, Lockwood, London. 341 pp.

Watt, K. E. F. 1968. "Ecology and Resource Management—A Quantitative Approach." McGraw-Hill, New York. 450 pp.

Watts, L. F. 1951. A program for range lands. *Unasylva* **5**, 49–54.

Weaver, J. E., and F. W. Albertson. 1956. "Grasslands of the Great Plains: Their Nature and Use." Johnsen Publ. Co., Lincoln, Nebraska. 395 pp.

Weaver, J. E., and W. W. Hansen. 1941. Native midwestern pastures: Their origin, composition and degeneration. *Nebraska Univ. Conserv. Surv. Div., Nebraska Conserv. Bull.* **22**, 1–93.

Wedel, W. R. 1961. "Prehistoric Man on the Great Plains." Univ. of Oklahoma Press, Norman, Oklahoma. 355 pp.

West, O. 1955. Veld management in the dry, summer-rainfall bush-veld. *In* "The Grasses and Pastures of South Africa" (D. Meredith, ed.), pp. 624–636. Central News Agency, Cape Times Ltd., Parow C. P., Union of South Africa.

West, O. 1965. Fire in vegetation and its use in pasture management with special reference to tropical and subtropical Africa. *Commonwealth Agr. Bur. Mimeo. Publ.* **1**, 1–53.

Wheeler, J. L., T. F. Reardon, and L. J. Lambourne. 1963. The effect of pasture availability and shearing stress on herbage intake of grazing sheep. *Australian J. Agr. Res.* **14**, 364–372.

White, E. M. 1961a. A possible relationship of little bluestem to soils. *J. Range Management* **14**, 243–247.

White, E. M. 1961b. Drainage alignment in western South Dakota. *Am. J. Sci.* **259**, 207–210.

White, E. M. 1964a. Morphological-chemical relationships of some thin A horizon solodized soils derived from moderately fine material on a well-drained slope. *Soil Sci.* **98**, 256–263.

White, E. M. 1964b. The morphological-chemical problem in solodized soils. *Soil Sci.* **98**, 187–191.

White, E. M., and J. K. Lewis. 1967. Recent gully formation in prairie areas of the Northern Great Plains. *Plains Anthropol.* **12-37**, 318–322.

White, E. M., and J. K. Lewis. 1969. The effect of a clay soil's structure on some native grass roots. *J. Range Management* **22** (in press).

White, L. A., ed. 1959. "The Indian Journals, 1859–1862, by Lewis Henry Morgan." Univ. of Michigan Press, Ann Arbor, Michigan. 229 pp.

Whittaker, R. H. 1953. A consideration of climax theory: The climax as a population and pattern. *"Ecol. Monographs"* **23**, 41–78.

Whittaker, R. H. 1962. Classification of natural communities. *Botan. Rev.* **28**, 1–239.

Williams, C. B. 1964. "Patterns in the Balance of Nature." Academic Press, New York. 324 pp.

Williams, O. B. 1961. Principles underlying the improvement of dryland country. *Wool Technol. Sheep Breeding* **8**, 51–58.

Williams, R. E. 1954. Modern methods of getting uniform use of ranges. *J. Range Management* **7**, 77–81.

Williams, R. E., B. W. Allred, R. M. Denio, and H. A. Paulson, Jr. 1968. Conservation, development and use of the world's rangelands. *J. Range Management* **21**, 355–360.

Williams, W. A. 1966. Range improvement as related to net productivity, energy flow and foliage configuration. *J. Range Management* **19**, 29–34.

Willoughby, W. M. 1959. Limitations to animal production imposed by seasonal fluctuations in pasture and by management procedures. *Australian J. Agr. Res.* **10**, 248–268.

Wills, G. F. 1946. Tree ring studies. *N. Dakota State College, Agr. Expt. Sta., Bull.* **338**, 1–24.

Wilson, A. D. 1969. A review of browse in the nutrition of grazing animals. *J. Range Management* **22**, 23–28.

Winch, J. E., G. W. Anderson, and T. L. Collins. 1966. Chemical renovation of roughland pastures. *Proc. 10th Intern. Grassland Congr., Helsinki, 1966* pp. 982–987. Finnish Grassland Assoc., Helsinki, Finland.

Workman, J. P., and J. F. Hooper. 1968. Economic evaluation of cattle distribution practices. *J. Range Management* **21**, 301–304.

Wynne-Edwards, V. C. 1965. Self-regulating systems in populations of animals. *Science* **147**, 1543–1548.

Chapter VII Forestry Viewed in an Ecosystem Perspective

EGOLFS V. BAKUZIS

I. INTRODUCTION

The question "What is a forest?" has been answered differently by different people at different times. From the sixth to the fifteenth century and even later forest was a legal term. Originally it meant land, whether covered with trees or not, reserved for hunting grounds for the royalty and protected by special laws and administration. The outmoded English terms "wald," "wold," "weeld," or "weald" referred to a natural concept.

The Society of American Foresters (1950) defined forest as "a plant association predominantly of trees and other vegetation." The Food and Agriculture Organization of the United Nations (1963) defined forest land as all land bearing vegetative associations dominated by trees of any size, exploited or not, capable of producing wood and other forest products, or exerting an influence on the climate or on the water regime, or providing shelter for livestock and wild life. Forest was defined (1963) as forest lands bearing a tree or bamboo cover, whether productive or not. The term woodland is frequently used interchangeably with forest, and relations of both to savanna are rather confused.

Forestry is defined as the scientific management of forests for the continuous production of goods and services (Society of American Foresters, 1950). Forest ecology studies forest ecosystems and provides the biological foundations for silvicultural planning. Silviculture attempts to understand biological processes for the purpose of control and increase of production of forest goods and services.

Fields such as forestry and range, wildlife, and watershed management deal with natural systems to a great extent and can be considered in part as applied ecology. Agriculture has departed relatively far from dealing with natural and seminatural conditions, but even there an ecological viewpoint is of great value.

One of the most fundamental elaborations on the nature of forest was made by Morosow in 1904. His ideas are known to the western world through the German translation of his book (Morosow, 1928). He defined forest as an aggregation of trees affecting each other and associated plants and animals, and interacting with soil and local climate. He viewed forest as a geographical and historical phenomenon.

Morosow's views substantially expressed the present notion of ecosystem. Similar views were held by Graves, Cajander, Sukachev, Dengler, Rubner, and have been recently restated by many others. Clements (1909) called the forest an organism. The context shows that he was in agreement with Graves, Zon, and others who had the notion of a system being involved.

A forest is an ecosystem involving the total physical environment and the whole of living organisms (the biocenose) according to Lutz (1959).

Sukachev, in 1954 and earlier, referred to forest as a biogeocenose (his preferred term for the ecosystem concept). A biogeocenose is any area on the surface of earth where some biocenose and the parts of the atmosphere, lithosphere, and pedosphere that correspond to it remain the same and react on one another in a uniform way, thus forming in the aggregate an indivisible, interdependent complex.

Ecosystem investigations involve two main approaches: One deals with pattern and process, the other with matter and energy. The latter dominates in contemporary modern ecology (Leeuwen, 1966). An exception to this statement is the interdisciplinary studies of forest ecosystems conducted for nearly two decades under the leadership of the late Sukachev (Sukachev and Dylis, 1968).

We must also recognize the contributions made by the classic schools of ecological thought. Forestry interests have been particularly well served by the schools of forest types by Cajander, Sukachev, Pogrebnyak, Hartmann, Scamoni, and others. Forestry has also benefited from the contributions by Clements, Braun-Blanquet, and representatives of the polyclimax school such as Daubenmire, Braun, and others. New workers have brought new ideas into ecology which may serve for better future understanding of forest ecosystems.

Unfortunately, much of the heritage is fragmentary and lacks a strong reference framework such as a classification system and thus presents great difficulties in integration.

This review will focus on special problems close to the central interests of forestry. Borderline problems such as the use of natural resources, of course, are essential; the Fifth World Forestry Congress held at Seattle, Washington, in 1960 was specially devoted to the multiple use of forest lands. Many studies and discussions have followed.

II. HISTORY OF FORESTS AND DEVELOPMENT OF FORESTRY

A. Postglacial Forests and Ancient Cultures

Diversity of the present flora of North America has been largely influenced by two processes in the history of the earth: first, mountain formation during the middle and late Tertiary and associated desiccation of large areas of the continent; and second, fluctuations of climate during the Pleistocene epoch (Knapp, 1965). Glaciation in North America did not lead to decimation of numerous tree species as in Europe. The history of forests in the United States has been reviewed in a book entitled "The Quaternary of the United States," edited by Wright and Frey (1965).

Recent climatic changes may be of help in explaining some of the processes occurring in contemporary forest ecosystems, particularly problems of reproduction, succession, changes in growth rates, migration of tree and forest line in the arctic and at high elevations, relations among forests, grasslands, and bogs, spread of diseases, and other problems.

Evaluation of the activities of prehistoric man and man's actions during historic times may provide explanations in cases where natural factors have failed to explain. Early man was a gatherer and hunter. Some 12,000 years ago neolithic man shifted from food-collecting to food-producing, developing an agriculture and clearing land (Bates, 1961). Relict steppe with open forests were the first places for settlement in Europe. Burning converted mesophytic hardwood forests into agricultural land. Warm dry forests were converted to grape plantations (Tischler, 1965). In this process, forests were gradually removed from the fertile land; cultivation of conifers additionally strengthened the northern appearance of forests over wide areas of Europe.

Man, probably, reached America via the Bering Straits about 30,000 years ago, or by some reports, 50,000–70,000 years ago (McKee, 1966). Before he learned to farm some 7000 years ago he hunted, fished, and gathered wild plants. Altogether American Indians used more than 2000 plant species (Driver, 1961). Indians learned to domesticate plants but not animals. Fire was a tool for hunting, primarily the bison. By 1500 A.D. Indians had greatly expanded the grasslands (Dix, 1964).

B. Forestry in Europe

The Middle Ages in Europe are sometimes called the Wood Age, and during this period multiple use of forests reached its peak. However, many forests were grazed and other forests with low growth rates and poor wood quality under a regime of shifting agriculture were unable to supply the developing cities with desired wood products. Forest litter was an important fertilizer for agricultural land. This resulted in depletion of forest soils and stagnation of forest stands. Forestry gradually gained partial independence of agriculture and the great forest reconstruction work started. Conifer cultivation started in the eighteenth century.

The eighteenth century marks the start of scientific development of forestry. Duhamel du Monceau (1700–1782) of France is regarded as the first forest botanist and silviculturist. The first book on silviculture, "Silvicultura Oeconomica" (1713), by H. von Carlowitz, presented information on climate, soils, site, and humus; emphasized the significance of ground cover; and discussed problems of drainage and irrigation (Fiedler and Reissig, 1964). The book compared yields from forestry and agricultural crops. Reaumur and Buffon in 1742 called attention to losses of production due to excessively long rotations (Endres, 1923).

The importance of light was recognized early. However, opinions were sharply divided on how to make the best use of light in forest production. Two schools of thought developed: One, which originated from H. Cotta (1763–1844), advocated intensive thinning such that each tree should be fully exposed to light; the other school, initiated by G. L. Hartig (1764–

1837), believed that high production could be achieved by keeping the forest dense. The argument has lasted over one hundred years, the views have been somewhat modified, long-term experiments have been carried out, and yet the problem has not been adequately solved. Hopefully, intensively instrumented ecosystem measurements made during the current International Biological Programme will contribute something new to this problem.

The deductive theoretical approach was very strong at the start of scientific forestry. The idea of normal forest developed in the eighteenth century. It served as a basis for the development of forest statics, a branch of forestry dealing with efficiency of forestry management. The equilibrium equation by Faustmann was added around 1840, and different theories of soil and forest rent which had great impact on selection of species and rotation in European forests followed. Cotta as early as 1816 (Cotta, 1816, 1835), sounded a warning of "overlooking the forest behind the trees." Another pioneer, F. W. L. Pfeil (1783–1859), called attention to studies of local conditions. The desire to reconstruct depleted forests in Germany was so great that the work was done before conditions were sufficiently investigated.

Ebermayer (1876) conducted his classic studies of forest litter and forest productivity from 1866 to 1876 in Bavaria. He also published on water circulation and carbon cycling in the forest. As fundamental studies on matter-energy exchange in forest ecosystems they have been surpassed only by the later work under the supervision of Sukachev.

Ebermayer's work may have stimulated the call by Karl Gayer in the 1880's to return to nature as the source of knowledge. The back-to-nature movement gained momentum and reached its peak with the ideas about *Dauerwald* or the continuous forest by A. Möller (1922). During the heated arguments, the primeval forest was pictured as a goal for management, being highly productive, healthy, of mixed species, and uneven-aged. Few West European foresters had seen it. American (Schenck, 1924) and Russian (Tkatschenko, 1930) experience helped to clarify this misconception. The dispute about the natural forest versus artificial has mellowed but certain aspects of this dispute are raised again from time to time in different form.

One of the problems that contributed to the original argument was the decline in growth of planted Norway spruce on former hardwood sites in Saxony. The problem was investigated by Wiedemann (1925) and his main conclusion was that selection of species was not proper. Decline in productivity of second generation spruce on raw humus sites in Finland was investigated by Siren in 1955. Holmsgaard *et al.* (1961) noted that in Denmark Norway spruce are planted on former hardwood sites and there is no decline in productivity in the second generation. These three studies are based on very intensive field investigations and thus are good examples of the multitude of problems involved in productivity studies.

Ecological studies in Europe are closely tied with forest classification systems, which then provide a framework for integration of the obtained knowledge and are useful in the discovery of contradictions in research findings. Ecological type maps aid in application of research findings to the proper field locations. Forestry practices are closely associated with ecological type mapping. Forest types serve for reforestation, thinning protection, and in the Finnish classification system also for yield prediction. In other systems yield tables are modified to adapt them to individual forest types. It is impossible for a forester to even think of management of forests without an ecological classification system.

C. Forestry in the United States

The early English settlers in Virginia in the seventeenth century adopted the shifting agriculture already practiced by the Indians. Shifting agriculture was still in use in the middle of the nineteenth century (Nye and Greenland, 1960). Concern for a consistent timber supply was strikingly apparent in early colonial times.

Of all the pioneers in United States forestry—Pinchot, Fernow, Schenck, Hough, and Graves—the latter had the viewpoint closest to the notion of the present-day ecosystem concept. His definition of forest types (Graves, 1899) contains all essential elements of forest ecosystems and resembles Morosow's ideas (1928) which were expressed later. Clements (1909) noted that there is good agreement between foresters and ecologists: "The application of the methods of forest types to forestry brings it into harmony with the fundamental principles of ecology. . . . Reproduction, development, and succession are at the bottom of the same great processes and that this process can be accurately and thoroughly understood and made of use only in so far as it is directly connected with the habitat factors which are its causes, just as the functional structures are its results."

However, conditions for the development of an ecological system of forest types as a basis for forest management and research were not favorable. A symposium on forest types was arranged by the Society of American Foresters in 1913. Greeley (1913) emphasized use of forest cover types and studies of life histories of individual species. This approach to ecological problems has been followed by the United States Forest Service, in general, to the present. The development of the forest cover type system was initiated by the Society of American Foresters (1954) and they have worked for nearly thirty years on improvement of the forest cover type system. Abundant ecological information, concerning forest stand relations to soil moisture, nutrients, successional status of the cover types, and other features, have been introduced into type descriptions. Local modifications and refinements have been added for the New England states (Westveld, 1956). Further improvements are possible by subdividing the extensive regions, by subdividing cover types which occur over a great variety of heterogeneous sites, and by further clarification of the nature of successional status.

The application of a soil classification system to forestry needs has been difficult (Kittredge, 1938). These difficulties are partially due to the fact that soil classification system was developed to serve the needs of agriculture. The new classification system (which is still in preparation) is expected to bring considerable improvements. The difficulties with soil classification systems are rather general, because morphological soil characteristics have to be interpreted in physiological terms.

Forest hydrological studies in the United States differ markedly from European studies. Most of the European work deals with problems of water surplus, while American studies are dominated by problems of water conservation. Forests once credited as "being the oceans of the continent" are frequently responsible for local water shortages. On the other hand, fire, which was recognized as the greatest menace to forests, has become a tool of silviculture.

The transition from virgin to second-growth forests has produced many ecological problems. Numerous short-term studies have been made; however, careful integration of research findings on a large scale is still wanting.

D. Recent Developments in World Forestry

According to the Food and Agriculture Organization of the United Nations (1963) nearly 4126 million hectares or nearly one third of the whole world land surface is covered with forests, ranging from the arctic tundra to equatorial swamp forests, from the desert scrub to mountain rain forests, from homogeneous plantations to luxuriant jungles. Of the world's forests about one third is coniferous. The Soviet Union possesses 46% and North America 36% of the coniferous forests of the world. The total growing stock is 340,000 million m^3 of roundwood or 1300 million tons by weight as compared with the world's production of 800 million tons of grain and 290 million tons of steel (Glesinger, 1962).

The total acreage in the United States is 1903.8 million acres (771 million ha) of which cropland comprises 465 million acres (188 million ha), or 25%; grasslands, 699 million acres (283 million ha), or 33%; forest cover, 648 million acres (262 million ha), or 32% (Bennet *et al.*, 1958). It is predicted that more forest land will be converted to more intensive uses by agriculture, urban developments, etc. In the United States it is predicted that by the year 2000, 73 million acres (30 million ha) of forest land will have been converted to cropland, and 50 million acres (20.2 million ha) will have been converted to urban development industrial, transportation, and other facilities (Wenger, 1967). Similar predictions have been made for other parts of the world. At present reduction of forest land is occurring in developing countries, but forest land is increasing in western Europe, the United States, and in China. In the United States, there was a net gain of 3.2 million ha (7.9 million acres) of forest land during the decade to 1963, and a 10 million ha (24.7 million acres) increase during the previous decade.

Tree farms and monocultures are no more deplored in forestry than in agriculture, provided species are used in their proper ranges of climates and sites, attention is given to pests, fire, wind, damage by other factors, etc. (Scott, 1966).

Such orchard forestry depends more on physiology and autecology, while naturalistic silviculture depends on synecological knowledge. Naturalistic does not mean following nature which is undisturbed by man in regard to species composition and structure. In naturalistic silviculture, species composition and structure can be altered, but it is nevertheless adapted to its site. It adheres as closely as possible to natural conditions under sound economic planning (Leibundgut, 1962). The ecosystem concept is valid for both orchard and naturalistic forestry; only emphasis on different subsystems changes.

The contemporary process of urbanization raises some forestry problems: What role will forests play in the developments of the megalopolis? Forestry problems in relation to urbanization were discussed in a special symposium arranged by the Connecticut Agricultural Experiment Station in 1962 (Waggoner and Ovington, 1962). In the United States, of most interest is the megalopolis complex stretching from Boston to New York, Philadelphia, Baltimore, and Washington for some 500 miles (800 km). There are 16.2 million acres (6.6 million ha) of forests comprising 48% of the megalopolis area. Most of the forests have developed on abandoned farmland, and they serve primarily for recreation, conservation of the landscape, and wildlife (Gottmann, 1961).

Problems of dryland and wetland forestry, amelioration of depleted sites, mechanization of forestry operations, environmental pollution, and economic problems caused by increasing labor costs and competition with other products are very acute in highly developed countries. To establish a balance in the economy, objective evaluation of forest amenities and influences is needed.

III. FOREST AS AN ECOSYSTEM

According to Klir and Valach (1967), we can study only finite parts of the totality of the matter in a finite time. The definition of system within a distant object represents the standpoint from which we want to study the object. Every material object in nature is unique. However, if we define a certain system in this object in such a manner that it possesses only properties which exist from a certain viewpoint, several objects may exist in which we can find systems satisfying this definition. This statement may explain the multitude of approaches to the problems of ecosystems; it may also indicate some of the basic problems in applying classification systems.

In this discussion forest will mean the whole ecosystem, while forest cover or forest vegetation will refer to the vegetational subsystem or phytocenose. Forest site or site will be used for the other subsystem or the totality of the physical environment.

A. General Features of Organisms and Environment

1. FOREST ORGANISMS

The organismic part of an ecosystem can be considered as represented by biocenose in the restricted sense of this term. According to Sukachev (1960), biocenose consists of phytocenose, zoocenose, and microbocenose, which can be further subdivided into different populations. The organismic part of an ecosystem also can be subdivided to consist of producers, consumers, and decomposers. Forestry is primarily interested in producers, especially trees and their utilitarian parts.

Forest ecosystems cannot be patterned as typical food chains; trees are not usually palatable. Under natural conditions, they are decomposed by microorganisms or destroyed by fire. In managed forests, they are removed to great extent from the system.

A forest stand is an aggregate of trees and other associated plants. There are many characteristics for classification of forest stands, such as species composition, age structure, density, etc. A forest stand may consist of many layers.

Animals play an important role affecting forest reproduction, growth, and succession. They affect the health status of trees by causing injuries and by opening wounds for penetration of fungi. There are a number of insect pests and diseases which have changed the composition of forest stands permanently.

Soil organisms are essential to the existence of plants and animal life as they are a vital part of the cyclic pattern of matter in nature.

2. FOREST ENVIRONMENT

There is a set of concepts—environment, site, habitat, biotope, eco-tope, niche, and others—which frequently are poorly defined and are used interchangeably. Of the multitude of possible definitions, a few are given below:

The forest environment is a complex of atmospheric, topographic, and soil influences, complicated by the interaction of the forest and its inhabitants, plant and animal, with the environment, and by competition between individuals of the forest (Braun, 1950).

Site is an area, considered by its ecological factors with reference to its capacity to produce forests and other vegetation; and combination of biotic, climatic, and soil conditions of an area (Society of American Foresters, 1950).

Site originally meant a place where a tree or other organism lived, but with time its meaning has been changed and it refers now to the totality of external factors acting upon an organism or community (Braun-Blanquet, 1964). Thus the meaning of site and environment is substantially the same. Site is more commonly the preferred term in forestry literature, while ecological literature in general uses the term environment.

Livingston and Shreve (1921) noted that environmental factors can be classified in two fundamental ways: according to their source of origin, such as climatic, edaphic, and biotic factors, or according to their mode of action, such as moisture, nutrients, heat, light, other radiation, and mechanical force.

Factors, according to their mode of action, cannot be substituted or compensated. Moisture cannot replace nutrients; heat cannot replace light. However, organisms can tolerate a range of variation in factor intensities.

Space and time provide coordinates for object identification, while energy is a coordinate for system identification (Theobald, 1966). Since there is no substantial difference between matter and energy, we can expand the statement by Theobald and use matter as an additional coordinate for identification of ecosystems. Moisture, nutrients, heat, and light are the four major coordinates for ecosystem identification.

The four coordinates are still ill-defined and difficult to measure quantitatively. Morphological features are easily observable and many are measurable; however, they have no direct relationship to physiological properties. According to Braun-Blanquet (1964), soil morphology and morphology of vegetation cannot be directly related because the relationships exist on a physiological basis. Relationships between plant communities and soils depend on the physiological conditions in terms of moisture, nutrients, aeration, and heat balance. Identical communities have identical soils in a physiological sense if the climates are similar. Morphologically, soils of the same communities may only be analogous. This is essentially the principle of bioecological equivalence, the most essential part of the definition of forest types by Cajander. Bioecological equivalence is recognized by other schools of ecological thought. It is the directive correlation as defined by Sommerhoff (1950) in his theoretical study of analytical biology, and explained as the mechanism of homeostasis in terms of the set theory by Ashby (1964). The principle of equifinality by Bertalanffy (1951) refers to similar problems.

In recent years, emphasis has been placed on the study of particular functional or operational environment and its complexity and variability rather than a generalized environment as it appears from the studies by Mason and Langenheim (1957) and Platt and Griffiths (1964).

In order to be classed as an environmental phenomenon a given phenomenon must be (1) operationally significant to an organism, (2) directly effective at some time during the life of the organism, and (3) effective as to sequence as ordered by the ontogeny of the organism. We thus have a concept of environment that is organism-directed, organism-timed, organism-ordered, and organism-spaced (Mason and Langenheim, 1957). This is in agreement with the principle of bioecological equivalence by Cajander and other classic schools of ecological thought (Braun-Blanquet, Sukachev, Pogrebnyak). The problem is how to measure the operational environment. It might appear that measurement should be done at the contact zone between organism and environment; however, organisms possess a power to be selective. A transformation matrix is necessary, as shown by Lange (1965), to relate the input state to the output state of an element of a system; the coupling of elements of a system, and the coupling of subsystems should be known. Jenny (1961) dealt substantially with this problem. Earlier Jenny (1941) attempted to solve the difficulties in searching for independent environmental factors in his study of soil formation. This method provides an opportunity to solve specific cases, predominantly on a broad geographical scale. The method can provide insight into profiles within ecosystem space.

B. Energy Flow and Cycling of Matter

Energy budgets for forest ecosystems have been prepared very rarely. Separate cycles of matter—water, carbon, nitrogen, and different mineral cycles—have been studied rather frequently. Lately cycling of pollutants and radioactive substances have gained particular attention. Unfortunately, these investigations frequently lack description of general conditions of the ecosystems involved. Thus, collection of such information does not allow one to arrive at generalizations.

Generalizations about cycling patterns even in coordinated studies, such as those under the leadership of Sukachev (Sukachev and Dylis, 1964), are very difficult. The theory of systems analysis still has not progressed very far, and biologists are frequently unaware of what is already accomplished. More intensive work is needed in the future.

1. ENERGY BUDGET

According to Baumgartner (1965), the conversion and exchange of matter are caused and accompanied by energy exchanges and transport. Since the conversion and exchange of mass follows the budget of energy, plant life can be described best by measuring constituents of this budget. Investigations of the heat budget of plant cover are as useful as investigations of matter.

Exchange of matter refers to the reception, or release of solid, liquid, and gaseous materials within, into, and from the immediate plant environment; in addition, the conversion of matter from one aggregate form to another, and its temporary fixation are often implied. Budget of matter refers to the actual presumed recording of specific exchanges and storage variations within the immediate plant environment. By analogy, equivalent definitions are valid for the energy budget (Baumgartner, 1965).

The energy budget for plant cover must be based ultimately on the principle of conservation of energy. Since solar radiation is the primary energy source, it is reasonable to relate the other heat fluxes to radiation. All heat fluxes which increase the energy of a stand are assumed to be positive while heat losses are negative.

The energy budget at a reference level above the stand was given by Baumgartner (1965) as follows:

$$S + L + V + B + P + K + R = 0$$

where S = radiation flux; L = air heat flux, convection; V = evaporation heat flux; R = precipitation heat flux; B = soil heat flux; K = assimilation heat flux; and P = heat flux within and between plant bodies and air, conduction.

It is easy to estimate S and B individually, but the separate evaluation of L and V by the flux method is difficult. Transpiration requires 60% of the net radiation received; photosynthesis, 1.5%; respiration, 0.5%; and flux of sap, 0.001% (Baumgartner, 1967a).

Energetics of soil formation have been studied by Volobuev (1964). The most important items of energy expenditure in soil formation are (1) expenditures on biological transformation of substances, (2) physical and chemical weathering, (3) hydrothermic cycle, and (4) migration of substances within the soil profile. Much radiant energy is stored in soil

humus. Biological processes are linked with transpiration and evaporation in absorbing part of the solar radiation. Part of the energy is dissipated as heat. The energy balance in soil formation is:

$$Q = W_1 + W_2 + b_1 + b_2 + i_1 + i_2 + g + v$$

where Q = quantity of energy participating in soil formation; W_1 = energy from physical destruction of parent rock; W_2 = energy from chemical destruction of parent rock; b_1 = energy accumulating in humic substances; b_2 = energy dissipated in biological reactions and transformation
of organic and inorganic substances (expended in part as heat); i_1 = energy used for evaporation from soil and plants; i_2 = energy used in transpiration; g = losses of energy in mechanical migration of salts and particles in soil; and v = energy expended in process of heat exchange with
atmosphere.

About 0.2–0.5 cal/cm^2 per year in tundra and 10–15 cal/cm^2 in the
humid tropics are used for weathering. A 250 m^3/ha internal runoff of
water with shifting of substances to 2 m depth requires 2 cal/cm^2 per year.
Energy used for evapotranspiration in tundra and desert is 3000–6000
cal/cm^2 per year; in tropic forests, 60,000 cal/cm^2 per year. We have no
direct determination of expenditures of energy on biological processes in
the soil. As an indirect measurement, the accumulation of organic matter
can be used. In tundra, 2.5–25 cal/cm^2 per year are needed to accumulate organic matter; in forests of the humid tropics, 2000 cal/cm^2. In
forests of temperate regions, steppes, and savannas, 100–400 cal/cm^2 are
needed per year to accumulate organic matter (Volubuev, 1964).

2. WATER BUDGET

The water budget was stated by Baumgartner (1967b) as follows:

$$N = A + V + R - B$$

where N = precipitation; V = evaporation; R = storage in soil; B = utilization in soil; B = utilization; and A = runoff.

Energy balance observations are valuable aids in the study of plant
water relationships, especially in connection with vaporization processes.
They permit an insight into the physical causes of the phenomena, and
provide an independent check on hydrometrically obtained vaporization
values (Baumgartner, 1965).

Annual use of water by forests has been estimated to range from 3
inches (75 mm) in the boreal forests to 125 inches (3000 mm) in the tropical forests; in the United States from 10 to 55 inches (250–1375 mm). For
upland soil in temperate latitudes, the climax vegetation generally evapo-
rates most of the growing season rainfall. Xeric succession increases
evapotranspiration; hydric succession decreases it. Secondary succession

TABLE 1
Transpiration of Forest Stands in Germany[a]

Species	Foliage (kg/ha)	Average daily leaf transpiration (H_2O/g fresh weight)	Average daily stand transpiration	
			Liter/ha	mm
Birch	4,940	9.50	47,000	4.7
Beech	7,900	4.83	38,000	3.8
Douglas-fir	40,000	1.33	53,000	5.3
Larch	13,950	3.24	47,000	4.7
Spruce	31,000	1.39	43,000	4.3
Pine	12,550	1.88	23,500	2.35

[a] Bauer and Weinitschke (1967), after H. Polster.

tends to increase evapotranspiration (Lull, 1964). Certainly it is not succession by itself but the different requirements of various species involved in successional changes which change the evaporation rates, as can be visualized in Table 1.

Forests of Central Europe transpire 250–400 mm of annual precipitation. Evaporation also has to be added. As a result, streamflow from forests is 5–15% less than from open land. However, there is less erosion in forest regions (Bauer and Weinitschke, 1967).

3. Nutrient Cycling

The most important partial processes of nutrient cycling in the forest are (1) nutrient uptake, (2) nutrient storage, (3) nutrient return, and (4) nutrient losses. Nutrients can be added to the system by fertilization, by entering through the atmosphere, especially with rain, by release in weathering, or by entering through groundwater and surface water movements. Nutrients can be lost through leaching and by removal of crops (Ehwald, 1957).

Dengler (1944), using data supplied by Albert, presented an illustration of nutrient cycling in pine and beech stands in Germany. Ehwald (1957) showed a similar picture, using data by Mina concerning oak stands in forest steppe in the U.S.S.R. This information is compiled in Table 2.

Nutrient cycling actually is more complex than is implied in Table 2. There are several minor cycles involved such as leaching from the foliage, consumption by animals, decomposition of litter and humus, etc.

Nutrient requirement is a physiological concept, it refers to specific performances, mostly requirements for growth. Nutrient status is a measure of the degree to which the nutrient requirements for maximum

TABLE 2

ANNUAL CYCLING OF NUTRIENTS IN FOREST STANDS[a,b]

Process	Species	Nitrogen	Phosphorus	Potassium	Calcium
Uptake	Pine	45	5	7	29
	Beech	50	13	15	96
	Oak	87.7	6.5	79	95.1
Return	Pine	35	4	5	19
	Beech	40	10	10	82
	Oak	55.6	3.1	58.6	82.8
Storage	Pine	10	1	2	10
	Beech	10	3	5	14
	Oak	27	3	13.9	0.8
Removed by thinning	Oak	5.1	0.4	6.5	10.5

[a] Dengler (1944) (originally from Albert), for pine and beech; Ehwald (1957) (originally Mina), for oak.

[b] In kg/ha.

growth is satisfied (Tamm, 1964). Nutrient requirements of forest trees have been studied for 80 years. Table 3 shows the requirements for three important species on medium sites, as based on European experience according to Loycke (1963).

On the best sites, there is a luxury uptake: for pine 150%, spruce 140%, and beech 145%; on poor sites there is a limited uptake: for pine 35%, spruce 50%, and for beech 33% of the values shown in Table 3 (Loycke, 1963). As noted by Ehwald (1957), nutrient uptake of forest stands show very different values when obtained by different authors. According to Ehwald (1957), for an average production of dry matter forest stands require annually 25–45 kg/ha of nitrogen, 20–40 kg/ha or even 65 kg/ha of calcium; and depending on the computation procedures, either 11–33 kg/ha or only 4–11 kg/ha of potassium, 3–6 kg/ha of phosphorus, and the

TABLE 3

NUTRIENT REQUIREMENTS OF MAIN SPECIES ON MEDIUM SITES[a,b]

Species	Calcium	Magnesium	Phosphorus	Potassium
Pine	90	21	17	56
Spruce	75	19	21	58
Beech	86	17	20	62

[a] After Loycke (1963).

[b] In kg/ha/year.

TABLE 4

CRITICAL CONTENTS OF NUTRIENTS IN FOLIAGE OF SCOTS PINE AND
NORWAY SPRUCE FOR ASSESSMENT OF FERTILIZATION NEEDS[a,b]

Species	Nitrogen	Phosphorus	Potassium	Calcium	Magnesium
Pine	1.2	0.12	0.42	0.29	0.05
Spruce	1.3	0.13	0.50	0.36	0.06

[a] Baule and Fricker (1967).
[b] In percent of dry matter.

same amount of magnesium. Nutrient uptake may be 50% higher than the above values if the forest stands occur on the best site class.

Lately foliar analysis has been used frequently for examining the nutrition status of forest stands. Baule and Fricker (1967) presented a table of critical values for Scots pine and Norway spruce, as shown in Table 4.

Sampling cannot be done during the growing season; the preferred time for Scots pine is from late September to late November, for Norway spruce from October to January. Needles should be collected from 8 to 15 trees, near the top, preferrably from codominants (Baule and Fricker, 1967).

The production of dry matter on the best forest soils of central Europe compares well with the production on the best agricultural soils. Considerably less mineral nutrient, except for calcium is required for production of 1 ton of dry matter forest crop than required for 1 ton of agricultural crops (see Table 5).

A considerable amount of nutrient is stored in older stands. In a 115-year-old Norway spruce stand there were 0.75 tons of nitrogen, 1 ton of calcium, 0.5 tons of potassium, 100 kg of magnesium, and 100 kg of phosphorus stored per hectare. In addition a layer of 3–8 cm thick raw humus stored 300–1000 kg nitrogen, 100–500 kg calcium, 50–200 kg potassium, and 100 kg phosphorus per hectare (Ehwald, 1957).

TABLE 5

AMOUNT OF MINERAL NUTRIENTS REQUIRED TO PRODUCE
1 TON OF FOREST AND AGRICULTURAL CROPS[a,b]

Crops	Nitrogen	Phosphorus	Potassium	Calcium
Forest	4–7	0.3–0.6	1–5	3–9
Field cultures	10–17	2–3	8–26	3–8

[a] Ehwald (1957).
[b] In kg/ha.

TABLE 6

STANDING CROP AND ANNUAL INCREASE OF DWARF SHRUBS
IN DIFFERENT FOREST TYPES IN NORWAY[a]

Forest types	Standing crop (kg/ha)				Annual increase (kg/ha)			
	Dry matter	CaO	K_2O	P_2O_5	Dry matter	CaO	K_2O	P_2O_5
Myrtillus MT	5400	43.1	25.7	10.7	800	10.2	8.8	2.7
Vaccinium VT	6200	53.6	31.3	12.7	1,400	12.5	11.9	3.4
Vaccinium uliginosum–								
Calluna bog	7700	44.5	19.1	17.1	1,500	15.6	8.6	3.6
Empetrum ET	9200	58.0	33.8	19.5	2,100	15.1	14.6	6.0
Calluna CT	14,400	68.1	55.9	28.4	2,600	17.9	16.9	6.4

[a] Mork (1946).

There are few comprehensive investigations about the production and nutrient status of secondary forest vegetation. Mork (1946) analyzed dwarf shrub vegetation in different forest types in Norway. The results of his study are shown in Table 6.

As the increment of wood decreases with decreasing site quality from *Myrtillus* type to *Calluna* type, the amount of dwarf shrub production increases. After burning, a considerable amount of nutrients can be expected to be added to the soil, particularly to the most oligotrophic sites (Mork, 1946). According to Ehwald (1957), the production of secondary vegetation under pines exceeds considerably the production of secondary vegetation under spruce on comparable sites. This may compensate fully for the lesser wood production of pine stands as compared with wood production of spruce stands (see also Fig. 13).

Nutrient status in tropical and subtropical forests have rarely been investigated. A comprehensive report on this situation was prepared by Bazilevich and Rodin (1966) and Rodin and Bazilevich (1966). According to their data, ash elements and nitrogen in the leaves of tropical woody plants range from 3 to 10%, in savannas they may reach 14% of dry weight. In wood and roots, ash contents are considerably lower but still exceed that of temperate regions. Accumulation of nutrients in tropical forest ranges from 2000 to 5300 kg/ha. In equatorial savannas accumulation of nutrients is 700 kg/ha; in dry savannas, 1000 kg/ha. The aerial parts of the bamboo synnusia alone contains 2300–4300 kg/ha of nutrients. In subtropical forests, the accumulation of chemical elements reaches 5300 kg/ha. The uptake per hectare per year was calculated by

Bazilevich and Rodin (1966) to be 2000 kg/ha, of which Si comprised 780 kg; N, 430 kg; and Ca and K, 200 kg per hectare each. Compared with temperate hardwood forests, tropical forests draw 3 to 4 times the quantity of chemical elements into the biological cycle every year. In dry savannas the uptake is 320 kg/ha, and Si is the main element involved because of the predominance of grasses. Annual uptake of mangroves is 370 kg/ha; bamboo, 560 kg/ha.

Bazilevich and Rodin (1966) calculated that the total return of chemical elements in tropical forests exceeds 1500 kg/ha per year; the amount of nitrogen (250–260 kg/ha) is less than that of silicon (750 kg/ha) but more than that of the other elements (Ca, Mg, Fe, Al, K, S, Mn, P, Na, Cl, listed in decreasing order). The moist tropical forest thus shows a greater return of chemical elements in the annual litter than any other type of ecosystems. It was calculated that about 500 kg/ha of nutrients were retained in the net increment (of which nitrogen contributes 170 kg/ha) in the moist tropical forests. In subtropical forests, 150 kg/ha of ash substances and 51 kg/ha of nitrogen were retained in the net annual increment. The amount of leaching from crowns by rain in moist tropical forests of Ghana (1500–1800 mm rainfall annually) was 286 kg/ha, of which 222 kg/ha was potassium.

In spite of rapid decomposition in tropical forests, the amount of litter and surface humus fluctuates from 2000 to 10,000 kg/ha, or about half to one third of the amount of the litter in temperate hardwood forests. In montane tropical forests, litter and surface humus sometimes reach 60,000 kg/ha. Chemical elements accumulated in moist tropical forest litter amount to 80–300 kg/ha, which is less than in temperate forests. In savannas accumulation of surface litter and humus is only 1000–3000 kg/ha. In subtropical forests there are 5000–30,000 kg/ha of litter and surface humus (Bazilevich and Rodin, 1966).

C. Classification of Forest Ecosystems and Ecosystem Models

Ecological classification systems have made great contributions to the development of ecological foundations of forestry. An immense amount of ecological knowledge has been incorporated in the development of classification systems, and is in an organized form. It is impossible to review here even the basic features of the major schools. In general, they have all attempted to classify ecosystems, and each has developed methods particularly suited for the area where the specific type of approach first originated. Thus the Scandinavian schools prefer dominance methods, because they work in conditions where environment limits the number of species, and where differences in vegetation appear as changing patterns of dominant species and growth rate. The presence of a multitude of species does not lead to development of strong dominants in the more favorable conditions of central and western Europe.

The recognition of the principle of bioecological equivalence is common to the major European schools; it enables considerable reduction in the number of site types and has greatly facilitated mapping of vegetation. The systems are convertible in terms of other systems and [as has been shown by Wiedemann (1929), Wagner (1954), Sokolov, as cited by Pogrebnyak (1955)], Cajander's Braun-Blanquet's and Sukachev's classification systems can be transformed into multidimensional coordinate systems where moisture and nutrients are the two major axes. The system by Pogrebnyak (1930) was developed as a multidimensional classification system. The methods of ecological groups are essentially multidimensional coordinate systems, which can be used either as classification systems or for ordination. Ellenberg (1950) introduced a method for computation of community coordinates as simple unweighted averages of species present. There are two axes in the outline of Clementsian primary succession schemes: a moisture axis, giving the position of xeric and hydric succession, and [as used by Cooper (1913) and others] having intermediate conditions already at the beginning of the succession. The other axis of Clements was time. However, it was stated that "fertility increases with time" and thus time can be substituted by a correlated nutrient axis. Actually Clements observed differences in fertility and concluded that these represented development stages in time.

Thus it appears that the classification systems of all major schools of thought can be linked with multidimensional coordinate systems. This, of course, is merely a way to reorganize the systems and facilitates comparisons and better understanding of the systems.

A coordinate method, called method of synecological coordinates, has been used in studies of forest ecosystems since 1957 in Minnesota (Bakuzis, 1959). It is called synecological to emphasize that the performance of individual species is evaluated with respect to conditions of competition as they occur in the area under investigation. Four major coordinate axes were used—moisture, nutrients, heat, and light—signifying the intensity of the biotically effective parts of the essential environmental factors, or, in other words, referring to the operational or functional environment. The scales used are on a relative basis from 1 to 5; thus they are convenient for estimation and avoid the use of zero. The use of zero can be criticized because there is no species which can persist with zero moisture or zero light. Species growing on the driest forest sites are evaluated as having moisture coordinate 1, species growing on wettest forest sites are assigned moisture coordinate 5. A preliminary assignment of coordinates was made from literature data and adjusted to local conditions after an extensive reconnaissance. Such a reconnaissance was performed in Minnesota in 1957 and it yielded 356 plant lists obtained from surveying the apparent full range of variability of moisture, nutrient, heat, and light conditions. Field work for an intensive investigation was undertaken in 1960 and 1961 in the central pine section of Minnesota; it included 55 sample areas, about 4000 m² in size, representing the apparent local moisture-nutrient range. In selection of sample areas homogeneity was sought in environmental conditions allowing the density of the stand to vary with the provision that some of the eight subplots measured would be rather undisturbed. Subplots were measured with concentric circles, 1, 2, 4, 8, 16, 32, 64, and 128 m² in size for the purpose of studying the development of the spatiotemporal pattern. On all subplots 8 m² and less, plants were listed according to species presence and dominance. Reproduction and shrubs were counted and their ages estimated for the same subplots. Trees above 1-inch diameter at breast height were measured on subplots 32 m² and less, and those above 4-inch diameter on two largest subplots.

Synecological coordinates provided a set of ecosystem submodels from the species present in the ecosystem, which serve primarily to estimate the operational environment.

Several sets of soil triangular coordinate systems have been developed. The same approach is used for analysis of data from chemical analysis of plants and their parts. The basic rule for the use of triangular coordinates requires that the sums of the three elements are equal for all sets of those three elements. This restricts the use of triangular coordinates considerably. However, there are several methods of data transformation. The standard method used in the Minnesota studies is to express the original measurements in terms of their means. The second step is to compute percentages for each of the three elements involved based on these transformed values. The distribution patterns in triangular coordinates are centered by this procedure. To obtain better insight into the effectiveness of the triangular coordinates selected (e.g., nitrogen-phosphorus-potassium or simply N-P-K coordinates), it is desirable to plot the original values in the triangular coordinates.

The illustrations presented here can be roughly subdivided into two parts: Figs. 1–6 deal with submodels of ecosystem space and Figs. 7–13 analyze special features of forest vegetation.

Figure 1 presents geographical features of ecosystem space. It uses the classic drawing by Vysotsky as redrawn from Pogrebnyak (1963), showing the broad geographical changes of vegetation from the arctic to the southern desert. Three major processes are going on: paludification, mesification and xerification. Paludification has been intensively investigated in Europe, but mesification has only recently been reported (Zonn, 1955). Xerification of forest vegetation was reported by Kalela (1941) in eastern Patagonia. The cover-type system of the Society of American Foresters (1950) is used to illustrate the differences between the boreal forest region and the northern region after the latter has been subdivided into a western and an eastern part. Only permanent forest cover types are shown. Figure 1 further shows a provisional outline of Minnesota forests and prairie by geographical sections together with the site complexes in moisture-nutrient coordinates (edaphic fields) for individual sections. It indicates that a geographical subdivision has a particular pattern which it occupies in the total ecosystem space.

The geographical nature of forest ecosystems is further explored in Fig. 2, which shows bivariate combinations of all four major synecological coordinates (combinations of moisture, nutrient, heat, and light coordinate axes). Information was obtained from Minnesota studies and from outside sources as indicated by the figure. Methods used by different authors are very different. However, the results, in general, show similar patterns. Figure 2 also illustrates what happens with bivariate synecological fields when dominance weights are used for computation of synecological coordinates. The fields expand considerably. Correlation coefficients between the corresponding coordinates based on presence and dominance are high (Bakuzis, 1967), ranging from 0.6 to 0.9, all being highly significant. However, there is a strong concentration of independent transformations within the center of the edaphic field, as shown in the last diagram of Fig. 2. Plot size and number affects the extent of synecological fields.

Figure 3 shows an approach to a three-dimensional ecosystem space. The top series of diagrams show the third dimension plotted on bivariate fields. Viewed in pairs, the diagrams allow for an insight into the mutual relations of the four coordinates. A double-grouping system demonstrates another attempt to view four-dimensional patterns. There are six possible combinations to cover all groups of double classification; only two of them are shown. Triangular coordinates with transformations can be used to present four major coordinates on one graph. However, only the fourth coordinate is presented in original values, and the others have been transformed. The four triangular diagrams shown do not exhaust all possible combinations in application of triangular coordinates to study the structure of ecosystem space.

Figure 4 shows the performance of three major species in the six bivariate combinations of the synecological fields. It also shows soil groups plotted in synecological coordinates. A comparison with the species distributions reveals that the models are in agreement with previous knowledge.

Figure 5 deals with only the upland part of the central Minnesota pine section site complex. It shows soil triangular coordinate systems plotted within the synecological moisture-nutrient system and shows moisture, nutrient, heat and light coordinates plotted in nitrogen-phosphorus-potassium (N-P-K) coordinates and in soil organic matter–sand-gravel–silt-clay (CM-SG-SC) coordinate systems. In addition, site index and basal area of forest stands are shown in all three sets of coordinates. The association between moisture and nutrient coordinates on upland sites is close. Both coordinates also are closely related to light coordinates; only the correlation is negative. Heat coordinates, however, show a distinctly different pattern. Of practical importance is the different performance of site index and stand basal area in corresponding pairs of coordinate systems. After comparisons with Fig. 6, it can be concluded that stand basal area is more closely associated with nitrogen than is site index. Site index tends to decline at the highest silt-clay content, while basal area continues to increase with increasing silt-clay contents. These differences in response of site index and basal area to the same environmental factors have been noted by Mader (1963) and earlier in European literature.

Figure 6 also shows the expanding triangular field moving from topsoil to the parent material in N-P-K coordinates. The same also happens in OM-SG-SC coordinates with the exception that the A_0-horizon cannot be presented in these coordinates and that the A_2-horizon shows a very different pattern of distribution, evidently due to the leaching effect. An attempt was made to use a different transformation (first transformation) by dividing nitrogen analysis data by 100 instead of dividing nitrogen, phosphorus, and potassium data by their own averages. Figure 6 also shows distribution patterns based on 32 groundcover species in N-P-K coordinates using the chemical analyses data from foliage samples. Note that the distribution outline is oriented differently than that of the soil analysis. In addition, nitrogen distribution in N-P-K coordinates show no distinct pattern as compared with the pattern in topsoil. These patterns were studied to establish a possible transformation rule which could link soil analyses data with plant chemical analyses and thus contribute to the general study of the structure of forest ecosystem space.

IV. POPULATION DYNAMICS IN FOREST ECOSYSTEMS

A. Competition

Vegetation dynamics are closely associated with competition and pattern development. With the exception of the absolute limits of species boundaries, distribution of plants only indirectly depends on physico-chemical environmental factors. These factors mostly influence the competitive ability of plants. Ellenberg (1956) called attention to the fact

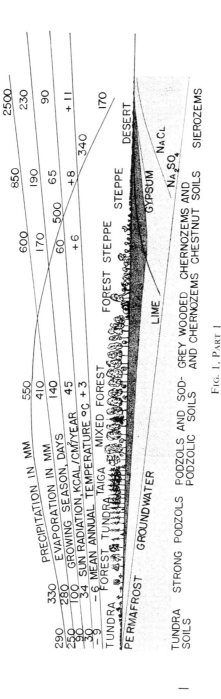

FIG. 1, PART I

PRECIPITATION IN MM
EVAPORATION IN MM
GROWING SEASON, DAYS
SUN RADIATION, KCAL/CM²/YEAR
MEAN ANNUAL TEMPERATURE °C

290	330	550	600	850	2500
250	280	410	170	190	230
90	100	140	60	65	90
30	34	45	500	+8	+11
−9	−6	+3	+6	340	170

TUNDRA FOREST TUNDRA TAIGA MIXED FOREST FOREST STEPPE STEPPE DESERT

PERMAFROST GROUNDWATER LIME GYPSUM NaCl Na₂SO₄

TUNDRA   STRONG PODZOLS   PODZOLS AND SOD-   GREY WOODED   CHERNOZEMS AND   SIEROZEMS
SOILS   PODZOLIC SOILS   AND CHERNOZEMS CHESTNUT SOILS

211

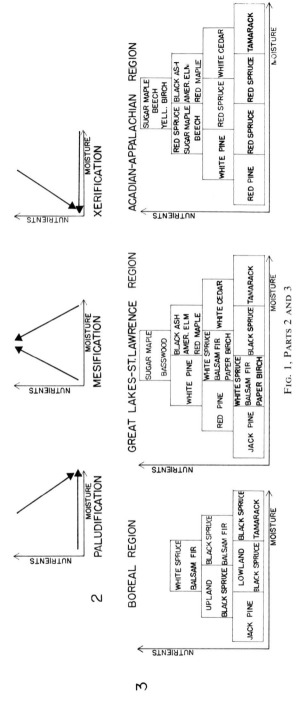

FIG. 1, PARTS 2 AND 3

212

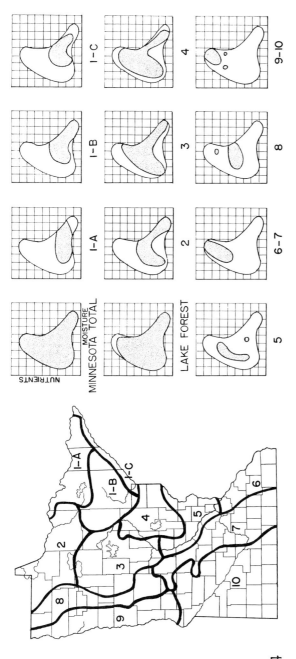

FIG. 1. Geographical nature of forest ecosystem space. (1) North–south profile through European part of USSR from tundra, forest-tundra, taiga, mixed forest, forest-steppe, steppe to desert with corresponding climatic and soil developments (redrawn from Pogrebnyak, 1963; original by Vysotsky). (2) Principal vegetation and soil processes: paludification, mesification, and xerification. (3) Modification of the forest cover type system of eastern North America (data from *Society of American Foresters*, 1954, and additional sources) presented in analogy to Clementsian primary succession; time axis is substituted by nutrient axis, scale is relative. (4) Provisional geographical subdivisions of Minnesota forests. Sections 1 to 5 belong to the lake forest formation, subdivision 1-A is an igneous rock outcrop area; sections 6 and 7 belong to the deciduous forest formation; section 8 is the lake forest-prairie transition or forest-prairie; sections 9 and 10 represent forest groves in the prairie region. Edaphic fields show distribution patterns of local forests in relative moisture-nutrient coordinates (Bakuzis, 1959).

213

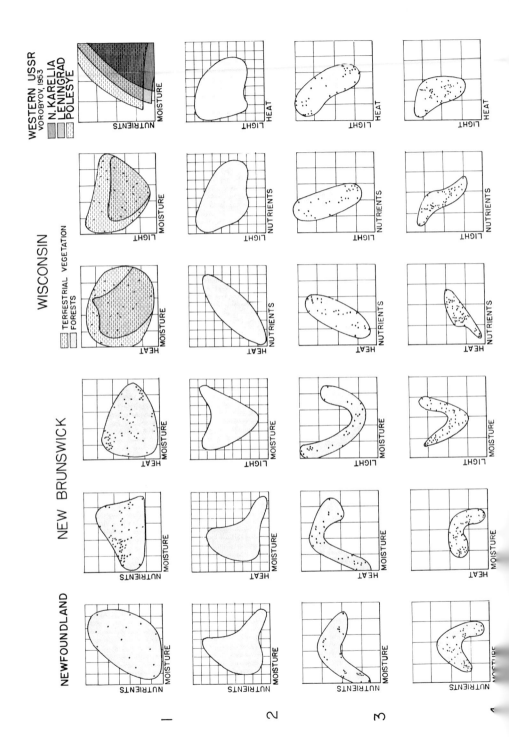

214

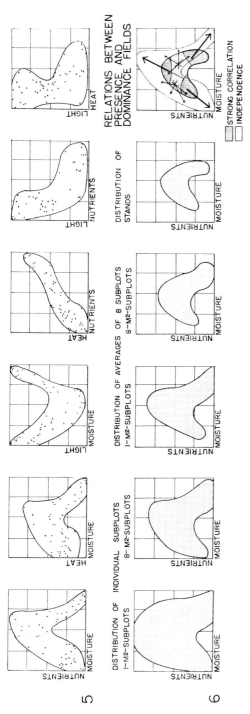

FIG. 2. Forest ecosystem space in different regions. (1) Subsystems with moisture as the main axis (environmental factors directly assessed): Newfoundland (after Damman, 1964), New Brunswick (after Loucks, 1962), Wisconsin (after Curtis, 1959), Western regions of USSR (Vorobyov, 1953). (2) Distribution of 356 Minnesota forest communities in six bivariate combinations of moisture, nutrient, heat, and light axes (environmental factors assessed by evaluation from species presence in the communities: Bakuzis, 1959). (3) Distribution of 30 forest communities of California coastal redwood region (environmental factors directly measured; Waring and Major, 1964). (4) Distribution of 55 forest communities of central Minnesota pine section in six bivariate fields (environmental factors assessed from species presence; Bakuzis, 1967). (5) Distribution of 55 forest communities of central Minnesota pine section in six bivariate fields environmental factors estimated from species by weighting them according to dominance; Bakuzis, 1967). (6) Effect of plot size and number of plots on the extent of the edaphic field of 55 forest communities of the central Minnesota pine section in comparison with edaphic field determined from total species survey and weighing species by dominance (environmental factors assessed from species presence or dominance, respectively; Bakuzis, 1967).

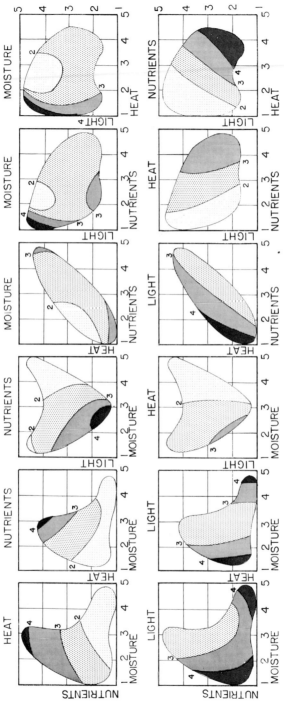

FIG. 3. Multidimensional characteristics of forest ecosystem space in Minnesota based on distribution of 356 forest communities of moisture, nutrient, heat, and light coordinates on a relative scale from 1 to 5. Original values are obtained for individual species from literature data and they represent intensity of operational environment; species are evaluated under conditions of competition; community coordinates assessed as unweighted averages of all species present. (1) Third coordinate mapped on bivariate distributions of community coordinates and shown with the aid of horizontales. Viewed in pairs, four coordinates can be considered simultaneously. (2) Fourth dimension approached with the aid of double classification; only two of six possible combinations are shown. (3) Fourth dimension of ecosystem space shown with mapping of the fourth dimension on triangular coordinates. Use of triangular coordinates requires transformation of three of the original coordinate values.

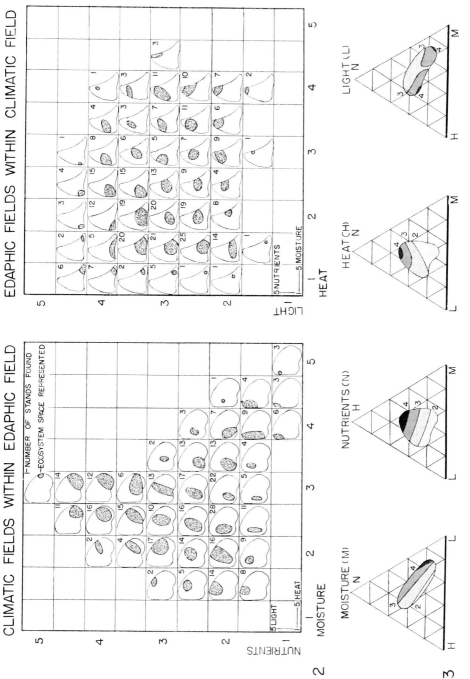

CLIMATIC FIELDS WITHIN EDAPHIC FIELD

EDAPHIC FIELDS WITHIN CLIMATIC FIELD

Fig. 3, Parts 2 and 3

FREQUENCY OF OCCURRENCE IN PERCENT: ▦ 1-40 ▦ 41-70 ■ 71-100

Pinus banksiana

Larix laricina

218

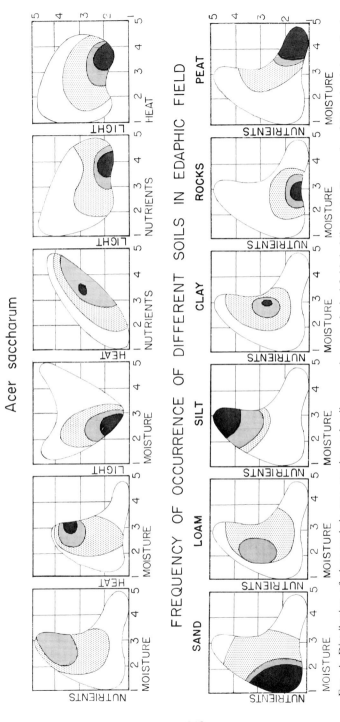

Acer saccharum

FREQUENCY OF OCCURRENCE OF DIFFERENT SOILS IN EDAPHIC FIELD

Fig. 4. Distribution of characteristic tree species and soil groups in synecological fields in Minnesota. Ecographs of *Pinus banksiana, Larix laricina,* and *Acer saccharum* show their frequency of occurrence in individual subunits of the bivariate coordinate systems in moisture, nutrient, heat, and light axes, on a relative scale from 1 to 5. Note the correspondence between distribution patterns of species and soil groups. Basis 356 plots.

219

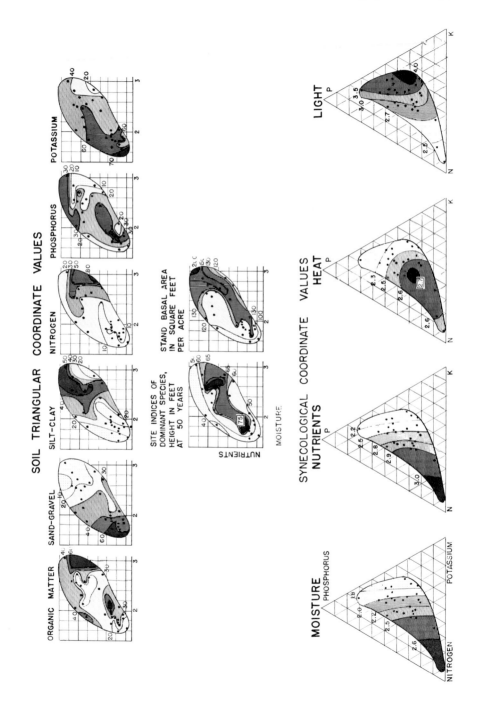

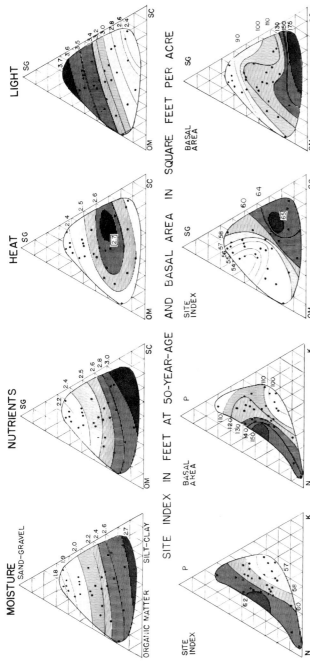

Fig. 5. Relation between synecological coordinates and soil coordinates of the 6-inch topsoil in upland forests of central pine section of Minnesota based on 30 forest community and soil analyses; not all combinations are shown. Two sets of soil coordinates are shown: nitrogen-phosphorus-potassium (N-P-K), and organic matter–sand-gravel–silt-clay (OM-SG-SC). The original values of N, P, and K (or OM, SG, and SC) are divided by the corresponding means of all 30 plots, and the coordinate percentages computed from these relative values. Synecological coordinates of moisture, nutrients, heat, and light are on a scale from 1 to 5 and represent the intensity of operational environment. Site indices and stand basal area are shown in all three sets of coordinates; note the differences of their patterns in different sets of coordinates.

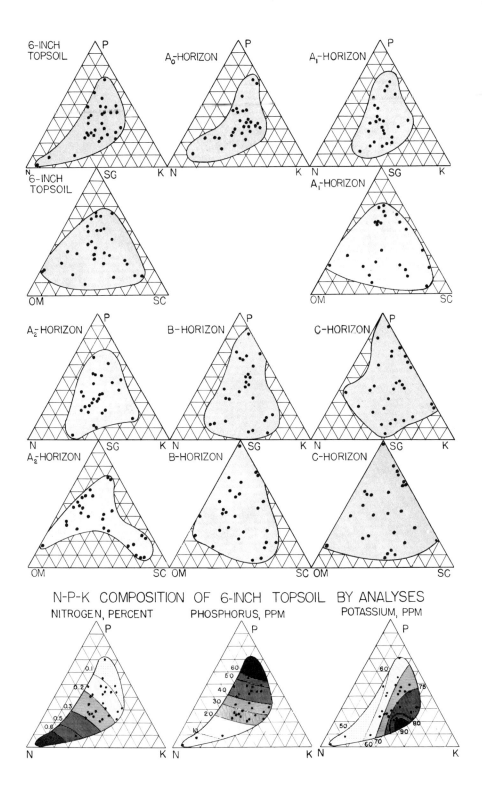

N-P-K COMPOSITION OF 6-INCH TOPSOIL BY ANALYSES

NITROGEN, PERCENT   PHOSPHORUS, PPM   POTASSIUM, PPM

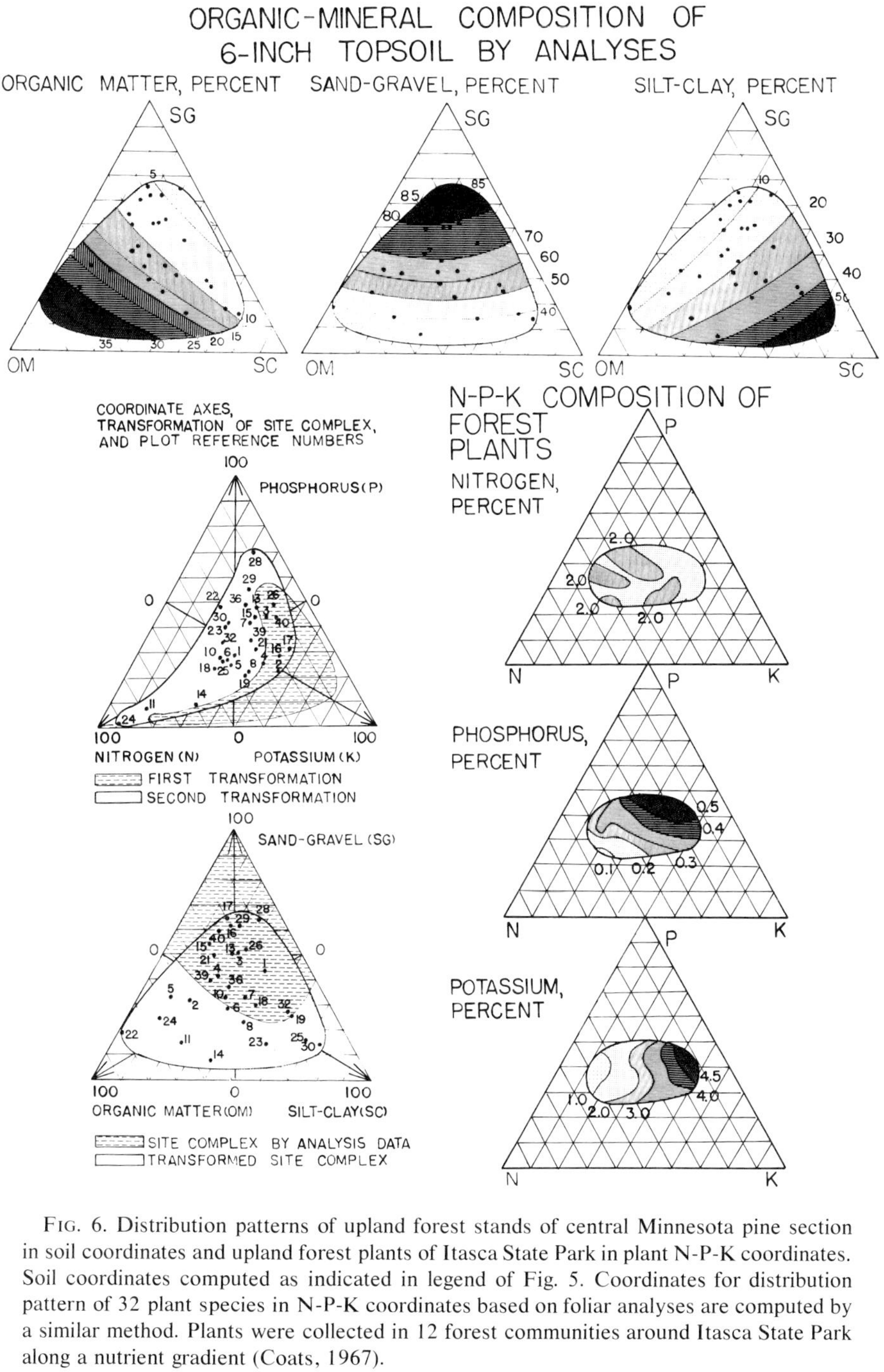

ORGANIC-MINERAL COMPOSITION OF 6-INCH TOPSOIL BY ANALYSES

FIG. 6. Distribution patterns of upland forest stands of central Minnesota pine section in soil coordinates and upland forest plants of Itasca State Park in plant N-P-K coordinates. Soil coordinates computed as indicated in legend of Fig. 5. Coordinates for distribution pattern of 32 plant species in N-P-K coordinates based on foliar analyses are computed by a similar method. Plants were collected in 12 forest communities around Itasca State Park along a nutrient gradient (Coats, 1967).

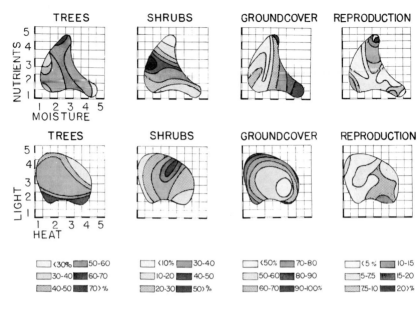

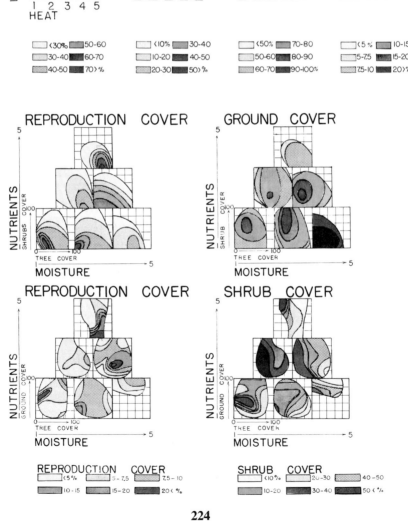

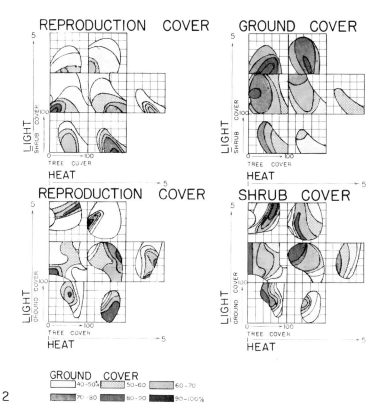

FIG. 7. Forest cover relations in Minnesota, based on estimates in 356 communities. (1) Distribution of tree, shrub, ground, and reproduction covers in moisture-nutrient and heat-light coordinates. (2) Cover interrelations within moisture-nutrient (edaphic) and heat-light (climatic) coordinates. Coordinates represent intensity of the operational environment on a relative scale from 1 to 5 and are assessed on the basis of species present in a community. They refer to conditions of competition. Original data for determination of species synecological coordinates are from literature. Covers are estimated occularly and expressed as percentages of ground covered by their horizontal projections.

that a species responds differently (physiological performance) to environmental factors in the absence of competition from what it does under conditions of competition (synecological performance).

Competition is usually defined as a situation where several species or individuals have similar requirements for materials which are scarce. Competition first becomes operative at higher densities and on better sites. There is a significantly greater number of stems per unit area on poor sites than on good sites when the ages of stands are the same (see Fig. 12). Competition results in the development of vertical and horizontal structure in stands.

Figure 7 shows estimated cover relationships in Minnesota forests. Tree cover has the greatest closure along the mesic moisture line, reaching its peak at the highest nutrient level. This might be an explanation for widespread reports in ecological literature that climax communities are the most productive. This might be true in some conditions. As shown in Fig. 13, this is not true for the Minnesota central pine section or for the northern forest of Wisconsin.

Figure 7 indicates that the distribution of shrubs follows the average nutrient level with a maximum under dry conditions. This shrub maximum occurs at minimum forest cover and low volumes of standing crop as shown in Fig. 13. The information presented in Table 6 refers to dwarf shrubs only. Maximum density of dwarf shrubs, both in Norway and Minnesota, occurs at the lowest nutrient level, while tall shrubs respond to medium levels of nutrients.

Reproduction cover shows three major centers: (1) pine and oak reproduction on dry-poor sites, (2) sugar maple reproduction on mesic-rich sites, and (3) spruce-fir reproduction on moist sites. These are centers for advance reproduction. As shown in Fig. 10, the optimum sites for advance and subsequent reproduction may disagree. This has led foresters rather frequently to erroneous predictions about subsequent natural reproduction. Ground cover responds to positions where there is neither a strong tree cover nor dense shrub cover. However, some groups of ground cover tend to respond to special positions. As noted earlier, dwarf shrubs occupy low nutrient level sites, while mosses under Minnesota conditions are abundant only on wet-low nutrient sites, ferns and their allies on moist-low nutrient sites; while vines are complementary for dwarf shrubs by occupying rich sites.

Figure 7 further illustrates cover interaction in greater detail in a series of five-dimensional graphs which allow a trained mind to comprehend seven-dimensional relations. These diagrams outline reproduction and shrub relationships, pointing out conditions of sharp antagonisms and special cases of coexistence. On the best sites, reproduction cover reaches a maximum under relatively dense tree cover and thus shrubs and reproduction do not coexist. On medium sites shrubs are vigorous, and tree reproduction survives temporarily under a denser tree cover than shrubs. Partial cutting stimulates shrub invasion on these sites. On poor-dry sites, shrubs and reproduction coexist under rather open tree cover. Partial cutting could be allowed on these sites. On poor-wet sites, there is no reproduction and shrub coexistence, and partial cutting favors shrub invasion. Interpretation of relationships in the heat-light coordinates may lead to a better understanding of the complex of multiple competitions. However, broad regional differences should always be considered; conclusions from Minnesota conditions may not hold elsewhere.

TABLE 7

RANK CORRELATION COEFFICIENTS BETWEEN RED PINE OVERSTORY
AND BALSAM FIR UNDERSTORY BASAL AREA, DEPENDING ON
SAMPLE PLOT SIZE IN MINNESOTA[a,b]

Location	Size of understory plots (m²)	Size of overstory plots (m²)			
		10	40	70	100
Chippewa plots					
	10	+0.25	+0.06	−0.02	−0.01
n = 30	40	+0.15	−0.15	−0.08	−0.32
r = 0.463, 1% sign.	70	−0.04	−0.26	−0.24	−0.41
r = 0.361, 5% sign.	100	−0.10	−0.39	−0.38	−0.42
Itasca plots					
	10	+0.39	−0.12	−0.02	+0.03
n = 40	40	+0.25	−0.27	−0.38	−0.34
r = 0.418, 1% sign.	70	+0.18	−0.38	−0.45	−0.51
r = 0.325, 5% sign.	100	+0.15	−0.41	−0.47	−0.55
Superior plots					
	10	+0.43	+0.38	+0.12	+0.22
n = 30	40	+0.49	+0.26	−0.04	−0.05
r = 0.463, 1% sign.	70	+0.44	+0.05	−0.11	−0.15
r = 0.361, 5% sign.	100	+0.39	+0.02	−0.25	−0.32

[a] Bakuzis (1954).
[b] Underlined numbers indicate correlations on plots of equal sizes for overstory and understory.

The average number of shrubs for the Minnesota central pine section is 24,000 stems per acre (60,000/ha), hardwood reproduction 20,000 stems per acre (50,000/ha) and conifer reproduction 5000 per acre (12,500/ha). From these totals 21,000 shrubs, 19,000 hardwoods, and 4000 conifers per acre were less than 5 years old. Conifer reproduction consisted primarily of balsam fir invading sites which are inadequate for later growth. Frequently ecologists and foresters have made judgment errors about succession and reproduction because of lack of familiarity with the mechanics of competition involved.

B. Pattern of Distribution

Pattern is defined here as the spatiotemporal distribution of the vegetational subsystem in the ecosystem structure. Understanding pattern requires an understanding of concepts such as diversity and similarity. Usually the number of different species is assumed to be responsible for the entire diversity. This is only taxonomic diversity. Species may differ taxonomically but belong to the same ecological group; a community may consist of representatives of very different ecological groups. Microsite variability may be more or less responsible for vegetation patterns. The assertion that in climax (mesic-rich conditions) diversity is at a maximum is true for ecological diversity in Minnesota, but not true for taxonomic diversity or for microsite diversity. Taxonomic and microsite diversity in Minnesota forests is greatest near the center of the edaphic field and tends to follow ecotone lines.

The pattern of distribution affects forestry at many different scales. At the largest scale it affects management problems of administrative-economic nature. At a medium level pattern of distribution affects forest protection, in terms of spread of insects and diseases, and spread of fire. At the smallest scale, it affects forest reproduction and production, both in terms of quantity and quality.

Pattern depends on environmental conditions, species morphological characteristics, interaction between species, and between species and the environment. Zinke (1962) investigated the effect of a single tree of *Pinus contorta* on soil properties. Soil pH, nitrogen content, exchangeable bases, exchange capacity, and soil volume weights changed considerably with increasing distance from the tree.

Kershaw (1960) investigated species association in grasslands and reported a change in correlation from negative to positive with increasing plot size. Ghent (1963), measuring balsam fir reproduction with different sizes of nested plots, discovered systematic changes in computed means which tended to decrease with increasing plot size. Different silvicultural systems introduce interesting problems of spatial distribution. Regular clear-cutting in strips provides a simple example of spatial distribution pattern and its change with time (Bakuzis and Brown, 1962).

The change of relationships between overstory of pines and a balsam fir-spruce understory depending on plot size is shown in Table 7.

The Itasca plots were further analyzed to discover the stability of the correlation. For 10 years prior to the measurement, correlation on the 100 m² plot size was $r = -0.50$, while 20 years earlier, it was $r = -0.41$, indicating that competition has increased during the last 20 years, at least with respect to this particular plot size.

The changing sign of the correlation coefficients with changing plot size indicates that each plot size measures a selected set of effects and leaves another set of effects unexplained. The maximum distance between trees within a plot is equal to the diameter of the plot, or the average distance roughly could be equal to the radius of the plot. Thus, small plot sizes explain relations between trees which are close to each other; large plot sizes focus attention on relations among trees at a greater distance.

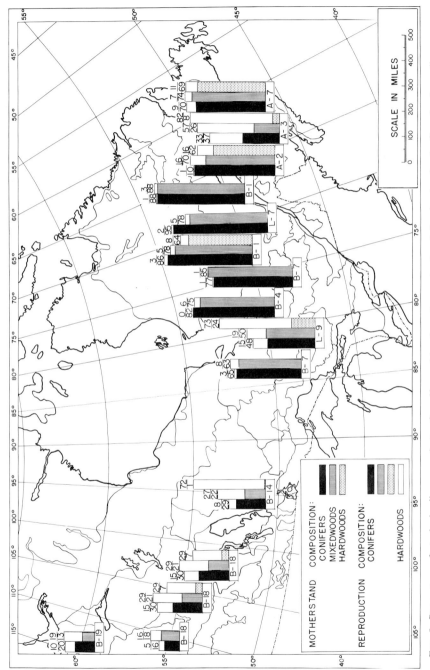

FIG. 8. Percentage of stocked milacre quadrats of conifer and hardwood reproduction following logging in Canada, depending on composition of motherstands (after Candy, 1951).

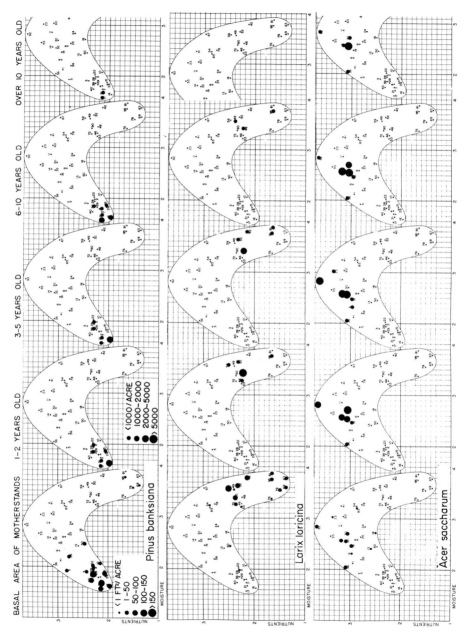

It frequently happens that foresters conclude that trees standing side-by-side should be the first to be thinned out without consideration as to whether or not they truly affect each other negatively. A second-story balsam fir close to the base of a large red pine has its active root system in an area where there are no active pine roots. Stemflow may provide some special sources of moisture; litter fall and leaching from the pine crown may provide more nutrients. Extending the plot size to 500 m² in another area indicated that the undulating topography overshadowed competition effects.

C. Reproduction

Coordinated reproduction surveys over large areas covering different sites, especially when a classification system of forest types is used, provide deep insight into fundamental problems of process which cannot be easily detected from study of a single locality. Unfortunately such surveys are very rare. In 1951 Candy compiled results from a Canadian survey but the results were only grouped; there was scarcely any analysis but the valuable material. Bakuzis and Hansen (1965) analyzed this material with respect to reproduction of balsam fir. Figure 8 shows the general pattern of total reproduction and conifer reproduction which survived logging operations. Most of the conifer reproduction is balsam fir, especially in the eastern part of the surveyed area. The amount of conifer reproduction ranged from 5000 stems per acre in the east to 500 stems per acre in the west. Fire in standing timber resulted in reproduction of 100–2000 conifer stems per acre. Fire after logging was disastrous; sometimes a whole rotation is needed for the forest to return (Hosie, 1953).

Some insight into the status of advance reproduction in Minnesota can be obtained from Fig. 7. An intensive study in the central pine section of Minnesota provided more detailed data, examples of which are shown in Fig. 9. Advance reproduction of jack pine and tamarack is retrogressive: It can only persist on all sites where motherstands are growing for a short time. These motherlands developed following fire; both species are very intolerant to shade and suited for subsequent and not for advance repro-

FIG. 9. Advance reproduction in relation to motherstand basal area in moisture-nutrient coordinates (edaphic field) of central Minnesota pine section, based on 55 stands. Basal area in square feet per acre (1 m²/ha = 4.355 ft²/acre) and number of reproduction per acre (1 ha = 2.47 acres). Stand basal area, and reproduction by age groups is given for *Pinus banksiana, Larix laricina,* and *Acer saccharum.* Coordinates are on a relative scale from 1 to 5; they represent the intensity of operational environment. Coordinates for communities computed as average from the coordinates of all species present. Information about species coordinates is obtained from literature; the values refer to conditions under competition.

duction. Sugar maple reproduction persists under its own motherstands for long periods. Balsam fir and black ash reproduction, especially the 1–2 years old, appear over nearly the whole edaphic field in the central pine section except on clearcut areas. These species, however, show only good growth on very restricted sites in the section, from which they vainly try to spread over new sites.

Reports about subsequent reproduction to clearcutting operations are very scattered and because of a lack of sufficient background material cannot be interpreted. Extensive investigations were carried out in Latvia from 1940 to 1944, started by Dr. R. Markus in 1939. Clearcutting in strips was practiced there for nearly a hundred years, following strong regulations and with excellent historical records. Forest-type classification was done by a system which resembled closely Sukachev's approach, but leaning toward Cajander's idea in recognition of permanent forest types. Investigations were done over the whole site complex, excluding rare oak-ash types. Figures 10 and 11 show some of the highlights of this investigation. The type list is given in Fig. 11. The edaphic field is provisionally outlined as a regular triangle which reflects the neighborhoods of the types and not their ecological amplitudes. This outline was developed in 1954 (Bakuzis, 1959), and a similar triangular outline was reported from that area by Heikurainen (1964).

Figure 10 shows the relationships between the final yield of motherstands and numbers of advance and subsequent reproduction as determined on 5 to 22-year-old clearcut strips. To avoid shrub and grass invasion forest stands were kept rather dense, and this in part explains the small numbers of advance reproduction. After logging operations, only the well-developed and healthy advance reproduction was retained. The width of strips was 40–80 m in a 5-year sequence. The total number of strips investigated was 441. The whole site complex is well subdivided by the main species, each having its own pattern and optima for production, advance reproduction, and subsequent reproduction.

Figure 11 shows the average maximum reproduction capacity in each forest type, and the average maximum reproduction intensity of annual reseeding rate, and for each type summer temperature (May–August) at which the maximum rate occurred. The patterns of capacity and intensity are interrelated. Maximum reproduction capacity was reached on the average in 6–7 years (range from 3 to 13 years). Aspen reached its maximum first, followed by birch, pine, and finally by spruce which was the slowest and which on some forest types tended to develop another cycle of seeding about 15–20 years after the cut. Maximum reproduction intensity was reached by aspen in the first year, by birch in 2–3 years, by pine in 3 years, and by spruce in 8–10 years. Dry sites reproduced better in

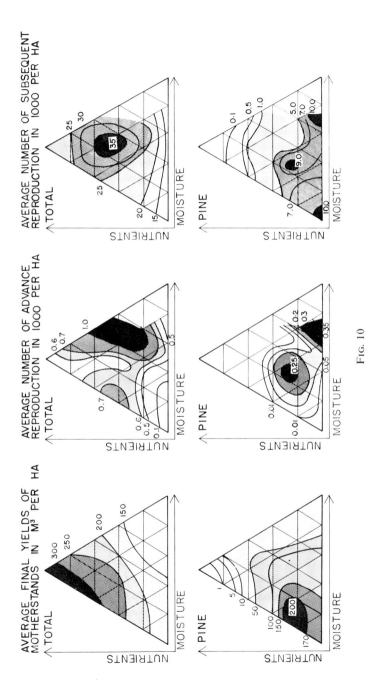

Fig. 10

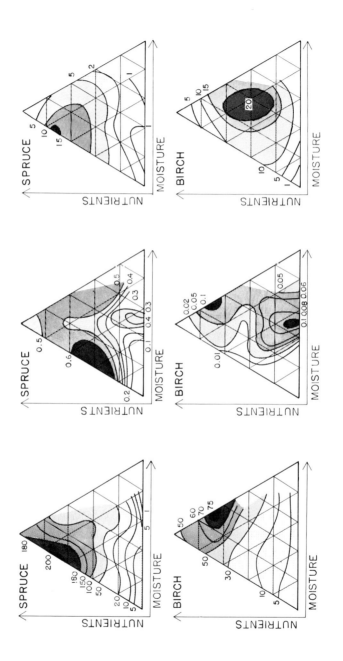

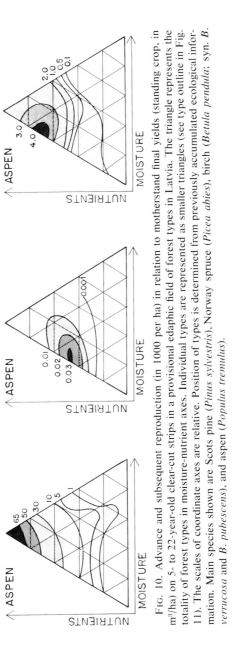

Fig. 10. Advance and subsequent reproduction (in 1000 per ha) in relation to motherstand final yields (standing crop, in m³/ha) on 5- to 22-year-old clear-cut strips in a provisional edaphic field of forest types in Latvia. The triangle represents the totality of forest types in moisture-nutrient axes. Individual types are represented as smaller triangles (see type outline in Fig. 11). The scales of coordinate axes are relative. Position of types is determined from previously accumulated ecological information. Main species shown are Scots pine (*Pinus sylvestris*), Norway spruce (*Picea abies*), birch (*Betula pendula*; syn. *B. verrucosa* and *B. pubescens*), and aspen (*Populus tremulus*).

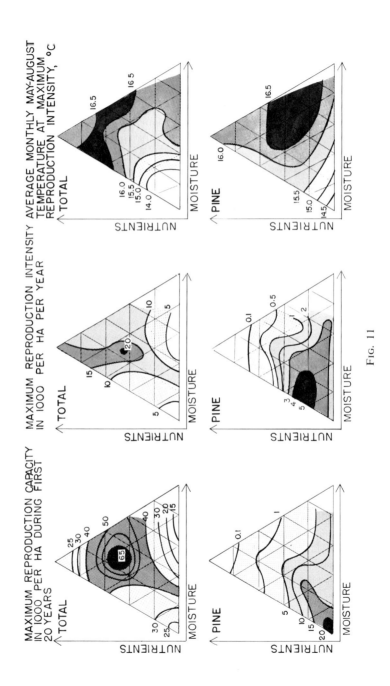

Fig. 11

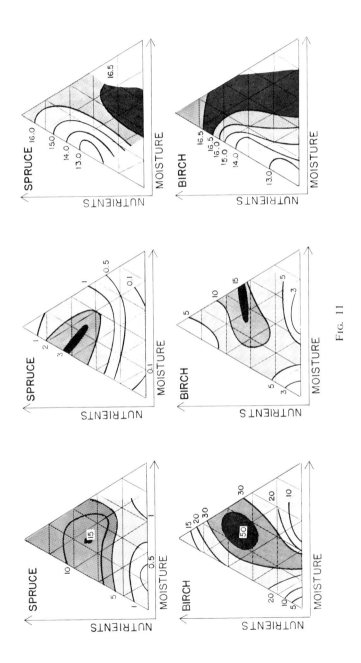

Fig. 11

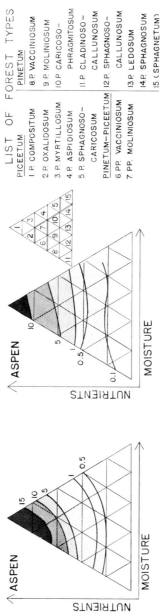

LIST OF FOREST TYPES

PICEETUM	PINETUM
1. P. COMPOSITUM	8. P. VACCINIOSUM
2. P. OXALIDOSUM	9. P. MOLINIOSUM
3. P. MYRTILLOSUM	10. P. CARICOSO–
4. P. ASPIDIOSUM	PHRAGMITOSUM
5. P. SPHAGNOSO–	11. P. CLADINOSO–
CARICOSUM	CALLUNOSUM
PINETUM–PICEETUM	12. P. SPHAGNOSO–
6. PP. VACCINIOSUM	CALLUNOSUM
7. PP. MOLINIOSUM	13. P. LEDOSUM
	14. P. SPHAGNOSUM
	15. (SPHAGNETUM)

FIG. 11. Reproduction capability of forest types after clear-cutting in strips in a moisture–nutrient coordinates) in the forests of Latvia, based on 441 sampled areas. Forest types are arranged according to their relative position to moisture and nutrient conditions as evaluated from ecological studies of these types. The outline does not include some rare hardwood types (oak, ash, black alder). Maximum reproduction capacity represents the average of individual maxima within individual forest types, and is expressed as the total number of seedlings and sprouts of all ages per hectare. Maximum reproduction intensity refers to the number of individuals of 1-year age classes and represents reproduction rate within a certain forest type. It is computed as the average from all stands within a type. Average monthly May–August temperature during maximum reproduction intensity.

years of low summer temperatures; on the better sites and especially on peat soils, warm summers were most favorable for reproduction. The effect of precipitation is not shown in Fig. 11, but it has a bimodal effect, with both dry and wet sites reproducing better in wet years or wet summers (Bakuzis, 1959).

V. PRODUCTION IN FOREST ECOSYSTEMS

The concept of productivity is sometimes confused with the concept of production or the concept of fertility. The concept of fertility can be considered from a purely natural science point of view or in combination with additional economic considerations. Soil fertility refers to a number of soil characteristics and processes occurring within soils which are important to plant growth. Site fertility also includes climatic conditions. Inclusion of the effect of vegetation leads to the concept of ecosystem fertility. The amount of energy in calories fixed as organic matter by the natural vegetation during a unit time serves as a natural measure of fertility. Productivity is the result of interaction of fertility and management measures (Hoffmann, 1963).

There are different types of production systems, each with special characteristics. Peat bogs provide a special case in matter-energy exchange in which nearly all net production accumulates in a similar manner to the fossil fuels of earlier geological times. Wood is the next natural product in which the energy of the sun can be stored for a considerable period of time.

A. Total Production

Total production is a more recent concept of forestry than is the production of merchantable wood. The term merchantable is somewhat confusing because it implies marketing. The essence of the term is that merchantable wood has qualities suitable for utilization. Although the early pioneers were aware of the importance of total production in the forest and a considerable amount of knowledge has been gradually accumulated since their day, the modern concepts of production in forestry came into being with the contributions by Boysen-Jensen starting in 1910. His work, "Die Stoffproduktion der Pflanzen" (Boysen-Jensen, 1932), belongs to the classic contributions in forestry literature and is a significant contribution to science in general.

The early belief was that production of wood was determined by the amount of foliage. Hartig in 1896 had already proved that the same amount of foliage on the best sites produces twice as much wood as a similar amount of foliage on poor sites (Assmann, 1961). Boysen-Jensen tried to explain the differing productivity of light-demanding and shade-tolerant species, and developed techniques for measurement of dry-matter production, loss of matter through respiration, and net production. He formulated a working hypothesis for general production studies as follows: "Net production is equal to gross production minus losses through respiration in roots, stems, branches, and leaves."

In addition to Boysen-Jensen, the Danish school of thought has other important contributors, such as Möller (Mar:Möller), Müller, Holmsgaard, and others. Burger in Switzerland is another well-known contributor to fundamental studies of forest production. The ecosystem concept has produced interest in the fundamental problems of primary production on a worldwide basis, and is the central theme of the International Biological Program within which several large forestry projects are included (Ellenberg, 1967).

Primary production is defined as the total mass produced by an autotrophic plant community per time unit per unit soil area. There are several measures of production: (1) dry-matter weight per unit area and time, (2) elementary analysis as weight measures per unit area and unit time within the biomass, (3) energetically as the amount of energy fixed per unit area and unit time (Lieth, 1965). A summary on measurement of dry-matter production by plant cover was published by Woodwell and Bordeau (1965).

Certainly the newer methods do not eliminate the use of methods of wood measurement developed in forestry and constantly improved over several generations. The newer methods are complementary to the older ones in many situations.

The old forestry methods had made some progress in estimation of branches and stumpwood. Erteld and Hengst (1966), reviewing the knowledge about branchiness of different species, concluded that merchantable branchwood can be estimated as percentage of merchantable stemwood dependent upon tree height but independent of species as shown in the following tabulation:

Tree height (m)	Branchwood (%)
5–15	30–25
15–20	18–12
20–30	16–8

Determination of twig volume is more complex. It depends primarily on species and age. Before the trees reach 5–7 m height all their volume may be in this category. At maturity of 90–100 years, this percentage has decreased to about 10% of the total volume of the tree. The decrease is very rapid for the first 30–40 years, reaching about 30% of the total tree volume, but further decrease in the amount of unmerchantable twigwood is slow (Erteld and Hengst, 1966).

In older German literature it was assumed that stumps and roots comprise 20–25% of the volume above the ground for pine, 25–34% for spruce, and 20–30% for beech. Recent studies in Denmark have reduced these figures to 17% for spruce, 20% for beech, and 15% for oak (Erteld and Hengst, 1966).

Ovington (1962) reported that the forests in England utilize 1.3% of the total incident energy for dry-matter production. According to the information from C. M. Möller et al. (1954), only a small part of gross production is channeled into wood. A beech stand in Denmark at 8 years of age accumulated 34% of gross production as wood increment, at 25 years of age it reached its maximum with 43%, and at 85 years it was down again to 35%.

Polster (1961) presented information on the use of assimilated substances as shown in the following tabulation:

Item	Percent
Total assimilated matter	100
Lost in respiration	45 (25–60)
Lost as litter	16
Not harvested roots	3
Seeds	1
Lost in logging and transport	3
Actual wood production	32

Under American conditions the loss in logging and transport far exceeds these figures.

According to Weck (1955) silvicultural practices have been aimed at increasing photosynthesis, and little further progress in this direction can be expected. Silviculture should try to develop some techniques aimed at decreasing respiration losses. Careful evaluation of fundamental research findings might provide some hints into this or other directions in which possibilities could be explored.

Jahnke and Lawrence (1965) demonstrated that a geometrical evaluation of tree form is important in making judgments about photosynthetic efficiency.

Trendelenburg and Mayer-Wegelin (1955) compiled numerous studies by Burger made from 1929 to 1953 concerning the efficiency of foliage

TABLE 8

EFFICIENCY OF FOLIAGE OF DIFFERENT SPECIES
IN WOOD PRODUCTION[a]

Species	Fresh weight of foliage (kg) to produce 1 m³ solid merchantable wood	Number of leaves or needles, and surface area (m²) of 1 kg foliage	
		Number	Area (m²)
Scots pine	1200	38,000	5.5
White pine	1050	85,000	9.6
Larch	1230	450,000	10.5
Douglas-fir	1300	150,000	6.8
Spruce, even-aged	3000	130,000	5.5
Spruce, uneven-aged	4000	180,000	5.2
Fir	2400	95,000	5.6
Oak	1100	2100	12.4
Beech	880	7000	16.0

[a] Trendelenburg and Mayer-Wegelin (1955); original data from Burger (1929–1953).

of different species in wood production in Switzerland. The data are given in Table 8. Table 8 indicates that pine needles produce more efficiently than fir and spruce. Efficiency decreases with age.

Different tree dominance classes show differences in contribution to wood production, as indicated by Table 9.

Ebermayer (1876) was quite convinced that different species of trees produce nearly the same amount of wood by weight if the sites are comparable. Differences among species are considerable if production is mea-

TABLE 9

CONTRIBUTIONS OF TREE DOMINANCE CLASSES
IN STAND STRUCTURE AND GROWTH[a]

Tree classes	Percent in stand by number	Percentage contributed to volume growth of stand
Dominants	25	58
Codominants	45	37
Intermediates	20	4
Suppressed	8	1
Oppressed	2	+

[a] Weck (1955).

sured by volume; they are less if weight is used as a basis. The calorific values can also be different, and this may change the total evaluation of the species. Of course, industry is frequently interested in special properties which may favor some species more than others. Erteld and Hengst (1966) gave volume and weight ratios for fir, spruce, and beech. Adding the calorific values supplied by R. Müller (1959) and assuming the values for fir equal to 100 we arrive at some ratios, as shown in the following tabulation:

Species	Volume	Weight	Energy
Fir	100	100	100
Spruce	89	94	96
Beech	58	89	80

Table 6 shows the production of dwarf shrubs in Norway. This production increased with decreasing site quality, which may appear paradoxical. Figure 13 shows additional paradoxes from Minnesota and northern Wisconsin. Competition and selection of species are the main causes. Contributions of secondary forest vegetation to productivity have been reported rather frequently, but these reports are very diversified and frequently lack basic information which allows them to be put into a perspective. Their use is rather casual and restricted.

There are some general estimates of total terrestrial production. Newbould (1963) reported that it is generally agreed that the total amount of dry-matter produced is on the order of $4–6 \times 10^{10}$ tons per year. Annual production in dry and short-grass prairie range from about 1.6 tons/ha in short-grass prairie to 5 tons/ha in tall-grass prairie; hardwoods produce 6–7 tons/ha, conifers about 6.4 tons/ha; tundra produces less than 0.9 tons/ha (G. Müller, 1965).

Net productivity may vary from 0.11 g/m^2 per day in desert to 18 g/m^2 for sugar cane; most ecosystems vary from 1–6 g/m^2 per day. On a global scale, perhaps half of the gross primary production is derived from microorganisms, mainly in the sea (Brock, 1966).

B. Production of Merchantable Wood

There have been several attempts to estimate the total growth potential of forests of the world. Table 10 shows data as presented by Weck and Wiebecke (1961).

Of the total potential production of 4430 million tons, 3488 million tons are hardwoods or 5000 million m^3, and 942 million tons are conifers

TABLE 10

ESTIMATED FOREST AREA AND ESTIMATED GROWTH POTENTIAL[a] FOR DIFFERENT
FORMATION CLASSES OF THE FORESTS OF THE WORLD[b]

	Estimated area		Estimated growth		
Formation class	Million hectares	Percent	Tons/hectare year	Total million tons/year	Percent
Equatorial rain forest, lower range	440	18	3.5	1540	35
Equatorial rain forest, mountain range	48	2	3.0	144	3
Monsoon forests and humid savanna	263	11	1.8	474	11
Dry savanna and dry mountain forests in tropics	530	21	1.0	530	12
Temperate rain forests and laurel; precipitation below 1000 mm	20	1	7.2	143	3
Sclerophyllous forests	177.5	7	1.0	178	4
Summergreen forests and mountain conifers	393	16	2.2	865	19.5
Boreal conifers	605.5	24	0.9	556	12.5
Total	2477	100	1.8	4430	100

[a] In tons of dry matter production per hectare and per year.
[b] Weck and Wiebecke (1961).

or 2090 million m³. Of the annual growth potential of 4430 million tons, the present annual harvest is 1600 million tons (Weck and Wiebecke, 1961). The standing crop in European forests is 75 m³/ha; in the United States the average of old-growth standing crop is 86 m³/ha; in the U.S.S.R., 110 m³/ha. Annual increment in Europe is 2.3 m³/ha; in the United States, 2.2 m³/ha, and in the U.S.S.R. 1.2 m³/ha. The highest standing crops in the United States are those on the Pacific Coast, 257 m³/ha; Rocky Mountain area, 106 m³/ha; northern region, 56 m³/ha; and southern region, 47 m³/ha. The annual increment, in cubic meters per hectare, is 3.0 in the Pacific Coast, 1.0 in the Rocky Mountains, 2.0 in the north, and 2.6 m³/ha in the south (Food and Agriculture Organization of the United Nations, 1967).

An important problem since the very beginnings of forestry has been the determination of site class for productivity studies. Up to 1888, yields were used to determine site class. Height as a measure of site class was suggested in 1765 by Oettelt, and was introduced in practice in 1879 by Franz and Baur (Kramer, 1965). Lately attempts have been made to use average increment computed over total age as a measure of site quality. There are many difficulties involved in the use of this measure because it cannot be directly determined as height.

Determinations of site class (or site index) are very difficult in the selection forest. Theoretically this would require very many age determinations and height measurements. The problem is complicated because of the long-time suppression of advance growth. An early system by Flury used the diameter-height relation for site determination. A satisfactory measure still has not been found.

Figure 12 shows a three-dimensional interpretation of a yield table prepared for Scots pine in Germany by Wiedemann (1949). This figure indicates that although final standing crop and total yield (includes all thinnings during the rotation and the final yield) show some tendency to be parallel to stand average heights; considerable errors would follow such an assumption. The slopes of the curves showing the number of stems per hectare and mean diameters at breast height (dbh) are rather highly negatively correlated. Some predictions could be made considering several age groups separately.

Figure 13 shows some relationships of site classes and yields in moisture-nutrient coordinates (in the edaphic field) using different sources of reference. Pogrebnyak (1963) commented that the yield classes for dry-rich ecotopes are extrapolated because of lack of field data. These sites are outside the adaptation limits for forests and they are occupied by grasslands. Thus if there is no additional information, a triangular model for the edaphic field is a better approximation to reality than is a rectangular system. This is related to the triangular form of Clementsian primary succession schemes.

Depending on the moisture-nutrient combination, the site class or site index of each species may vary independently.

Distribution of yields in moisture-nutrient coordinates show different patterns in different regions. In the northern areas such as Newfoundland and Latvia, the highest final yields are in mesic-rich conditions. The main species contributing to maximum yield in both situations is spruce. The central pine section in Minnesota and the northern forest of Wisconsin are areas of the lake forest formation where mesic-rich positions are occupied by tolerant hardwoods with occasional scattered white pine. Pines and spruce fir occur at medium and lower nutrient levels. Production of secondary forest vegetation under the closed canopy of tolerant hardwoods is rarely significant unless the stand is disturbed. This leads to the paradox that the most fertile sites are not the highest producers. In the central pine section, wood production at medium nutrient levels in relatively dry situations is considerably reduced by shrub competition. Using

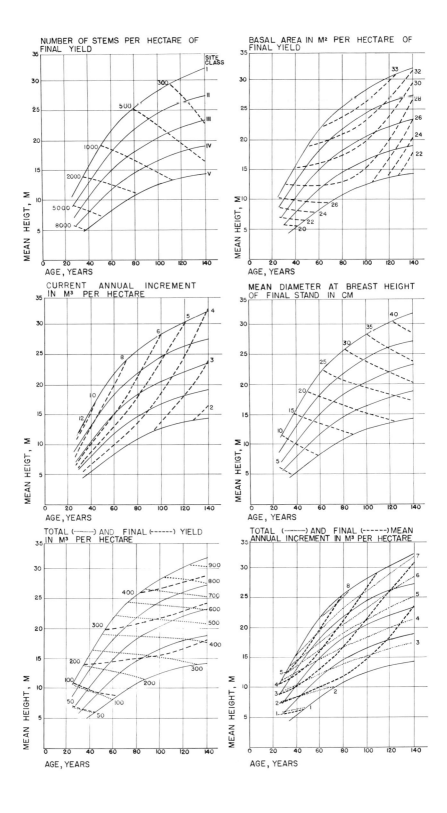

data from Buckman (1966) for hazel shrub (*Corylus* sp.) it appears that the shrub layer may reduce final yield on a 100-year rotation by about 50–100 m³/ha.

Interpretation of yield data, as with interpretation of other ecological information, could be much more fruitful if an agreement could be reached about reference frameworks. As discussed earlier the moisture-nutrient coordinate system is particularly suited and is already in wide application.

The application of mathematical models to problems of yields is 200 years old. Substituting time for environmental factors, Mitscherlich transformed his yield law into a growth law. Weck (1955) stated that Mitscherlich's growth curve fitted rather well to information accumulated in German yield tables. Weck (1955) and Blanckmeister (1957) considered that of all growth curves suggested, the curve by Backman is best suited to this fit.

Climatic indices for prediction of forest growth have been rather popular. However, they worked best in areas from which the original data used in their development came. Paterson prepared a climatically determined vegetation productivity index (CVP) in 1956 and improved it in 1961. This has been successfully tested in some areas.

Czarnowski (1964) considered that the index by Paterson neglected air humidity and grossly exaggerated the role of solar radiation. He used Mitscherlich's yield equation as a starting point, collected numerous functions about climatic and soil relations and the response of vegetation, and developed a complex formula with 11 parameters for determination of the productive capacity of a forest site. The parameters are of three categories: (1) species indigenous properties, (2) climatic conditions, and (3) soil properties. The study needs further testing.

VI. REEVALUATION OF FOREST ECOSYSTEMS

A. Forestry and Natural Resources Viewpoint

The ecosystem concept is a remote idea which can be approached asymptotically but is never fully reached. Fragments of this concept can be traced back to the beginnings of forestry practice and to the common-sense knowledge of hunters and farmers in different parts of the world.

FIG. 12. Development of normal stands of Scots pine (*Pinus sylvestris*) by site classes in Germany (data by Wiedemann, 1949). Site classes are determined as average height of trees at a given stand age. Number of stems per hectare, basal area in m²/ha, mean diameter at breast height in centimeters, current annual increment in m³/ha, total (cumulated) and final (standing crop) yield in m³/ha, total and final mean annual increment in m³/ha are shown.

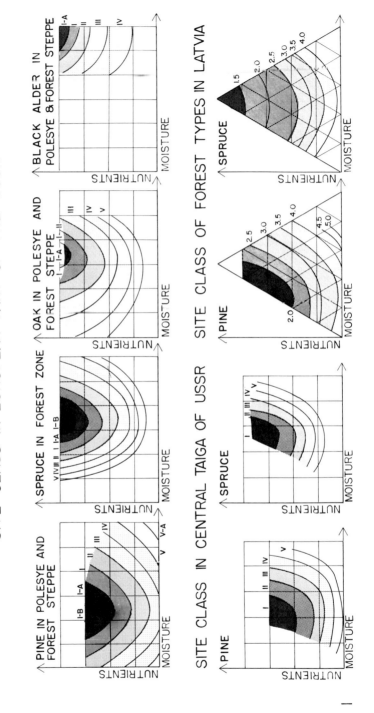

SITE CLASS IN EUROPEAN PART OF THE USSR

BLACK ALDER IN POLESYE & FOREST STEPPE

OAK IN POLESYE AND FOREST STEPPE

SPRUCE IN FOREST ZONE

PINE IN POLESYE AND FOREST STEPPE

SITE CLASS OF FOREST TYPES IN LATVIA

SPRUCE

PINE

SITE CLASS IN CENTRAL TAIGA OF USSR

SPRUCE

PINE

248

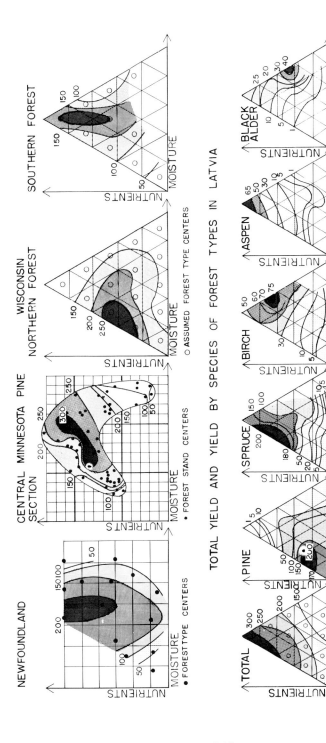

Fig. 13. Site class and yield of forest stands in moisture-nutrient coordinates (edaphic field) of forest ecosystems. (1) Site classes in relation to relative moisture-nutrient coordinate systems are shown for several areas of the European part of the USSR (Pogrebnyak, 1963; Vorobyov, 1953) and for Latvia (Bakuzis, 1959). (2) Yield of mature and nearly mature forest stands in m³/ha is given for Newfoundland (Damman, 1964), central pine section of Minnesota, northern and southern forests of Wisconsin (Wilde et al., 1949), and for Latvia by individual species (Bakuzis, 1959).

The vocabulary has changed, thoughts have deepened, technical solutions have developed, and theories have been tried to clarify the concept; this work will continue in the future.

The ecosystem concept can be considered at different levels and from different viewpoints. Each point of view develops its operational level, a principal level which is used most frequently. However, not all problems can be solved at the standard level. The standard level for forestry needs is based on delineation of ecosystems at a level of forest communities which respond homogeneously to treatment. A physical environmental subsystem ancillary to this criterion is determined. This level is essential for developing of forest classification systems. The final stage of series of ecosystems which can develop on similar physical environments comprise the permanent forest type, frequently called forest site type, as with Cajander's forest types in English translations. This series of ecosystems can also be considered as a group of forest types as done by Sukachev's school of thought. There are no principal differences involved.

The next higher level is the level of landscapes. A landscape is defined as a part of the earth's surface with a certain physiognomy, or a characteristic appearance with causal relations and structure. It is a material and spatial structure, a complex of effects, and with a historical development. Each landscape consists of landscape cells or ecotopes, which are fundamental elements of landscape with considerable homogeneity. The cells of neighboring landscapes differ (Bauer and Weinitschke, 1967). Forest ecosystems are a part of landscape ecosystems.

The fundamental forest ecosystem of forest type agrees with a certain level of a hierarchic classification of plant ecology. However, forest types may not be the best unit for wildlife management, range management, watershed management, farm and suburban forestry, or park and wilderness maintenance needs. The problem is similar to that which forestry has in the application of soil classification systems. There is much useful information, although the systems involved have been treated with different approaches, but redundant information is introduced, and essential information is missing.

Actually watershed management, park and wilderness maintenance, and probably wildlife and range management are considerably more interested in the landscape level than is forestry.

On the other hand, the principal unit of forest ecosystems is too large for many problems of animal ecology and for microbial ecology. Smaller and smaller ecosystems can be defined from specific points of view.

The problem is whether or not it is possible to serve, at least partially, all systems of different sizes and systems constructed from different points of view.

According to Theobald (1966) energy is a coordinate for system identification. Space and time serve as coordinates for object identification. Energy by itself cannot supply a physical explanation of experimental results. Energy, however, does allow us to elicit behavioral laws for the substantive concepts of theory.

Thus energy is a valuable aspect of ecosystems but it is not enough. Matter is physically related to energy and it supplies the missing coordinates for a multidimensional ecosystem framework. As already reviewed and demonstrated by examples, the different approaches to ecosystems should yield enough information to place a system in energy (e.g., heat, light, and mechanical force) and matter (moisture, nutrients, air) coordinates. Most of the ecological work already done in the past supplies enough information to allow estimates of at least a couple of ecosystem coordinates.

All branches of science interested in natural resources appear to have a definite interest in the structure of ecosystem space as outlined by this coordinate system.

An ecological survey or a preliminary reconnaissance for such survey should use a framework of wide interest.

As outlined by Schultz (1967), the ecosystem concept embraces a number of unique subconcepts which provide models for research in research fields and provides a frame for questioning the efficacy of existing practices and policies.

The functional-factorial approach, developed by Jenny (1941) for soil studies and applied to vegetation studies by Major (1951) and many others thereafter, has produced valuable profiles for understanding of ecosystem space properties.

Reevaluation and systematization of the inheritance of the past is one of the basic needs. This material is indispensable to establish scales for an ecosystem survey in matter-energy coordinates.

Much experimental and theoretical work is needed to establish proper quantitative definitions for moisture, nutrient, heat, light, and other regimes. Further detailed work is needed on the coupling of ecosystem elements and development of transformation matrices for inputs and outputs. The work by Lange (1965) seems to supply a mathematical program needed for such an undertaking.

B. General Systems Analysis

Systems analysis provides a rigorous approach to a very large complex of interrelated phenomena. The first books specially designated to serve the needs of ecology and resource management have been published

(Watt, 1966, 1968). General texts already are numerous. These methods require direct application of mathematics to biological relations, while conventional statistical methods have an essential role in preliminary investigations and subsequent testing of mathematical models.

Dimensional analysis offers an opportunity to obtain mathematical expressions for unknown functions from rather general considerations. Results, of course, always should be checked by experimental data. Application of dimensional analysis to problems of forest ecology and management was demonstrated by Khilmi (1962) in 1957. His book, originally written in Russian, was translated to English in 1962. So far there have been no followers in English and German language publications. At least this approach requires a careful examination.

Classic forestry literature provides many useful models which can be adopted to the contemporary use of systems analysis. The models of normal forest, yield models, economic production balance (Faustmann's formula), the period-plan according to Ostwald's relative forest rent theory (Markus, 1967), the classic forest statics, and others supply valuable information and ideas for application of modern operations research methods and simulation.

The Clementsian models of different kinds of climaxes and types of succession require a careful reexamination. Substituting nutrient axis for Clementsian time removes many of the difficulties in understanding his regional vegetation complexes. Time is not a factor that can be used for system identification as concluded by systems analysis (Ashby, 1956; Theobald, 1966). This does not affect investigation of dynamics, succession, and evolution, but merely removes time from a position where it does not conceptually belong.

We must be careful in judging the work of contributors by their use of metaphors like organism and machine. Forest was an organism according to Clements (1909); it is a machine with input according to Ashby (1956). Conceptually it was and is a system, more specifically, an ecosystem.

The entropy and information concepts may provide some possibilities to objectively approach some aspects of resource management where adequate measures still are missing (Lindsay, 1959; Moles, 1966).

Problems of stability, adaptation, regulation, and self-organization are among those problems of ecology which can be approached with the aid of cybernetic principles. The role of humus in forests is particularly interesting as a mediator between the living tree and the purely mineral soil and is part of a feedback system. A question is how useful is the cybernetical theory of relations by Leeuwen (1966) in the investigation of problems such as spatiotemporal distributions and others? It was developed concurrently with studies in succession. Probably in combination with a fresh look at Clementsian succession patterns, it will provide a basis for development of models for succession studies in national parks, wilderness, and other areas of interest as outlined by Stone (1965).

VII. SUMMARY AND CONCLUSIONS

In delineation of ecosystems, forestry puts special emphasis on forest trees and their utilitarian parts. A notion of the ecosystem concept can be traced far back in the history of forestry. Problems of pattern and process and problems of matter-energy exchange have been of equal importance in forestry. Spatiotemporal order is the basis of silvicultural systems aimed at utilizing efficiently the energy supplied by the sun and the matter of the earth.

Forestry recognizes the contributions made by the classic schools of ecological thought.

The related ideas of bioecological equivalence, on the one hand, and the idea of classifying environmental factors according to their mode of action, on the other, have been extremely helpful in overcoming the difficulties of relating ecological requirements of vegetation to the morphology of the environment. The same idea has lately been expressed as the organism-directed approach to the environment.

As a result, any kind of information on ecosystems can be referred to a general and fundamental matter-energy coordinate system of multidimensional ecosystem space. The most important of these ecosystem coordinates are the regimes of moisture, nutrients, air, heat, light, and mechanical energy with all their components.

The classic ecological classification systems, such as those by Clements, Cajander, Sukachev, Braun-Blanquet, and others, can be transformed in terms of ecosystem coordinates. Systems such as those by Pogrebnyak, Ramensky, Ellenberg, and gradient systems already are constructed in terms of ecosystem coordinates. Any other information providing enough background to estimate at least some of the coordinates can be helpful for further development of a more or less general ecosystem framework.

Studies of subsystems have made the greatest contributions in advancing the total ecosystem concept in the past. They will play important roles in the near future as well. The problems involved include collection of ecological functions, transformations from input to output states, and coupling of subsystem elements and coupling with other subsystems.

Institutions with large resources should organize permanent interdisciplinary fundamental ecosystem studies. A permanent ecological survey should be established nationally, preceded by extensive ecological reconnaissance studies.

ACKNOWLEDGMENT

The study was supported by the Graduate School, University of Minnesota, and the Minnesota Agricultural Experiment Station.

REFERENCES

Ashby, W. R. 1956. "An Introduction to Cybernetics." Wiley, New York. 295 pp.
Ashby, W. R. 1964. The set theory of mechanism and homeostasis. *Gen. Systems* **9**, 83–97.
Assmann, E. 1961. "Waldertragskunde." Bayerischer Landwirtschaftsverlag, Munich. 490 pp.
Bakuzis, E. V. 1954. Unpublished report.
Bakuzis, E. V. 1959. Synecological coordinates in forest classification and in reproduction studies. Ph.D. thesis, University of Minnesota, St. Paul, Minnesota. 244 pp.
Bakuzis, E. V. 1967. Some characteristics of forest ecosystem space in Minnesota. *Papers 14th Congr. Intern. Union Forest Res. Organ., Munich. 1967* Vol. 2, pp. 107–125.
Bakuzis, E. V., and R. M. Brown. 1962. Elements of model construction and the use of triangular models in forestry research. *Forest Sci.* **8**, 119–131.
Bakuzis, E. V., and H. L. Hansen. 1965. "Balsam Fir: A Monographic Review." Univ. of Minnesota Press, Minneapolis, Minnesota. 445 pp.
Bates, M. 1961. "Man in Nature." Prentice-Hall, Englewood Cliffs, New Jersey. 116 pp.
Bauer, L., and H. Weinitschke. 1967. "Landschaftspflege und Naturschutz." Fischer, Jena. 302 pp.
Baule, H., and C. Fricker. 1967. "Die Düngung von Waldbäumen." Bayerischer Landwirtschaftsverlag, Munich. 259 pp.
Baumgartner, A. 1965. The heat, water and carbon dioxide budget of plant cover: Methods and measurements. *In* "Methodology of Plant Ecophysiology" (P. E. Eckardt, ed.), pp. 495–512. UNESCO, Paris.
Baumgartner, A. 1967a. The balance of radiation in the forest and its biological function. *Biometeorol.* **2**(II), 743–754.
Baumgartner, A. 1967b. Entwicklungslinien der forstlichen Meteorologie. *Forstwiss. Centr.* **86**, 156–175 and 201–300.
Bazilevich, N. I., and L. E. Rodin. 1966. The biological cycle of nitrogen and ash elements in plant communities of the tropical and subtropical zones. *Forestry Abstr.* **27**, 357–368.
Bennett, J. B., H. R. Josephson, and H. H. Weston. 1958. The heritage of our public lands. *Yearbook Agr. (U.S. Dept. Agr.)*, pp. 42–62.
Bertalanffy, L. 1951. Problems of general systems theory. *Human Biol.* **23**, 302–312.
Blanckmeister, H. 1957. "Mathematischer und physischer Grundriss für Forstwirte." Neumann, Radebeul. 524 pp.
Boysen-Jensen, P. 1932. "Die Produktion der Pflanzen." Fischer, Jena. 108 pp.
Braun, E. L. 1950. "Deciduous Forests of Eastern North America." Blakiston, Philadelphia. 596 pp.
Braun-Blanquet, J. 1964. "Pflanzensoziologie." Springer, Vienna. 865 pp.
Brock, T. D. 1966. "Principles of Microbial Ecology." Prentice-Hall, Englewood Cliffs, New Jersey. 306 pp.
Buckman, R. E. 1966. Estimation of cubic volume of shrubs (*Corylus* spp.). *Ecology* **47**, 858–865.
Candy, R. H. 1951. Reproduction on Cut-Over and Burned-Over Land in Canada, Res. Note. No. 92. Forestry Branch, Canada. 224 pp
Clements, F. E. 1909. Plant formations and forest types. *Proc. Soc. Am. Foresters* **4**, 50–63.

Coats, R.N. 1967. Evaluation of an ecological nutrient gradient in the central pine section of Minnesota. M.S. thesis. Univ. of Minnesota, St. Paul, Minnesota. 120 pp.

Cooper, W. S. 1913. The climax of Isle Royal, Lake Superior, and its development. *Botan. Gaz.* **55**, 1–44, 115–140, and 189–235.

Cotta, H. 1835 (first ed., 1816). "Anweisung zur Waldbau." Arnold, Dresden. 394 pp.

Curtis, J. T. 1959. "The Vegetation of Wisconsin." Univ. of Wisconsin Press, Madison, Wisconsin. 675 pp.

Czarnowski, M. S. 1964. "Productive Capacity of Locality as a Function of Soil and Climate with Particular Reference to Forest Land." Louisiana State Univ. Press, Baton Rouge, Louisiana. 174 pp.

Damman, A. W. H. 1964. Some forest types of central Newfoundland and their relation to environmental factors. *Forest Sci. Monographs* **8**, 1–62.

Dengler, A. 1944. "Waldbau auf ökologischer Grundlage." Springer, Berlin. 596 pp.

Dix, R. L. 1964. A history of biotic and climatic changes within the North American grasslands. *In* "Grazing in Terrestrial and Marine Ecosystems" (D. J. Crisp, ed.), pp. 71–89. Blackwell, Oxford.

Driver, E. 1961. "Indians of North America." Univ. of Chicago Press, Chicago, Illinois. 668 pp.

Ebermayer, E. 1876. "Die gesamte Lehre der Waldstreu mit Rücksicht auf die chemische Statik des Waldbaues." Springer, Berlin. 416 pp.

Ehwald, E. 1957. Über den Nährstoffkreislauf des Waldes. *Sitzber., Deut. Akad. Landwirtschaftswiss. Berlin* **6**, No. 1, 1–55.

Ellenberg, H. 1950. "Unkrautgesellschaften als Zeiger für Klima und Boden." Ulmer. Stuttgart. 141 pp.

Ellenberg, H. 1956. "Aufgaben und Methoden der Vegetationsgliederung." Ulmer, Stuttgart. 136 pp.

Ellenberg, H. 1967. "Internationales biologisches Programm." Deutsche Forschungsgemeinschaft, Bad Godesberg. 28 pp.

Endres, M. 1923. "Lehrbuch der Waldwertrechnung und Forststatik." Springer, Berlin. 326 pp

Erteld, W., and E. Hengst. 1966. "Walderstragslehre." Neumann, Radebeul. 332 pp.

Fiedler, H. J., and J. Reissig 1964. "Lehrbuck der Bodenkunde." Fischer, Jena. 544 pp.

Food and Agriculture Organization of the United Nations. 1963. "World Forest Inventory." FAO, Rome. 113 pp

Food and Agriculture Organization of the United Nations. 1967. "Wood: World Trends and Prospect." FAO, Rome. 130 pp.

Ghent, A. W. 1963. Studies of regeneration in forest stands devastated by the spruce budworm. *Forest Sci.* **9**, 295–310.

Glesinger, E. 1962. The role of forestry in world economic development. *Proc. 5th World Forestry Congr., Seattle, Wash., 1960* Vol. 1, pp. 191–195. Univ. of Washington Press, Seattle, Washington.

Gottmann, J. 1961. "Megalopolis." Twentieth Century Fund, New York. 810 pp.

Graves, H. S. 1899. Practical forestry in the Adirondacks. *U.S. Dept. Agr., Bull.* **26**, 1–84.

Greeley, W. B. 1913. Classification of forest types. *Proc. Soc. Am. Foresters* **8**, 76–78.

Heikurainen, L. 1964. Improvement of forest growth on poorly drained peat soils. *Intern. Rev. Forestry Res.* **1**, 40–113.

Hoffmann, F. 1963. Betrachtung zum Begriffe der Bodenfruchtbarkeit. *Wiss. Z. Tech. Univ. Dresden* **12**, 217–226.

Holmsgaard, E., H. Holstener-Jørgensen, and A. Yde-Andersen. 1961. Bodenbildung, Zuwachs und Gesundheitszustand von Fichten Beständen erster und zweiter Generation. I. Nord-Seeland. *Forstl. Forsoksva. Danmark* **27**, 1–167.

Hosie, R. C. 1953. Forest regeneration in Ontario. *Forestry Bull. Univ. Toronto* **2**, 1–134.

Jahnke, L. S., and D. B. Lawrence. 1965. Influence of photosynthetic crown structure on potential productivity of vegetation based primarily on mathematical models. *Ecology* **46**, 319–326.

Jenny, H. 1941. "Factors of Soil Formation." McGraw-Hill, New York. 281 pp.

Jenny, H. 1961. Derivation of state factor equations of soils and ecosystems. *Soil Sci. Soc. Am. Proc.* **25**, 385–388.

Kalela, E. K. 1941. Über die Holzarten und die durch die klimatischen Verhältnisse verursachten Holzartenwechsel in den Wäldern Ostpatagoniens. *Ann. Acad. Sci. Fennicae, Ser. A IV* **2**, 1–151.

Kershaw, K. A. 1960. The detection of pattern and association. *J. Ecol.* **48**, 223–242.

Khilmi, G. F. 1962. "Theoretical Forest Biogeophysics." Natl. Sci. Found., Israel Program Sci. Transl., Jerusalem. 155 pp.

Kittredge, J. 1938. The interrelations of habitat, growth rate, and associated vegetation in the Aspen community of Minnesota and Wisconsin. *Ecol. Monographs* **8**, 151–246.

Klir, J., and M. Valach. 1967. "Cybernetic Modeling." Iliffe, London. 437 pp.

Knapp, R. 1965. "Die Vegetation von Nord-und Mittelamerika und der Hawaii-Inseln." Fischer, Stuttgart. 373 pp.

Kramer, H. 1965. Bonitierungsmasstäbe in der Forstwirtschaft. *Schriftenreihe Forstl. Fak. Univ. Goettingen* **33**, 79–90.

Lange, O. 1965. "Wholes and Parts." Pergamon Press, Oxford. 74 pp.

Leeuwen, C. G. 1966. A relation theoretical approach to pattern and process in vegetation. *Wentia* **15**, 25–46.

Leibundgut, H. 1962. Orchard versus "naturalistic" silviculture. *Proc. 5th World Forestry Congr., Seattle, Wash., 1960* Vol. 1, pp. 404–408. Univ. of Washington Press, Seattle, Washington.

Lieth, H. 1965. Indirect methods of measurement of dry matter production. *In* "Methodology of Plant Ecophysiology" (P. E. Eckardt, ed.), pp. 513–518. UNESCO, Paris.

Lindsay, R. B. 1959. Entropy consumption and values in physical science. *Am. Sci.* **47**, 376–385.

Livingston, B. C., and F. Shreve. 1921. The distribution of vegetation in the United States, as related to climatic conditions. *Carnegie Inst. Wash. Publ.* **284**, 1–585.

Loucks, O. L. 1962. Ordinating forest communities of environmental scalars and phytosociological indices. *Ecol. Monographs* **32**, 392–412.

Loycke, H. J. 1963. "Die Technik der Forstkultur." Bayerischer Landwirtschaftsverlag, Munich. 480 pp.

Lull, H. W. 1964. Ecological and silvicultural aspects. *In* "Handbook of Applied Hydrology" (V. T. Chow, ed.), Sect. 6, pp. 1–30. McGraw-Hill, New York.

Lutz, H. J. 1959. "Forest Ecology: The Biological Basis of Silviculture," Publ. Univ. British Columbia, Vancouver. 8 pp.

McKee, R. 1966. "Great Lakes country." Crowell, New York. 242 pp.

Mader, D. L. 1963. Volume growth measurement—an analysis of function and characteristics in site evaluation. *J. Forestry* **61**, 193–198.

Major, J. 1951. A functional factorial approach to plant ecology. *Ecology* **32**, 392–412.

Markus, R. 1967. "Ostwald's Relative Forest Rent Theory." Bayerischer Landwirtschaftsverlag, Munich. 128 pp.

Mason, L. H., and J. H. Langenheim. 1957. Language analysis and the concept of environment. *Ecology* **38**, 325–340.

Moles, A. 1966. "Information Theory and Esthetic Perception." Univ. of Illinois Press, Urbana, Illinois. 217 pp.

Möller, A. 1922. "Der Dauerwaldgedanke." Springer, Berlin. 84 pp.

Möller, C. M., D. Müller, and J. Nielsen. 1954. Graphic presentation of dry matter production of European beech. *Forstl. Forsøksv. Danmark* **21**, 327–335.

Mork, E. 1946. On the dwarf shrub vegetation on forest ground. *Medd. Norske Skogforsøksv.* **33**, 269–356.

Morosow, G. F. 1928. "Die Lehre vom Walde." Neumann, Neudamm. 375 pp.

Müller, G. 1965. "Bodenbiologie." Fischer, Jena. 889 pp.

Müller, R. 1959. "Grundlagen der Forstwirtschaft." Schaper, Hannover. 1257 pp.

Newbould, P. J. 1963. Production ecology. *Sci. Progr.* **5**, 91–104.

Nye, P. H., and D. J. Greenland. 1960. The soil under shifting cultivation. *Commonwealth Bur. Soil Sci. (Gt. Brit.), Tech. Commun.* **51**, 1–156.

Ovington, J. D. 1962. Quantitative ecology and the woodland ecosystem concept. *Advan. Ecol. Res.* **1**, 103–192.

Paterson, S. S. 1961. Introduction to phyochorology of Norden. *Medd. Statens Skogsforskningsinst.* **50**, No. 5, 1–145.

Platt, R. B., and J. F. Griffiths. 1964. "Environmental Measurement and Interpretation." Reinhold, New York. 227 pp.

Pogrebnyak, P. S. 1930. Über die Methodik der Standortsuntersuchungen in Verbindung mit den Waldtypen. *Proc. Intern. Union Forestry Expt. Sta., Stockholm, 1929* pp. 455–471.

Pogrebnyak, P. S. 1955. "Foundations of Forest Typology. Acad. Sci. Ukrainian S.S.R., Kiew (in Russian). 455 pp.

Pogrebnyak, P. S. 1963. "General Silviculture." Izd. Selskokhoz. Lit., Moscow (in Russian). 399 pp.

Polster, H. 1961. Neuere Ergebnisse auf dem Gebiet der Standortsökologischen Assimilations-und Transpirations Forschung an Forstgewächse. *Sitzber. Deut. Akad. Landwirtschaftswiss. Berlin* **10**, No. 1, 1–43.

Rodin, L. E., and N. I. Bazilevich. 1966. The biological productivity of the main vegetation types in the Northern Hemisphere of the Old World. *Forestry Abstr.* **27**, 369–372.

Schenck, C. A. 1924. Der Waldbau des Urwaldes. *Allgem. Forst- u.Jadgztg.* **100**, 377–388.

Schultz, A. M. 1967. The ecosystem as a conceptual tool in the management of natural resources. *In* "Natural Resources: Quantity and Quality" (S. V. Ciriacy-Wantrup and J. J. Parsons, eds.), pp. 139–141. Univ. of California Press, Berkeley, California.

Scott, C. W. 1966. The changing aims of forestry. *Forestry* **39**, 10–16.

Siren, G. 1955. The development of spruce forest on raw humus sites in northern Finland and its ecology. *Acta Forestalia Fennica* **62**, No. 4, 1–408.

Society of American Foresters. 1950. "Forestry Terminology." Soc. Am. Foresters, Washington, D.C. 93 pp.

Society of American Foresters. 1954. "Forest Cover Types of North America (Exclusive of Mexico)." Soc. Am. Foresters, Washington, D.C. 67 pp.

Sommerhoff, G. 1950. "Analytical Biology." Oxford Univ. Press, London and New York. 207 pp.

Stone, E. C. 1965. Preserving vegetation in parks and wilderness. *Science* **150**, 1261–1267.

Sukachev, V. N. 1954. Die Grundlagen der Waldtypen. *In* "Festschrift für Erwin Aichinger" (E. Janchen, ed.), Vol. 2, pp. 956–964.

Sukachev, V. N. 1960. Forest biogeocenology as a theoretical basis for silviculture and forestry. *In* "Questions of Forestry and Forest Management" (A. B. Zhukov, ed.), pp. 41–50. Acad. Sci., Moscow.

Sukachev, V. N., and N. Dylis. 1968 (Russian ed. 1964). "Fundamentals of Forest Biogeocoenology." Oliver & Boyd, Edinburgh and London. 672 pp.

Tamm, C. O. 1964. Determination of nutrient requirements of forest stands. *Intern. Rev. Forestry Res.* **1**, 115–170.

Theobald, D. W. 1966. "The Concepts of Energy." Spon, London. 192 pp.

Tischler, W. 1965. "Agraokölogie." Fischer, Jena. 499 pp.

Tkatschenko, M, 1930. Urwald und Plenterwald in Nordrussland. *Proc. Intern. Union Forestry Expt. Sta., Stockholm, 1929* pp. 124–128.

Trendelenburg, R., and H. Mayer-Wegelin. 1955. "Das Holtz als Rohstoff." Hanser, Munich. 541 pp.

Volobuev, V. R. 1964. "Ecology of Soils." Natl. Sci. Found., Israel Program Sci. Transl., Jerusalem. 260 pp.

Vorobyov, D. V. 1953. "Forest Types of the European Part of the USSR." Acad. Sci. Ukrainian S.S.R., Kiew (in Russian). 452 pp.

Waggoner, P. E., and J. D. Ovington. 1962. Proceedings Lockwood conference on the suburban forest ecology. *Conn. Agr. Expt. Sta., New Haven, Bull.* **652**, 1–102.

Wagner, H. 1954. Gedanken zur Berücksichtigung der mehrdimensionalen Beziehungen der Pflanzengesellschaften in der Vegetationssystematik. *8th Congr. Intern. Botan., Paris, 1954* Sect. 7, pp. 9–11.

Waring, R. H., and J. Major. 1964. Some vegetation of the California coastal redwood region in relation to gradients of moisture, nutrients, light, and temperature. *Ecol. Monographs* **34**, 167–215.

Watt, K. E. F., ed. 1966. "Systems Analysis in Ecology." Academic Press, New York. 276 pp.

Watt, K. E. F. 1968. "Ecology and Resource Management." McGraw-Hill, New York. 450 pp.

Weck, J. 1955. "Forstliche Zuwachs- und Ertragskunde." Neumann, Radebeul. 160 pp.

Weck, J., and C. Wiebecke. 1961. "Weltwirtschaft und Deutschlands Forst- und Holzwirtschaft." Bayerischer Landwirtschaftsverlag, Munich. 200 pp.

Wenger, K. F. 1967. Multiple-use silviculture in the United States. *Papers 14th Congr. Intern. Union Forest Res. Organ., Munich, 1967* Vol. 4, pp. 619–630.

Westveld, M 1956. Natural forest vegetation of New England. *J. Forestry* **54**, 332–338.

Wiedemann, E. 1925. "Zuwachsrückgang und Wuchsstockungen der Fichte in den mittleren und unteren Höhenlagen der sächsischen Staatsforsten." Laux, Tharandt. 190 pp.

Wiedemann, E. 1929. Die ertragskundliche und waldbauliche Brauchbarkeit der Waldtypen nach Cajander im sächsischen Erzgebirge. *Allgem. Forst- u. Jagdztg.* **105**, 247–254.

Wiedemann, E. 1949. "Ertragstafeln der wichtigen Holzarten." Schaper, Hanover. 100 pp.

Wilde, S. A. 1958. "Forest Soils." Ronald Press, New York. 537 pp.

Wilde, S. A., F. G. Wilson, and D. P. White. 1949. "Soils of Wisconsin," Publ. Wisconsin Conserv. Dept. 525-49. 171 pp.

Woodwell, G. M., and P. F. Bordeau. 1965. Measurement of dry matter production of the plant cover. *In* "Methodology of Plant Ecophysiology" (P. Eckardt, ed.). pp. 519–527. UNESCO. Paris.

Wright, H. E., and D. G. Frey. 1965. "The Quaternary of the United States." Princeton Univ. Press, Princeton, New Jersey. 922 pp.

Zinke, P. J. 1962. The pattern of influence of individual forest trees on soil properties. *Ecology* **43**, 130–133.

Zonn, S. W. 1955. Die biogezönotische Methode und ihre Bedeutung für die Erforschung der Rolle der biologischen Faktoren in der Bodengenese unter Wald *Arch. Forstwesen* **4**, 578–587.

Chapter VIII Ecosystem Concepts in Fish and Game Management

FREDERIC H. WAGNER

I. INTRODUCTION

There are essentially two aspects to the practice of fish and game management. (The term "wildlife" is used herein to embrace both aquatic and terrestrial wild animals; the terms "fish and game" are used to distinguish, respectively, the fish and the birds and mammals.) The first aspect is the direct manipulation of wildlife populations. Most commonly such manipulation involves exploitation or harvest for economic or sporting purposes. (Exploitation is used here roughly synonymous with use and without the necessarily detrimental connotation implicit in Chapter I.) Population manipulation may also include the protection of endangered species for esthetic, educational, and scientific purposes; and it may involve the control of noxious or economically undesirable species.

The second aspect of fish and game management is manipulation of the environment to enhance or reduce the species in question according to the need. More often than not, the wildlife we utilize is produced in ecosystems not specifically manipulated by man for its production although such ecosystems may be substantially altered by or for other human activities. In some cases, however, our wildlife harvests are derived from systems managed with varying degrees of intensity for wildlife production.

Both of these applied aspects have rested heavily upon the population level of ecological theory for their base. Proper exploitive practices require knowledge of demographic patterns, of responses to exploitation, and in general of the regulatory patterns employed by wild animal populations. Environmental manipulation for wildlife, too, presupposes a knowledge of ways in which various environmental factors operate on species of interest, and again of the general principles of population regulation.

Hence, both fish and game management have been heavily population-ecology oriented, often with the individual species focused upon and only those parts of the remainder of the ecosystem considered that may impinge upon the target species. Indeed much of wildlife management has been applied population ecology, and its research along with that of economic entomology has been among the major contributors (Southern, 1965) to what probably has been the most active of the integration levels of ecology.

A. Past Usage of Ecosystem Concepts: Fishery versus Wildlife Biology

Paralleling this trend of population emphasis, fish management has also been strongly based in ecosystem theory to a degree unknown in game management. This dichotomy seems to have developed for several reasons. First, much of fishery theory has been developed from work on commercially important species, particularly marine and the salmonids. In these species, weight is generally a more important parameter economically than numbers. With the emphasis thus on biomass, the concept of production in the ecological sense (the rate at which energy-bearing tissue is produced) has naturally assumed great importance. This focus on production has been sharpened further by a need to take into account the indeterminate body-growth pattern of fish and the variability of growth rates and adult size in different environments. Much thought has been given to the relative merits of different growth models. Consequently, formulas for fish production exist (cf. Ricker, 1946) and empirical values are available for a variety of species.

The game biologists' concept of production or productivity has not been as explicit in terms of contemporary ecosystem theory. A. Leopold's early definition (1933)—". . . the rate at which mature breeding stock produces other mature stock, or mature removable crop . . ."—with some additional consideration for growth, and for biomass or energy would have approached the mark. While this definition is occasionally quoted, the emphasis in game has been more with numbers than with biomass—apparently for two reasons: (1) Birds and mammals have more nearly deterministic growth patterns and adult size than do fish, and therefore vary less with nutrition or age once adulthood is reached. (2) The game technician is more commonly managing a resource for sport. It is more often the number of animals than the weight which determines the number of hunters whose sporting desires can be gratified.

A second probable reason for the greater emphasis on ecosystem theory in the fishery field lies in the different trophic status of most commercial and sport fish, on the one hand, and most sporting game species, on the other. Most of the fish species are carnivorous, are near or at the top of the aquatic pyramid, and hence are the recipients of the energy flow through most of the food chains in a system. In some cases, one or two species of top carnivores may constitute the convergence point for energy flow from nearly all of the pathways present. On the other hand, most game species are herbivores and share their trophic level with many other species of animals, both invertebrate and vertebrate. Much of the energy flow may pass them by, and their production is therefore not so clearly a function of the primary production of their systems.

A third reason for the greater use of ecosystem theory in the fishery field is closely related to the last, and derives from the difference between the producers of the aquatic and the terrestrial systems. In the aquatic, particularly under pelagic conditions, phytoplanktors of course represent the producer level. As such, the entire trophic level is available for utilization; and potentially, all of the energy fixed in primary production can be moved up the food chains of the system each year toward the top carnivores. This near-complete annual utilization of the primary production does in fact commonly occur, as shown in a number of studies reviewed by Raymont (1966). In the terrestrial situation, on the other hand, much of the production may go into root tissue or woody stems where it is not available even to browsing or grazing ungulates, much less to the more specialized seed, mast, and bud feeders in which category most upland gamebirds and many waterfowl fall. Here again, fish production is more clearly a function of the primary production of a system than is game production.

A fourth reason may lie in possible differences in the pattern of population regulation between fish, on the one hand, and birds and mammals, on the other. This will be discussed at greater length later; suffice it to say at this point that one could probably come nearer gaining a concensus among ecologists for the view that fish numbers or biomass are limited by food (and therefore energy) than one could get for the same generalization about birds and mammals.

These are the reasons underlying the more complete embrace of ecosystem concepts by fishery biologists. The most productive fisheries do tend to coincide with areas of higher primary production. The latter tend to occur in areas of higher inorganic nutrient concentrations due to seasonal thermal overturn at high latitudes, to upwelling wherever it occurs, and to the higher concentrations near the continental margins provided by the emptying of nutrient-laden rivers. Accordingly, fishery biology has been very absorbed with matters of trophic structure, production, and nutrient cycling. And the fishery manager has experimented with fertilizing ponds, lakes, and even portions of the ocean with the objective of increasing primary production and the amount of energy which can be transmitted up the food chains to the desired fish species.

The game biologist, on the other hand, may find that forested areas, although having higher primary production than grassland or savannah (Ovington *et al.,* 1963), produce lesser game crops than the latter (Bourlière, 1963) where more of the production is usable and available. Or he may find that it is the structure of the vegetation in terms of cover, interspersion, and other habitat elements which is a more important limiting influence than the amount of energy-bearing food. These are the reasons why game biologists have thought largely in terms of populations and specific limiting factors (often not food) which restrain the number of individuals (not biomass) in their resource.

B. Recent Increasing Attention to Ecosystem Concepts

In recent years a number of developments in game management are demanding interpretation and application in terms of more comprehensive ecosystem principles. The growing prevalence of environmental pollutants, which cycle through the community food webs and converge and concentrate in carnivores, is demanding more complete understanding of nutrient cycling patterns. We are clearly in need of broader perspectives because of the diverse ways in which man's activities can affect game resources. As Holling (1966, p. 197) has stated: "We are in a moment of history when a bomb exploded in one part of the world affects the food of Arctic caribou and when insecticides broadcast in the northern hemisphere appear as residues in Antarctic penguins."

A second reason for the increasing recourse to ecosystem perspectives in wildlife management lies in a need to understand and predict long-range effects of exploiting natural systems. A considerable part of the world's food supply comes from largely wild, undomesticated systems. Most notable, of course, is the ocean which undoubtedly will be exploited to an increasing degree in the future. In game biology there is growing interest in producing meat with wild animals in parts of the world where production by game is greater than that of livestock. Natural systems tend toward equilibria in their bioenergetic, biogeochemical, and interspecific (particularly competitive and predatory) processes. When major components are removed from such steady-state systems, some form of adjustment in pattern and process will inevitably occur. Some systems seem to absorb perturbation with a minimum of change, as Darling (1964) has stated of western Europe. But others react violently, as we shall observe later in this review. If we are to maintain the productivity of the world's ecosystems and derive a sustained yield, we must understand and be able to predict the effects of perturbation, exploit judiciously, and take counteractions to prevent violent change.

Finally, the growing interest in systems analysis in ecology and resource management (cf. Watt, 1966, 1968) suggests that this approach might become a means for unifying these fields. Watt (1968) envisions all the fields of natural resource management as potentially being unified by a common body of theory and methods, and having a common body of processes and mathematical properties. In this perspective, game population problems not only have a common theoretical base with fishery problems, but also become part of the broader subject of managing whole systems.

C. Objectives of the Review

The objectives of this chapter are twofold. The first is to consider several aspects of wildlife management, particularly but not exclusively game, in terms of ecosystem concepts. The emphasis here is on game because less ecosystem attention has been given this group. To attempt a general review of ecosystem concepts in fishery management would involve the assimilation of an already mountainous aquatic literature. The result could not be contained in a one-chapter review such as this. Many excellent review works and symposia already exist in various aspects of this field.

The second objective is to explore the bases for several divergences of view and concepts between the fish and game areas. Several points of disunity exist between these two fields, and one cannot avoid wondering whether the principles underlying the population processes of these two groups may not be more similar than present concepts imply. One such divergence is the greater consideration of ecosystem principles in fish than in game management.

A second dichotomy is in exploitation theory. With fishery population theory dating back at least a half century to Baranov (1918), and the greater mathematical rigor which has pervaded fishery work, this field seems to have taken on more theoretical precision and depth. Critical review of game exploitation theory and evidence vis-à-vis fishery theory might disclose more similarities to the latter than now seem to exist. According to Watt (1968), the various fields of resource management often have not familiarized themselves with each others' theory and techniques. This has tended to be true even in such closely related fields as fish and game exploitation. A detailed review of exploitation theory would be somewhat tangential to the present subject, but is touched upon briefly here as it relates to ecosystem implications of single-species exploitation.

A final area of disunity, both in basic and applied ecology, is in population regulation theory. Since the regulation and equilibria of entire trophic levels and ecosystems constitute the collective regulation of the constituent species, and since the principles involved are germane to proper wildlife population manipulation and environmental management, a consideration of this subject seems appropriate here.

Hence, the pattern of this review is a consideration of several aspects of wildlife management in an ecosystem perspective, and an attempt at confrontations where dichotomies exist.

II. IMPLICATIONS OF SINGLE-SPECIES EXPLOITATION

In the short fish and game management history of less than a century, most research on the exploitation aspect has understandably concentrated on the responses of individual species to exploitation. A great deal still remains to be learned on this subject, and undoubtedly much research emphasis will continue. However, investigations more and more are looking beyond the species in question to broader effects of exploitation on other facets of the ecosystem, biotic and physical. While concentrating their efforts on single-species problems in their classic work, Beverton and Holt (1957, p. 24) stated in their opening section that this is ". . . now perhaps the central problem of fisheries research: the investigation not merely of the reaction of particular populations to fishing, but also of the interactions between them and of the response of each marine community to man's activity." Several developments of recent years in this general topic merit attention here.

A. Population Responses to Exploitation

Before considering some of the side effects on the ecosystem of exploiting individual species within the system, it may be well to review briefly the effects of exploitation on fish and game populations. The two fields have traveled rather different paths on this subject, and one wonders whether more interchange between the two disciplines might not turn up a common set of principles underlying the two.

1. SIGMOID THEORY IN FISHERIES

Beverton and Holt (1957) have traced two separate, though related, lines of development in fishery population theory. One, which they termed "the analytical," dates back to Baranov (1918) and is based on estimating separate population parameters of recruitment, growth, natural mortality, and fishing mortality. These are integrated into mathematical models which express the response of a fish population to varying levels of exploitation, and which hopefully predict the maximum sustained yield. The analytical method has received a great deal of emphasis, with models varying in detail according to the assumptions different authors make about the nature of the relationship between breeding population size and recruitment, body-growth patterns, survival characteristics, and other parameters.

The second line of development, actually convergent with the first, is based, as a first and simplest approximation, on the well-known logistic population-growth curve of Verhulst (1838) and Pearl and Reed (1920). It assumes that any given species has a characteristic, potential rate constant of increase in a specified physical environment. Termed by Lotka (1956) "the instantaneous rate of increase" and by Andrewartha and Birch (1954) "the innate capacity for increase," we shall use the notation r_n for this parameter. (A subscript is used with this parameter to avoid confusion with r, the actual rate of population increase which will also be used.)

In an unlimited environment a species would increase exponentially according to this rate. No environment is unlimited however, and every population presumably stops its growth when it reaches some equilibrium density, K. The pattern of growth which a population thus undergoes is described by the logistic formula:

$$\frac{dN}{dt} = r_n N \frac{(K - N)}{K}$$

A population growing according to this formula describes a symmetrical sigmoid curve (Fig. 1A) in which the inflection point is at the midpoint of density between zero and K.

Two important implications of this formula derive (1) when the rate of increase per individual in the population, $dN/N\ dt$, and hereinafter designated by the notation r, is plotted as a function of density, N, (Fig. 1B); and (2) when the increment of growth, dN/dt, is plotted as a function of density (Fig. 1C). The first implication follows from the parenthetical term of the equation and implies that the actual growth rate per individual declines as a straight-line function of the density (Fig. 1B).

The second implication (Fig. 1C) is the important one for our discussion at this point. If the time intervals, t, are taken as years in species which reproduce seasonally, then the increments of growth could be taken as the annual recruitment of individuals into the population from reproduction. These increments represent the excess of births over deaths at different densities of a population.

The parabola in Fig. 1C thus indicates that the birth rate is equal to the death rate at K, and no excess exists—the obvious condition of equilibrium. As a population is progressively reduced, however, births numerically exceed deaths to an increasing degree up to the midpoint between zero and K. Below this density the increase increments decline, even though the increase per individual is higher (Fig. 1B), because the breeding population or capitol upon which the interest rate operates is too small.

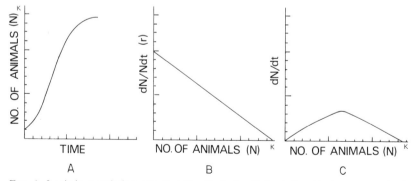

FIG. 1. Logistic population growth (A) with its implied relationship between r, the instantaneous rate of growth per individual, and population density (B), and between dN/dt, the increments of growth per unit of time, and population density (C).

The exploitation implication here is that at equilibrium the population has no margin which it can yield to exploitation. Any removal reduces the population by raising the decrement of natural mortality and harvest above the reproductive increment. At any density below equilibrium, an excess does exist and the population can be stabilized by the removal of exactly that excess. The parabola in Fig. 1C indicates the number of animals which could be removed on a sustained-yield basis at any given standing-crop level of the population. The highest sustained yield obviously could be removed from the density midpoint between zero and the natural equilibrium untampered by human exploitation.

The concept is used in a somewhat specialized sense in fishery problems. The additional dimension of weight is added so that the curve actually represents the biomass production of a population. Furthermore, as generally used in commercial fisheries, the model applies only to that segment of the population which has reached sufficient age to be taken by the fishing gear.

The validity of this model as a generalization about population behavior has been widely questioned (cf. F. E. Smith, 1952; Andrewartha and Birch, 1954; Slobodkin, 1954). Although these criticisms have all been well taken, the model and its implications are sufficiently close to population behavior to be useful conceptual tools and first approximations. Hutchinson (1957) pointed to ". . . the almost universal practice of animal demographers to start thinking by making some suitable, if almost unconscious, modification of this much abused function." Allee *et al.* (1949) and Dasmann (1964a) have reviewed several cases of approximate logistic fit to vertebrate populations. Tanner (1966) reviewed evidence from studies on 71 species reported in the literature and calculated r-density regressions. Of these, 47 were significant (negative) while 15 more were negative but short of significance suggesting the general existence in animal populations of a relationship similar to that in Fig. 1B and some type of sigmoid pattern.

In actuality, symmetrical sigmoid curves (logistics) may not be general. The more common curve applicable to the fisheries biomass situation may be a right-skewed curve with the inflection point to the left of mid-density (Gulland, 1962). Wagner *et al.* (1965) found the r-density regression in several pheasant populations to be curvilinear and approach a negative exponential, also implying a right-skewed growth curve. Similar curves may be inferred for a number of species in plots of reproductive rate on density (Kluijver, 1951), percentage spring-fall increase on density (Errington, 1945) and r on density in some laboratory populations (F. E. Smith, 1963). Such right-skewed sigmoids would imply a maximum, sustained-yield density somewhat less than half the unexploited, equilibrium density.

Beverton and Holt (1957) point out that the analytical and sigmoid approaches are actually convergent, leading to a similar general concept of population behavior. This concept basically holds that in a population at equilibrium, its productive processes (growth and reproduction) are equaled by dissipative processes (predation, respiration, and decomposition). There is no net excess of production over dissipation which could be taken as yield without altering the standing crop.

At densities below the equilibrium point, there is an excess of production over dissipation, the obvious condition underlying population growth. The excess per individual is highest at the very low densities, but because the producing population is small, the total production is small. At high densities near equilibrium, the producing population is large, but the production per individual and total production are small. In the intermediate ranges, production per individual is not at its maximum, but it is large enough that it can, with a moderate-sized, standing crop effect the largest total production.

The important point for our consideration here is that the imposition of a fishery on a balanced population temporarily increases the mortality rate above the reproductive rate and induces population decline. As it declines, reproductive and/or survival rates (the components of r) increase and eventually stabilize the population at some lower density, provided the level of exploitation does not exceed the density-dependent leeway present in the r-density relationship. The harvest thus reduces the population standing crop, the maximum sustained yield coming from no more than half, and quite possibly less than half, the pristine equilibrium level. In the process, a component of an interrelated system with homeostatic tendencies has been reduced, and one would expect some adjustment in the system.

2. Exploitation Theory in Game

With a decade or two less time to pursue what is in any event a young field, with a variety of species ranging from large ungulates with low reproductive rates to highly fecund upland game, and to migratory species which spend half the year away from their nesting range, and with generally less mathematical rigor, game management has not had the unifying benefit of a single, explicit theory or population model. The nearest thing to such a theory has been a general philosophy based on the work of Paul L. Errington. This philosophy, derived largely from his work with bobwhite quail (*Colinus virginianus*) and muskrat (*Ondatra zibetheca*) and with predation on these species, has two major facets which are generally assumed to operate, one or both, in most small-game species.

The first facet, here termed the winter threshold effect, visualizes game populations occupying environments with limited and generally well-fixed capacity to protect animals during the winter season. Each year the reproductive season produces a number of animals in excess of the winter threshold. This annual surplus inevitably disappears through predation, weather, or emigration because of the animals' intolerance to crowding into the limited habitat niches. The animals living within the security threshold experience little if any losses, barring catastrophic weather incidents. If pitched so as to remove no more than a number equivalent to the annual surplus, hunting can take a portion of animals without increasing the fall-spring mortality rate or affecting the population level (cf. Lauckhart and McKean, 1956; Uhlig, 1956).

The second facet, here termed the inversity principle, is based on spring and fall censuses in a number of populations which have shown higher rates of spring-fall increase in years when breeding densities were low than when high. Sufficient flexibility exists in this phenomenon that something approaching a summer or fall threshold level exists, and varying numbers of breeding adults are capable of producing the same crop of young by summer or fall. The implication here again is that even if the winter threshold effect does not operate, the population can be shot in fall and the reduced breeding population which results can produce virtually as large a crop of young as more breeders would have (Allen, 1956, 1966; Linder *et al.,* 1960).

The important point here is that hunting is assumed not to affect the standing-crop density of populations. Small-game hunting philosophy is based on this premise, two eminent spokesmen for the profession (Allen, 1947; Hickey, 1955) having stated that hunting is justifiable only as it does not affect standing-crop levels of harvested species. Sigmoid population-growth patterns have occasionally been alluded to in small-game publications, but it seems fair to say that the sigmoid model and its implications in terms of exploitation effects on standing-crop levels have neither served as a general conceptual base nor been fully perceived or embraced in small-game management.

Although critical review and resolution of this problem are needed, such an analysis is beyond the scope of this review. At this stage the evidence is conflicting. Several studies seem to show complete compensation for hunting loss with no effect on density (cf. numerous examples cited by Allen, 1954; Uhlig, 1956; Peterle and Fouch, 1959). Subjectively it does appear that many game populations can exist under heavy hunting pressure without obvious effects on their density. Yet Wagner *et al.* (1965) and Wagner and Stokes (1968) have pointed out certain population implications of Erringtonian theory and the fact that these do not hold in pheasant (*Phasianus colchicus*) populations. I have (Wagner, unpublished data) found a similar failure in bobwhite and ruffed grouse (*Bonasa umbellus*) data published respectively by Kozicky and Hendrickson (1952) and Edminster (1938).

Even less of a formal theory or population model exists in migratory waterfowl. Although migration largely obviates any assumption of a winter threshold effect, there has been at least some tendency in the field to apply approximate Erringtonian philosophy, perhaps dependent on the inversity phenomenon; and some workers (cf. Hickey, 1955) have been inclined to hold that hunting could not be justified if population levels were affected. However, since Hickey (1952) first showed that annual mortality rates in mallards (*Anas platyrhynchos*) are a straight-line function of the level of hunting kill, this relationship has been shown in a variety of species including the black duck (*Anas rubripes*) by Geis and Taber (1963), the green-winged teal (*Anas carolinensis*) by Moisan et al. (1967), and pheasant hen (Wagner et al., 1965). As Wagner et al. pointed out, such a correlation violates one implication that follows from a strict security-threshold phenomenon.

Although waterfowl populations clearly fluctuate with precipitation and the amount of habitat, Hochbaum (1947) some time ago signaled a concern for the possible effects of hunting on population levels. Today (1969) there is a wide, though by no means universal, suspicion that the shotgun is a significant depressant on waterfowl numbers.

The general exploitation philosophy in many large ungulates has been quite different from that in upland game and waterfowl. Some species, primarily those of climax communities, such as big-horned sheep (*Ovis* spp.) and barren-ground caribou (*Rangifer tarrandus*) in North America, have not fared well and restrictive exploitation philosophy is in order. However, the more common situation has been one of population increases occasioned by the elimination of large predators, improvement of habitat, and earlier restrictive hunting regulations (cf. A. Leopold et al., 1947). The need has often been to reduce populations and there has been little doubt of the additiveness of natural mortality and hunting kill. In some populations, there may be virtually no other source of mortality beside firearms once adulthood is reached (cf. Eberhart, 1960).

In summary, then, there is little doubt of the reduction effect of hunting on ungulate populations, the general goal being to reduce them to and hold them at levels where they maximize the energy flow from the vegetation without harming that vegetation. A. S. Leopold (1949) and Dasmann (1964a) have advocated an actual sigmoid management philosophy in big game. In waterfowl and upland game the picture is somewhat ambiguous. The suspicion persists that hunting influences waterfowl levels and perhaps upland game more often than we suspect. Perhaps a sigmoid view would be realistic in these species along with a goal to harvest at maximum sustained-yield densities rather than hope for a yield without depressing standing crops. Both Scott (1954) and Darling (1964) have advocated a sigmoid philosophy as a general approach to the management of wildlife.

As stated above, however, the manner in which populations respond to exploitation is relevant in this review to the effects on entire ecosystems which result from the exploitation-induced change in numbers of important species. Several responses of this kind have been documented.

B. Exploitation and Niche Competition

Two ecologically similar species in an ecosystem may compete for a common resource and thereby inhibit each other's population growth below that possible in the absence of such competition. A model exploring the consequences of such competition was early developed by Volterra (1926) and Gause (1934). This subject has elicited a great deal of interest among ecologists and has been neatly summarized by Slobodkin (1961) and Hutchinson (1965).

In brief, the formulas expressing the simultaneous growth of two competing species in Slobodkin's notation are:

$$\frac{dN_1}{dt} = \frac{r_1 N_1 (K_1 - N_1 - \beta N_2)}{K_1}$$

$$\frac{dN_2}{dt} = \frac{r_2 N_2 (K_2 - N_2 - \alpha N_1)}{K_2}$$

where N_1, r_1 and K_1 are the same values as in the logistic curve above for species 1; N_2, r_2, and K_2 are the logistic values for species 2; β is the depressive influence of one individual of species 2 on the growth of species 1; and α is the depressive influence of one individual of species 1 on the growth of species 2.

The outcome of the simultaneous growth of the two populations varies, depending on the magnitude of α and β, the K values for the two species, and in some cases on the relative numbers of the species at the time simultaneous population growth begins. If each species has a strong depressive influence on the other (α and β are relatively large), the two species cannot usually coexist, and the one which is more numerous when competition begins will eventually lead to the disappearance of the other. If the influences of the two species on each other are relatively inocuous, the two may coexist. If one influences the other more strongly than it is influenced, the species with the stronger depressive effect will exclude the weaker.

These consequences of the Volterra-Gause model seem to be borne out by a number of laboratory studies on two-species competition which Hutchinson (1965) summarized. And they seem to be borne out in nature by the fact that closely similar species usually display at least slight differences in niche requirements which reduce the reciprocal competitive influence and permit coexistence.

The consequences of the model can be changed by stopping growth of the two populations at some point below where they attain the alternatives described above. Thus in nature, a predator which limited population growth could prevent two species from increasing to the exclusion point, as Gause (1934) suggested. By the same token, the removal through exploitation of such a predator could release the restraint, allow the competing species to go to the Volterra-Gause end point, and conceivably lead to the extinction of one of the competitors.

Slobodkin (1964) demonstrated the first alternative with laboratory populations of two Hydrid species. By exploiting them, and thereby acting in the role of the predator, he enabled them to coexist when they could not do so without his intercession.

The recent work of Paine (1966) demonstrated the second alternative experimentally in a natural ecosystem. Along the west coast of North America, communities of sessile invertebrates develop on rocks of the intertidal zone. Some dozen or more of these species may be preyed upon by a single species of starfish (*Pisaster*). Paine postulated that their coexistence was made possible by *Pisaster;* when he systematically removed the latter from an experimental area, the number of prey species began to dwindle. He predicted that ultimately all but one would disappear.

Similar instances can be ascribed to the removal of large carnivorous mammals in North America and perhaps the influence of hunting by American Indians. These influences probably played a role in restraining population growth of the various species of ungulates on the continent and in holding their numbers within the carrying capacity of the vegetation under primeval conditions. While most of these ungulate species display marked differences in habitat and/or food preferences, a subject which will be pursued at greater length below, the North American elk or wapiti (*Cervus canadensis*) is perhaps the most versatile and nonspecific feeder of the 12–15 species extant on the continent (Murie, 1951). With the natural checks removed from this species, it has increased in many areas to densities at which it competes severely with species that it coexisted with prior to European settlement.

In Yellowstone National Park in western United States, the half-century–long excess of elk numbers has already led to the disappearance of two species which could not compete—the white-tailed deer (*Odocoileus virginianus*) and the beaver (*Castor canadensis*)—and it threatens some of the remaining competitors, particularly bighorn sheep (*Ovis canadensis*) (Anonymous, 1961). In Banff and Jasper National Parks in Canada, elk are similarly causing severe competition for mule deer (*Odocoileus hemionus*), moose (*Alces alces*), and bighorn sheep (Flook, 1964). In all of these cases, management needs now demand that humans artificially contain elk numbers, thereby acting in the capacity of equilibrium-maintaining influences of the primeval era.

In the case of successfully coexisting species, the consequences of the Volterra-Gause model can also be changed by altering the constants in the equations which have permitted coexistence. In actual cases, human exploitation can apply pressure to one or both species; this has the same effect as altering the constants. Larkin (1963) explored this possibility deductively through computer simulation by selecting arbitrary values for the equations and applying various levels of exploitation to each of two hypothetical fish species. Different combinations of exploitation levels altered the yield patterns for the two species. Under heavy enough pressure, either species could be eliminated while the other increased to numbers higher than its coexistent, equilibrium level.

These predictions seem to be borne out by several recent cases. In a fascinating analysis, Murphy (1966, 1967) has traced the changes in the Pacific sardine (*Sardinops caerulea*) population of the California Current system along the west coast of North America. In the early 1930's, fish 2 years and older in this resource were estimated to average 4×10^6 tons per year. Through fishing in excess of sustained-yield levels, reduction in the number of breeding age classes and consequent reproductive failures, the population declined to somewhere between one-tenth and one-twentieth of this level by the latter 1950's.

The sardine is a zooplankton feeder, and the calculated energy requirements of the resource at its 1930–1932 level constituted a major fraction of the total, calculated zooplankton production of the California Current system. As Murphy (1966) stated ". . . energy made available by the decline of the sardine should be reflected in an increase in other elements of the community. . . ." This expected response seems to have occurred with the increase of the anchovy (*Engraulis mordox*), a species with very similar ecology, including similar food habits. During the 1950's alone, the anchovy increase was more than fivefold, and at the end of the decade its annual spawning population was estimated at 4.8×10^6 tons, a value very similar to the annual biomass of sardines in 1930–1932. Murphy (1966) concludes that the reversion of these two species is not likely to occur unless man or nature acts to reduce the anchovy population.

A spectacular succession of changes in the deepwater fish of Lake Michigan, reviewed in a most interesting paper by S. H. Smith (1966, 1968) has evidently been brought about largely through heavy exploitation. Prior to the turn of the century, the geologically young and simple, deepwater fish fauna of the lake was dominated by a single top carnivore, the lake trout (*Salvelinus namaycush*), which fed mostly on seven species of chubs (*Leucichthys* spp.). In the early part of the century, the lake trout supported a commercial fishery which was fairly stable but which may have been somewhat overexploited, judging by the decline in yields by the 1930's. The exotic, parasitic lamprey eel (*Petromizon marinus*) arrived in the lake in the 1940's and immediately began parasitizing the lake trout. This added pressure apparently was a *coup de grâce* leading to the rapid decline which Ricker (1963) predicts will occur when a species, already at or beyond the allowable exploitation level, is subjected to even slight additional pressure. The lake trout fishery virtually disappeared by the 1950's.

A sequence of changes in the forage species of the deepwater ecosystem occurred concurrent with and after the decline of the lake trout. Smith ascribed these changes to the successive shifts of fishing pressure to, and population reductions of, the various forage species. However, one cannot help wondering whether the reduction of the lake trout may have invoked Paine's top-carnivore effect, at least in the later stages of the sequence, and may have been partly responsible for the drastic changes in the species composition of this system.

From 1898 to 1964, the commercial catch was dominated by the lake herring (*Leucichthys artedi*), lake whitefish (*Coregonus clupeaformis*), and the chubs. The combined annual catch of these species remained remarkably stable at somewhere between 11 and 20 million pounds.

The relative abundance of the species varied during this period. At the beginning of the century the herring was by far the dominant fish, alone producing a mean annual yield of about 17 million pounds. This take was apparently excessive, and the yield declined to where, by 1920–1929, the catch had dropped to one fourth the above level. The production of the system, previously used by herring, was now free for whitefish and chubs which were a rising component of the catch. Whitefish, which never increased to more than about a third of the combined catch, reached their peak in the late 1940's. But they too were apparently overexploited, and declined after 1950.

With both the herring and whitefish now reduced, the chubs rose rapidly, presumably making use of the energy freed by the decline of the other two. In the 1950's and early 1960's, the chub catch, which had totaled less than 4 million pounds prior to 1930, was now varying between 10 and 13 million, as much as the combined yield of all the forage species in the pre-1950 era.

Chubs, too, were apparently overexploited and began declining after 1962 setting the stage for the next and last major step in the sequence affecting the native species. This began in the late 1940's with the arrival of the exotic alewife (*Alosa pseudoharengus*). This species was first discovered in the lake in 1949 but it did not appear in the catch until the late 1950's, and then in small numbers. As late as 1962, the catch was still below 5 million pounds. But Smith concludes that the now-overexploited and declining chubs were at a competitive disadvantage, leaving a niche vacancy which resulted literally in an explosion of alewife. By 1965 the alewife catch was 14 million pounds, and a 30-million pound take was predicted for 1966.

Smith concludes that these changes were wrought primarily by the heavy species-specific exploitation. Although the upper Great Lakes (including Michigan) have been enriched during the period, eutrophication has not proceeded as far as the profound changes which have occurred in the eastern lakes.

Smith also concludes that these changes, which constitute the complete collapse of the deepwater fish community of the lake, are irreversible, as Murphy concluded of the sardine-anchovy reversal. Human intervention may be able to improve the present situation, but the near disappearance of some of the indigenes and the arrival of the alewife probably prohibit any return to the pristine condition. Recent management efforts have involved the introduction of several predatory species including attempts to return the lake trout to respectable numbers. The predator-free alewife has now increased to numbers which are detrimental to its own welfare, as suggested by the observed reduction in growth rate and weak year classes reported by Smith along with spectacular die-offs in the lake in recent years.

C. Conclusion

Our knowledge of the patterns and processes of ecological systems, while still in the early stages of development, is beginning to disclose some of the underlying principles on which these systems operate. Since the resources of any ecosystem are limited and the reproductive and growth tendencies of the biota constantly press it toward the limits of resources, we expect competitive interactions between organisms. The long-range, approximate stability of ecosystems suggests that the component species have achieved a degree of homeostatic coexistence. One would predict that alteration of the pattern of an ecosystem, as with exploitation for food or sport, would elicit adjustments of various types. We have examined a few examples of such adjustments.

The results of these exploitation-induced changes in the outcome of competitive situations point to a need for looking beyond single-species responses to the broader impact on other ecosystem components. Traditionally, we have weighed the values of predator control for wildlife management and livestock husbandry against the possibility that it could release incipient pest species. Now an entirely new consideration enters the picture: the possibility, as Paine suggests, that predation may promote community complexity. If this suggestion is correct—and it may yet not be past the hypothesis stage—then we must evaluate reduction of predatory species against the ecological, conservation, and esthetic values of community complexity. Parenthetically, one may wonder what the sale of dried starfish in the curio stores and tourist shops of coastal cities portends for the intertidal fauna of these regions.

Similarly, we need in exploitative situations to be aware of competitive species which could displace those exploited, and perhaps apply pressure to these competitors. Larkin suggests, as one implication of his work, that fishermen may need to guard against the use of highly selective gear which only takes a single, desired species. Murphy advocates applying "judicious" pressure to all of the ecologically similar species within a trophic level. In fact, he concludes more broadly: ". . . the intelligent use of living resources by man must be based on a thorough understanding of the total ecology of the communities involved, and it is unlikely that this would be dominated by any single environmental factor."

One could examine the effects of exploitation on other ecosystem processes beside the competitive. Foresters, for example, have devoted considerable attention to the effect of removing timber crops on the biogeochemical equilibrium of a forest system. Removal of an animal crop could also alter nutrient budgets and affect long-range productivity. For example, the annual removal of between a half million and a million cutthroat trout (*Salmo clarki*), mostly 12–16 inches in length, by tourists in Yellowstone National Park must effect a substantial nutrient transport. That transport might as easily be beneficial as inimical to the perpetuation of the pristine condition. With several million persons visiting the area each year, there may well be a long-range trend toward enrichment as is occurring in many lakes of western United States that are frequented by numbers of tourists. The fish removal could conceivably be retarding such a trend.

Our long-range goal should be to perpetuate the productivity and integrity of the world's ecosystems while utilizing them for human use. The goal of our ecological research is to understand these systems so that we can predict the effects of perturbation and avoid irrevocable changes.

III. VEGETATION COMPOSITION AND SECONDARY PRODUCTION IN UNGULATES

A. Recent Interest in Cropping Wild Ungulates for Food

The great interest of the past decade and a half among ecologists and conservationists in the potential of wild ungulates as a food source in Africa has undoubtedly been spurred, on the one hand, by the possibility that wild animals could be managed to contribute to the world food need, and on the other hand by the hope that such a land-use pattern would help perpetuate the wildlife resources of many parts of the world. Traditionally, wildlife resources have been unable to compete economically with other uses of the land, and have given way before land changes which often do not leave suitable or sufficient habitat. There have been exceptions where the sporting or trophy values of wild game have been sufficiently high to prompt affluent individuals to reimburse landowners for diverting land from the production of food and fiber. But such situations have hardly been the rule over the majority of any continent.

Origin of the suggestion that more biomass of edible meat could be produced via the wild ungulates in the bush, savannas, and plains of East and South Africa is difficult to trace back. According to Dasmann (1964b), many farmers in the Transvaal of South Africa had given up sheep raising "long ago" in favor of raising game for the market. Petrides (1956) introduced the idea into the American literature more than a decade ago, but much credit for really arousing interest in the idea should probably go to Dasmann and Mossman (1961), who actually worked out the economic advantage of the practice.

The rationale behind wild-game cropping as a land-use pattern has been well summarized by a number of authors in recent years (cf. Huxley, 1961; Talbot et al., 1965; Lambrecht, 1966; Watt, 1968). In brief, the two dozen or more species coexisting in these regions feed on the full vegetation spectrum, thereby utilizing the maximum possible amount of the primary production and making possible the maximum secondary production. Livestock use only part of the vegetation and produce less meat. The native species are also more disease resistant than livestock, have lower water requirements, wander over greater distances thereby spreading out the pressure on vegetation, and are more efficient converters of plant materials to flesh. [This latter point has been challenged by Steyn (1966). But data presented by Talbot (1963) and Dasmann (1964b) would seem to support the idea, and Nelson (1965) has suggested the same difference between the North American bison and cattle on southern Utah rangelands.]

This type of land use also appears to be the one most likely to ensure the long-range productivity of the ecosystems of many parts of East and South Africa. Darling (1964) and Pearsall (1964) have termed the habitat of this region "brittle" and "harsh." The soils and climate are not suitable for prolonged cultivation without serious damage to the soils, and a pastoral economy reduces the productivity of the vegetation as described above.

An interesting aspect of the soils problem (Huxley, 1961) lies in one possible effect of woodland eradication for tsetse fly (*Glossina moritans*) control (cattle are susceptible to sleeping sickness, the wild ungulates are not). Termites of this region carry large quantities of organic matter deep into the ground, thereby draining the surface nutrient pool. The drain is compensated for by deep-rooted woody plants which return nutrients to the surface. Woodland clearing in the interest of tsetse control would break the cycle and could conceivably result in a depletion of surface nutrients.

For these economic and ecological reasons, cropping of the native mammals appears to be the land-use pattern most likely to maximize productivity of the land in many areas of eastern and southern Africa for an indefinite period into the future.

B. Vegetation Responses to Herbivore Pressures

1. NICHE SPECIFICITY IN HERBIVORES

It has been well recognized for some time that the diet of a wide variety of vertebrates—whether terrestrial or aquatic, herbivorous or carnivorous—is influenced by two, general influences. The first is simply availability of what is present (cf. McAtee, 1932; Ricker, 1954a; J. J. Craighead and Craighead, 1956; Bartlett, 1958). The second is preference, purposeful selection, or what Ivlev (1961) has termed "electivity" (cf. Bartlett, 1958; Talbot, 1962).

An important conclusion of Elton's painstaking study (1966) is that herbivores have narrower niche boundaries than do carnivores, i.e., feeding specialization is more pronounced among the herbivores while carnivores tend to be more generalized in what they will consume. The degree of this specialization in ungulates has drawn a great deal of interest in recent years in connection with the complimentary food preferences of herbivorous species occupying savannah areas of Africa (cf. Huxley, 1961; Talbot and Talbot, 1963a,b; Darling, 1964). Any one herbivorous species thus applies pressure to a restricted portion of the collective plant species.

2. RESPONSES TO SINGLE-SPECIES PRESSURES

The effects of pressures from single, herbivorous species on vegetation composition have been reported in increasing numbers in recent years. The best information comes from the work of range scientists on the effects of livestock on vegetation, but there is growing knowledge on the influence of wild animals. Elton (1966) has reviewed a number of studies describing the influence of European rabbit (*Oryctolagus cuniculus*) and common vole (*Microtus agrestis*) on the composition of the British vegetation. A. Leopold *et al.* (1947) and numerous other authors have described the early disappearance of highly palatable browse species when deer populations increase.

Over much of North America and temperate Eurasia, the wildlife manager has but a single species of wild ungulate to work with in any one area. And in many parts of the world, a livestock or pastoral culture may emphasize a single domestic species.

Heavy, single-species pressures such as these can effect entire changes in the structure of the plant community. Thus, in the nineteenth century, after European settlement of the intermountain region of western United States, grazing pressure from livestock, particularly cattle, made heavy inroads into the bunch-grass vegetation of the foothills and valleys. With competition thus released, brushy species such as big sage (*Artemisia tridentata*), junipers (*Juniperus osteosperma* and *J. scopulorum*), bitter-brush (*Purshia tridentata*), and serviceberry (*Amelanchiar alnifolia*) increased markedly and turned the former grasslands into brushland. Within the present century, removal of large predators, restrictive hunting laws, and improved habitat conditions have led to widespread increases in mule deer. This browsing species is now placing heavy pressure on the shrubs which are slowly disappearing. With the livestock now removed from many areas, the vegetation is returning to the original bunch-grass type (A. D. Smith, 1949).

Similar effects have been reported from Africa. Elephants (*Loxodonta africana*), which include a great deal of browse in their diets, in particular have been reported to eliminate trees and shrubs, thereby converting woodlands and brush types to grasslands (Petrides and Swank, 1958; Buechner and Dawkins, 1961; Glover, 1963). Other changes have been wrought by hippopotami (*Hippopotamus amphibius*), by buffalo (*Syn-*

cerus coffer) and Uganda kob (*Adenota kob*) together, and by topi (*Damaliscus korrigum*) (Petrides and Swank, 1958).

Although Hairston *et al.* (1960) and Slobodkin *et al.* (1967) have postulated that herbivorous animals in general are held down by predators to densities at which they do not damage the vegetation, some species and groups of species evidently do overutilize that portion of the vegetation spectrum on which they feed. These plants decline, are replaced by others not similarly exploited, and the vegetation composition changes. The process is quite analogous to the cases discussed above of human exploitation on fish species which, placed at a competitive disadvantage, are replaced by unexploited forms.

The vegetation changes wrought by an herbivore may improve conditions for a second, as cattle in western United States have improved conditions for mule deer. But if the impact of the first herbivore is so great as to largely eliminate his required plants and lead to his demise, the benefit to the second herbivore may be short-lived. For the plant species eaten by the second herbivore no longer have the benefit of the pressure applied to the competing plants eaten by the first. Herbivore number two may now eat out its own plants and change the vegetation to some new pattern, or back to the original pattern as in the case of the western United States situation with the mule deer.

3. Responses to Interspecies Pressures

Avoidance of these alterations in plant communities and maintenance of productivity for the entire herbivore community would seem to lie in the simultaneous application of pressure by several herbivorous species, each with different food preferences. Range managers have recognized this principle by recommending common use of areas by sheep and cattle (cf. Cook, 1954; A. D. Smith, 1965). In recent years game management and range specialists Arthur D. Smith and George W. Scotter at Utah State University have experimented with the application of cattle grazing on winter deer ranges in Utah in an attempt to maintain productivity for both.

A striking example of the symbiotic effect herbivorous species can have in maintaining each other's portions of the vegetation and the level of production that can be maintained was reported by Holsworth (1960) for Elk Island National Park in Canada. The park comprises 49 square miles, predominantly forested, of which 45 are considered useful game range and 4 are termed an isolation area. The game populations of the 45 square miles in summer, 1959, were estimated at 650 bison (*Bison bison*), 600 elk, and 300 moose. The climax vegetation over much of the area was considered to be white spruce (*Picea glauca*), but over two-thirds of the area was covered with seral aspen-balm woodland (*Populus tremuloides* and *Populus balsamifera*). The remaining third was largely in meadows variously intermixed with damp areas and shrubs.

The three species showed substantially different habitat and food preferences. Moose were almost exclusively browsers, taking large quantities of willows (*Salix* spp.), aspen, balm, and hazel (*Corylus* sp.). Bison were equally specialized as grazers, taking grasses, sedges, and forbs. Elk were more versatile, largely grazing in summer (three-fourths of the diet was grass and forbs of which the latter constituted two-thirds), and taking more browse (mostly aspen and balm) in winter along with forbs, grasses, and ground litter. Holsworth concluded that nearly all of the seedlings and sprouts of deciduous species were utilized each year by moose and elk. This stabilized the succession, maintained openings in which the grasses and forbs that fed the bison could grow, and perpetuated the latter.

Although standing-crop biomass is a poor index of productivity, as will be discussed shortly and as many authors have pointed out (cf. Macfadyen, 1964; Odum, 1959; Petrides and Swank, 1965), a comparison of the biomass of this area converted to pounds per square mile (Table 1) is perhaps instructive. The bison were fed artificially during winter so that their production was not derived entirely from the primary production on the area. The standing crop produced naturally, therefore could be assumed at something less than the 26,544 lb total for all species, but more than the 12,131 lb for elk and moose alone—perhaps a value approaching 20,000 is realistic.

Petrides (1956) and Bourlière (1963) have listed standing-crop biomass values for ungulate herds in different parts of the world. While less than half the value given for one 11-square-mile study area in Nairobi Na-

TABLE 1

1959 STANDING-CROP BIOMASS PER SQUARE MILE OF LARGE UNGULATES, ELK ISLAND NATIONAL PARK, CANADA

Species	Total no. 45 square miles[a]	No./square mile	Mean weight[b] (pounds)	Pounds/ square mile
Bison	650	14.4	1000	14,400
Elk	600	13.3	530	7049
Moose	300	6.7	760	5092
Totals	1550	34.4	—	26,541
Totals w/o bison				12,131

[a] Density values from Holsworth (1960).

[b] Mean weights of bison and elk are those used by Petrides (1956). Mean moose weights obtained by assuming mean weight of bulls at 1000 pounds, of cows 700, of calves 400, and a 40 bulls:40 cows:20 calves population ratio.

tional Park, Kenya, by Petrides, 20,000 lb compares favorably with North American areas not grazed excessively. Since Elk Island Park is largely wooded, its values are even more significant in that they compare favorably with North American grassland areas where the entire annual production is available for use.

In conclusion, the complementary feeding habits of the species in Elk Island Park tend to maintain the vegetation in a form usable by all species, as George M. Scotter pointed out. The total secondary production is thereby greatly enhanced.

Probably the epitome of this phenomenon is the large, interspecies herds of herbivorous animals in East African plains and savannah. Talbot and Talbot (1963a,b) and other authors have emphasized the high degree of specialization and segregation in feeding habits of these species. Collectively their pressure leaves almost no facet of the vegetation uneaten. When livestock alone are grazed on these areas, the vegetation is impoverished and primary production materially reduced (Talbot and Talbot, 1963a). Although L. M. Talbot has informed the author (1967) that fire is an important influence in preventing woodland encroachment and probably in maintaining the vegetative diversity, one cannot help suspecting that the variety of herbivores is an important influence in maintaining that vegetative diversity and consequently the faunal diversity as well. In effect, a variety of herbivores may maintain vegetative diversity much as Paine (above) has suggested that predators maintain faunal diversity.

C. Secondary Production Levels in Ungulates

A number of authors have compiled data on, and discussed, carrying capacity of African areas based on standing-crop biomass values (cf. Petrides, 1956; Dasmann, 1962; Dasmann and Mossman, 1962; Bourlière, 1963; Talbot and Talbot, 1963a). However, as mentioned above, there is no necessary close relationship between standing crop and production. Since it is the production rate which determines the amount of energy-bearing tissue that can be passed on as yield on a sustained basis over a period of time, it is this parameter which needs to be explored to tell us more precisely the meat-production potential of wild ungulate herds.

Some authors (cf. Engelmann, 1966) use the term net secondary production to avoid any ambiguity. However, Odum (1959) has suggested that the term gross secondary production is not an appropriate one and that assimilation is the more apt word in this context. If we follow Odum's suggestion, there is then no problem of confusing gross and net secondary production and the adjectives can be dropped. Accordingly the term secondary production is used throughout this review in the sense of the amount of energy fixed in herbivore tissue per unit of time. The term productivity is used here in the more general sense of the capacity of a system to produce biotic material.

Secondary production is a function of reproduction, body growth, and survival of the growing and reproductive age classes in a population. These characteristics, of course, vary between different species of animals, and between populations of any one species, as we shall see. The data needed for calculating production values, then, are census figures on the numbers of animals per unit area, information on sex and age structure so that the growing individuals can be accounted for, knowledge of the birth rate, body-growth curves, and knowledge of survival patterns in the growing age classes.

Petrides and Swank (1965) contended in their most interesting paper on the African elephant with a species that did not reach mature weight until 25 years of age, and by its second year had still only gained 5% of its adult weight. The total growth increment of individuals in the first year of life was only 6–7% of the total, annual weight increment of the growing age classes. In such a species, survival and growth data based on 1-year intervals provide an adequate basis for reasonable production estimates.

However, in members of the Bovidae, Cervidae, and Antilocapridae, which comprise most of our large game species, approximate adult weight is usually reached in the second or third year of life (cf. Bannikov *et al.,* 1961; Talbot, 1963; Talbot and Talbot; 1963b; McEwan and Wood, 1966). Virtually all of that part of production involving growth takes place in the youngest two, or sometimes three, age classes. Detailed knowledge of growth rates and survival patterns are thus needed in this brief period subdivided by time intervals as short as 1 or 2 months. This is particularly needed in the vulnerable first year of life when, in different regions and species, intensified mortality may occur. Thus, accelerated loss may occur (1) shortly after birth, as a result of the prior nutritional state of the female as in the mule deer (Robinette, 1956), of separation from the female as in the white-bearded wildebeest (*Gorgon taurinus*) (Talbot and Talbot, 1963b), or of parasitism as in Texas white-tailed deer (Teer *et al.,* 1965); it may occur (2) at several months of age as a result of disease, again in the wildebeest (Talbot and Talbot, 1963b), of decline in nutritional content of the range vegetation for black-tailed deer (Taber and Dasmann, 1958), or of winter weather in temperate regions for white-tailed deer (Dahlberg and Guettinger, 1956) and saiga (*Saiga tatarica*) (Bannikov *et al.,* 1961); and finally accelerated loss may occur (3) at almost any of these ages as a result of predation.

To my knowledge, the only secondary production values published for large mammals in actual terms of energy fixed per unit area are those of Petrides and Swank (1965) for the elephant, for beef cattle in virgin North American grassland, and for white-tailed deer in the United States, previously published by Davis and Golley (1963). Petrides and Swank calculated production for elephants in a 28.5 square-mile area of Queen Elizabeth National Park, Uganda, at 0.34 kcal/m²/year from a standing crop of 7.1 kcal/m². Respective values for beef cattle were 0.86 and 7.5, and for white-tailed deer 0.64 and 1.3.

Davis and Golley's estimate (1963, p. 216) is based on synthetic data from the George Reserve deer herd in Michigan using an annual production and survival of 50 new animals per 100 in the population and 39 per square mile standing crop. The George Reserve herd is evidently a highly productive one, and while whitetail densities along the northern edge of the species distribution where winter range is limiting may typically be somewhat lower, 0.64 kcal provides an approximation of production levels possible in a highly productive herd. This probably is also a fair approximation of the level possible in a productive mule deer herd in western United States, a species which seems to be slightly less productive of young than whitetails but which average slightly larger in size.

In southern United States, where winter is not so seriously limiting and all of the range can be used throughout the year, Texas whitetail densities commonly approach or reach 100 per square mile (cf. Knowlton, 1964; Teer *et al.,* 1965). These deer are smaller than those of Michigan, possibly averaging no more than two-thirds the 130-lb liveweight allowed by Davis and Golley. However, this 30–40% lower weight is more than offset by the 2.5 times greater density. The production, assuming similar demography, is thus probably as high as or higher than the 0.64 value calculated by Davis and Golley. Similarly, black-tailed deer in Pacific coastal mountain ranges of North America, where climate is relatively mild, may reach densities in improved range of 100 per square mile (Taber and Dasmann, 1958). These animals, too, are small—actually about 10 lb smaller than Texas whitetails—and may have productivity levels slightly below, but comparable with, Texas deer at these densities.

Production can be calculated for the saiga of the Russian steppes based on data in the fine studies of Bannikov (1961) and Bannikov *et al.* (1961). If we adopt for density 5/km² (a middle value among the range given in their Table 7 for the Caspian area), May-June population structure given in their Table 36, and mean reproductive rate of 1.6 young per female of all ages (p. 213), weights of the respective sex and age classes given in their Tables 32 and 33, and the 1.5 kcal/g live weight used by Petrides and Swank (1965), we can calculate a standing crop of 0.26 kcal/m² on May 1 when all births are abstractly telescoped for simplicity of computation. Production during the year occurs entirely in the young and yearling age classes. Increments are calculated for fawns on July 1 following postnatal loss of 20%, on December 1 for fawns (75% surviving) and yearlings prior to winter cessation of growth, and on the following May 1 for the new fetuses added to the population. The year's increment is 0.20 kcal/m^2.

Production can also be calculated from the Talbot and Talbot (1963b) wildebeest data. Population structure and numbers for their 20,000 square mile area are given in their Table 18 and January 1, an average date of birth is used here as the start of the year (weights are provided in their Tables 1 and 2 and p. 24). The first weight increment is calculated here for September 1 following postnatal loss but prior to rinderpest loss. The second increment is calculated on the following January 1 for the calves surviving rinderpest, for the yearling and 2-year age classes, and for the neonates. The results provide a production value of 0.15 kcal/m^2 for the year from a January 1 standing crop of 1.10 kcal/m^2.

One other source of information can be used for very crude production estimates. Dasmann and Mossman (1961; and later published in Dasmann, 1964b) calculated potential annual yield values for 13 species of mammals on a 50–square-mile area of the Henderson and Sons, Ltd., ranch in Rhodesia. Their values were derived basically from the percentage of yearlings in the populations of each species, a statistic which is equivalent to the mortality rate from that date one year previous in a population at equilibrium. MacArthur and Connell (1966) suggest that the mortality rate, if expressed as energy or its biomass equivalent lost from the population per unit of time, can be taken as equivalent to the production rate since energy enters a population as fast as it is removed in the equilibrium situation.

Dasmann and Mossman calculated their yield values on the basis of adult animal weights. Where used here to calculate production, this will magnify the result because some of the animals dying will be young; a mean weight of the animals in the population, not given in these authors' data, would be preferable. This overestimate, however, is offset to some degree by the fact that the annual mortality indicated by the yearling percentage merely equates the young surviving in the population that replace those dying in the preceding year. Any young born but dying prior to the time the age structure is measured represent production not measured by the present method. Hence, the resulting values (Table 2) may roughly reflect the production level in each of these species.

All of these values are quite evidently crude. The mathematics have been oversimplified; birth by-products, such as the placenta and fluids, have not been included in the estimates, a deficiency in really precise production estimates; other sources of error are present. Hence, these can be considered no more than first approximations, but they do

TABLE 2

Approximate Secondary Production for Herbivore Populations[a]

Species	A: no on 50 miles[c]	B: % yearlings	C: no. yearlings (A × B)	D: pounds dressed carcass × 2[c]	E: total weight (lb) (C × D)	F: total weight (g)	G: g/m²	H: kcal/m²/yr (G × 1.5)
Impala	2100	28	588.0	130	76,440	34,627,320	0.267	0.40
Zebra	730	20	146.0	510	74,460	33,730,380	0.260	0.39
Steenbuck	200	35	70.0	24	1680	761,040	0.006	0.01
Warthog	170	60	102.0	140	14,280	6,468,840	0.050	0.08
Kudu	160	32	51.2	450	23,040	10,437,120	0.081	0.12
Wildebeest	160	22	35.2	520	18,304	8,291,712	0.064	0.09
Giraffe	90	20	18.0	2000	36,000	16,308,000	0.126	0.19
Duiker	80	42	33.6	40	1344	608,832	0.005	0.01
Waterbuck	35	20	7.0	400	2800	1,268,400	0.010	0.02
Buffalo	30	16.6[b]	5.0	1,140	5700	2,582,100	0.020	0.03
Eland	10	24	2.4	1,200	2880	1,304,640	0.010	0.02
Klipspringer	10	30[b]	3.0	28	84	38,052	0.000	T
Bush pig	10	50[b]	5.0	140	700	317,100	0.002	T
Totals	3785	—	1066.4	—	257,712	116,743,536	0.901	1.36

[a] Reported by Dasmann and Mossman (1961).

[b] Not given as percentage yearlings. These values are the recommended removal which in the other species approximated the yearling percentage.

[c] Weights were given as dressed. marketable carcass weights. They were multiplied by 2 to approximate the live weights.

perhaps provide some order-of-magnitude values on the levels of secondary production that can be expected in wild ungulate resources.

D. Discussion

Several tentative generalizations can be drawn from these observations of vegetation responses and ungulate production levels. First, the wildlife manager's problem in this area is often to maximize the production of ungulates while maintaining the productivity of the vegetation. Odum (1959, pp. 75–76) has shown that the total, primary production of any given area falls within fairly well-developed limits, and is more a function of the solar energy, water, and nutrients available on an area than of the kinds of plants present. In other words, plant monotypes (e.g., agriculture) and various combinations of natural vegetation may produce collectively at fairly comparable levels in any given area.

In temperate latitudes, the manager often has available only one or two species of ungulates, a situation also facing the range manager. Since each ungulate species is more or less specialized, it will often be true that part of the vegetation production will not be useful to the ungulate species present. Some of the primary production will go unused by them, and the level of secondary production therefore somewhat below the maximum possible if all primary production were usable.

Maximizing secondary production in the ungulates would seem to rest on two alternatives to the natural, temperate-latitude situation. One would be to change the vegetation to a form usable by the ungulates, the alternative widely used by range managers in western United States. Natural vegetation is removed and the terrain is seeded to exotic grasses palatable to livestock. This alternative has not been adopted by wildlife managers so far.

The second alternative has been advocated or experimented with by wildlife managers and other ecologists: This is the diversification of the ungulate resource to add species which will use most or all of the natural vegetation. Thus, a number of range managers are advocating common use by several species of livestock or livestock and wild game, as discussed above. Martin (1964) has advocated the judicious introduction of wild, exotic ungulates both to increase secondary production and to maintain the integrity of the vegetation. In southern United States, particularly in Texas, areas can be found in which the white-tailed deer shares its range with three or four species of livestock. In this same area, importation of Asiatic and African species has elicited considerable interest Craighead and Dasmann, 1966). In Africa some interest has been expressed in the domestication of wild species which could be added to the livestock list. Most often mentioned are eland (*Taurotragus oryx*) and buffalo as mentioned by Huxley (1961) and Darling (1964).

There are, of course, risks associated with the introduction of exotics as F. C. Craighead and Dasmann (1966) point out. These cannot be over-looked, but it would seem that research on the problem is in order.

At lower latitudes, combinations of ungulate species coexist, as in India (Schaller, 1967) and the African areas, discussed above. Even here, how-ever, it seems likely that the wildlife manager is faced with the challenge of learning what the vegetation responses are to different combinations and numbers of ungulates, what vegetation pattern is best suited to and most productive in an area, and what herds must be maintained to pro-duce that pattern. The problem evidently is a complex one, and it seems likely that we will see research on the subject moving toward the develop-ment of predictive models much like those being developed for predation (Holling, 1966) and spruce-budworm populations (Morris, 1963b).

A second generalization to be made from the above data, and discussed by other authors but once again demonstrated forcefully by those data, is the lack of any necessary relationships between standing crop and produc-tion. This was emphasized by Petrides and Swank (1965), who showed that in the elephant a given production value can only be achieved by a standing crop 20 times that value. The striking demonstration is the com-parison of the elephant values (0.34 $kcal/m^2/yr$ and production and 7.1 $kcal/m^2$ standing crop) with those for the saiga (0.20 and 0.26, respec-tively). The saiga production is more than half that for the elephant but from a standing crop only about 4% of the latter.

Those areas of eastern and southern Africa which carry the highest biomass of game are the savannas bordering the Congolese forest (Bour-lière, 1963). Some of these carry four to six times the biomass of areas farther east and south. Yet, elephants and hippos comprise up to 70% of the higher values. These areas, therefore, could conceivably have no higher secondary production than the ones of lesser biomass in Kenya, Tanzania, and Rhodesia.

The low production value for the Talbots' wildebeest also stands out in contrast to its rather considerable standing crop. The low calf survival and the consequent low recruitment of yearlings, despite the high repro-ductive rate for the herd, explains this seeming paradox.

A third generalization, not clearly answered by the data but bearing mention and future study, is the question of the comparative production of individual species in tropical and temperate zones. If niche specificity is greater or more restrictive among tropical species, as often suggested, and if the vegetation is more diverse with each plant species represented by fewer individuals, any one tropical herbivorous species may have less primary production available to it than a more widely feeding, temperate herbivore. The evidence tends in this direction. The deer values cited exceed all of the African values as does the beef-production value cited by Petrides and Swank. The saiga value exceeds 11 of the 13 Rhodesian values in Table 2 and the Talbots' wildebeest production.

In all cases the values range between 0.1 and 1.0 for individual species, in the approximate range for homoiotherms cited by Engelmann (1966), and well below the range cited for many poikilotherms. The collective herbivore population of the Henderson Ranch produced at a rate only slightly above 1.0 kcal/m^2/yr.

One final point needs to be touched upon here and can perhaps provide transition for the next section. When we analyze the trophic structure and production patterns of all or part of an ecosystem such as we have attempted here, we are inclined to think of the production of any one level being some function of that on the level below.

Yet available energy alone does not necessarily determine the production of any given species or level. Even where animal species appear to be limited primarily by food, the nutritive value of that food may be more limiting than its caloric content. Thus, the annual, late-summer decline in protein content of browse plants in California coastal ranges (Taber and Dasmann, 1958) effects a qualitative rather than quantitative dietary deficiency. The heavy fawn mortality that ensues doubtless plays an important role in limiting the blacktail deer population size. Burning the chaparral sets back the succession, markedly reduces the vegetation standing crop, and may reduce the total primary production (this last point is not clearly established in Taber and Dasmann's report). The protein content of the browse, however, increases two- or threefold, and the deer population responds almost immediately with a two- or threefold population increase in the first year.

A similar situation has been reported for the red grouse (*Lagopus scoticus*) in Scotland by Jenkins *et al.* (1963). Grouse densities are higher in areas where the heather (*Calluna*) is young but primary production is lower, than in areas of mature heather with higher production. Nutritive content of the young heather is higher.

The degree to which qualitative, nutritional deficiencies may limit the numbers of other wild animals is a largely unexplored subject.

IV. POPULATION REGULATION

A. General State of Population Theory

As fish and game workers have long understood, a knowledge of the principles by which animal populations are regulated or limited is a requisite for enlightened exploitation practice and environmental manipulation.

Indeed, the underlying philosophy of small game management in the United States is largely based on the indefatigable, early efforts of Paul L. Errington and his concepts of population regulation.

However, the subject of population regulation has probably been the most polarized area of ecology. As Elton (1966) and MacArthur and Connell (1966) have pointed out, generalizations exist which claim nearly every resource needed by animals or every environmental factor as the general cause of population limitation. Some authors argue that animal populations are limited exclusively by factors which modify their effects as density changes; others that all factors have a density-dependent effect and the distinction between density dependence and density independence is superfluous.

The problem is further confounded by a need to reconcile population and ecosystem principles. As wildlife management broadens its view to entire ecosystem perspectives, and particularly with problems where energy flow and yield must be maximized, we will need to relate population principles to those operating at trophic levels. The investigator working in energy-flow problems thinks in terms of efficiency rates and the degree to which trophic exchange is influenced by lateral energy loss from incomplete consumption of available energy, incomplete assimilation of what is consumed, and respiratory or maintenance loss. The general implication that the production and standing crop of any one trophic level (and by inference, the constituent species) are some function of those below, and that energy is therefore the significant limiting influence, is widespread.

Yet the population-oriented wildlife student thinks in terms of environmental factors, as discussed earlier. Food may at times be an important one, and there are a few advocates of food as the significant limitation; but there is no unanimity on this and clearly food is not the important source of limitation on many populations.

Nevertheless, there are numerous analogies and common concepts between these two levels of ecological integration. The concept of equilibrium is one such common concept, whether applied to the input-output equivalence of individuals in a population, or to the input-output equivalence of energy or materials in a population, a trophic level, or an entire ecosystem.

Another common concept is that of the effect of environmental factors, whether on population function or on energy and material exchange rates. Common, too, to expressing and predicting the effects of such factors is the emerging, conceptual tool of systems analysis. In this technique, functional relationships between various dynamic processes—e.g., reproductive and mortality rates at the population level, energy and material exchange rates at the ecosystem level—and the influential environmental factors are integrated into a computer model which permits expressing and predicting the magnitude of any particular population or ecosystem process, given specified values for one or more environmental factors (cf. Watt, 1966, 1968). The now classic achievement along these lines on a population problem is the spruce budworm (*Choristoneura fumiferana*) model (Morris, 1963b). This work will undoubtedly serve as a precedent for many future animal studies. The wildlife manager who can elaborate the patterns of influence upon, and predict changes in, a wildlife population as successfully as is now possible with the spruce budworm will indeed have achieved a triumph in the field.

Another influence which has prompted ecologists in recent years to ponder the extent of common ground between the population and ecosystem levels has been the papers of Hairston *et al.* (1960) and Slobodkin *et al.* (1967). These authors sought to generalize population regulation according to entire trophic levels.

Hence, as ecological theory continues to develop, we probably shall see continued integration of principles at the population and ecosystem levels. One problem affecting progress toward that integration is the disunity which currently exists in population theory. This disunity also poses a problem for wildlife management because our management policies are based on ecological principles, as discussed in the first section of this review. The effectiveness of those policies depends upon the correctness of the principles on which they are based.

This last section of the present review is an attempt to explore the bases for disunity in population theory in the hope of making some progress toward what Elton (1966) termed a need for "ecumenism."

B. Demographic Basis of Population Theory

1. DEMOGRAPHIC MODEL

Upon reading the population literature, I cannot escape the impression that the prevailing, theoretical disunity stems to a large extent from two sources: (1) semantic problems, and (2) a tendency to seek a single cause for the limitation of all or most populations. What is needed is more explicit statement in demographic terms of the questions for which we seek answers, and of such concepts as regulation, self-regulation, control, and limiting factor.

The purpose of this section is to explore these questions and terms in the context of a graphic model which hopefully will help clarify the subject. The basic ideas have been touched upon previously in relation to the ring-necked pheasant (*Phasianus colchicus*) populations (Wagner *et al.*,

1965) and elsewhere (Bohart and Wagner, 1967), but have not been elaborated at any length in the general case.

The model is clearly oversimplified. It would not be useful in an actual systems approach to a population, like that used for the spruce budworm, because the dependent variable—rate of population change—is too gross. For the latter approach, subdivision of population change into its natality and mortality components is more appropriate. Hence, the model is more in the nature of a conceptual construct enabling more explicit reference to the questions under consideration. It does permit one to visualize such practical problems as exploitation and predator control (its effect on the predator) along with the general pattern of influence of environmental factors. There undoubtedly are many species which cannot be viewed appropriately in this way, but it appears to the author that it is a realistic means, at least of formulating a preliminary concept for many species.

Two pertinent questions here can be asked in a general way with reference to Fig. 2. We may conceive of a population growing in an environment until it reaches some asymptotic level, K. Once it has reached this level, it remains there, on the average, barring any long-range change in the environment. It may fluctuate, but these changes tend to vary around a long-term, constant mean and between relatively constant limits. The first question is: Irrespective of the mean density at which this equilibrium is maintained, what extrinsic and/or intrinsic influences maintain this equilibrium?

The idea of equilibrium is not universally accepted, Andrewartha and Birch (1954) having been among the stronger critics of the idea. As entomologists, they have been impressed with the great range of variation in insect populations often caused by the random effects of weather. How-

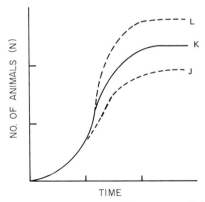

FIG. 2. Sigmoid population growth with equilibrium at population density K, and with alternative growth patterns and equilibrium densities (J and L).

ever, Ricker (1954b) has shown that populations fluctuating at random can eventually vary to zero or extremely high densities. And the idea of equilibrium has been embraced by a number of other entomologists (e.g., Nicholson, 1933; Morris, 1963a,b; Solomon, 1964; Klomp, 1966).

The second question posed by Fig. 2 is: What factors cause the population to go to equilibrium at K rather than at J or L? This, as we shall see shortly, is demographically related to, but nevertheless distinct from, the first.

The distinction between these two questions has been clearly embraced by several authors (cf. Solomon, 1957; Klomp, 1966). And as Wagner *et al.* (1965) have reviewed previously, Nicholson has embraced the two for more than 30 years. But other authors have not perceived the difference (cf. Bohart and Wagner, 1967) and this has, in my opinion, been one of the main semantic problems underlying the arguments in the literature. Consequently, the terms regulation, control, limit have often been used ambiguously. Only in recent years is the term regulation coming to be used in the case of the first question.

The questions need to be stated still more explicitly, however, if a model is to be developed as a basis for quantitative statements and measurement. To begin with, environmental factors do not operate on density, per se. The direct causal pathways involve the operation of environmental factors on birth rates and death rates, and consequently on rates of population change which represent the difference between birth and death rates. Population size changes as a secondary effect of the environmental influences on these rates.

This being the case, and because questions 1 and 2 actually are questions about rates of population change, they can be restated. If we use r as the instantaneous rate of population change, and take r as zero in a population at equilibrium (undergoing no change), we can restate the questions as follows: (1) Irrespective of the density at which it occurs, what influences maintain r at zero? (2) What influences reduce r to zero at density K? These questions can now be explored in the context of a graphic model (Fig. 3).

Relative to question (1), any population which can grow when placed in an environment must have an r value somewhere in the positive range on the ordinate in Fig. 3. Also, when a population goes to equilibrium, it must have a mean r value at zero. Consequently, a population that increases from low densities to or beyond its equilibrium density—either during its initial growth or in the course of its normal fluctuations once established—must have a negative correlation between r and density (N). Such correlations might be represented by the lines $r_b - r_5$ and $r_n - r_2$ in the case of initial growth, or by $r_3 - r_5$ and $r_0 - r_2$ in the case of established, fluctuating populations.

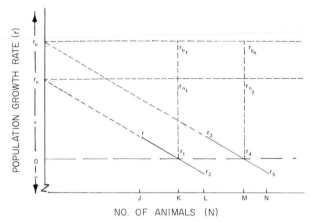

Fig. 3. Conceptual model showing relationships between *r*, rate of population growth per individual, and population density, given different patterns of environmental constraints. See text for explanation.

The general validity of this statement is borne out by Tanner's examination (1966) of the populations of 71 species. Of these, 47 had significant, negative correlations, and 15 more were negative but short of significance. Additional evidence of this sort has been provided by MacArthur (1960) for an ovenbird population, by Lack (1964) for a great tit population, and by Wagner *et al.* (1965) for six pheasant populations.

The points at which the population densities reach the $r = 0$ line become equilibrium points, and the corresponding values on the abscissa the equilibrium densities. For example, when a population exceeds K following an especially favorable year, r becomes negative and the population returns toward K. If the population should decline below K, r becomes positive and the population increases back toward K. In order for this pattern to hold, one or more influences, extrinsic or intrinsic, must increase mortality and/or reduce natality when a population is high. Conversely, one or more influences must ease their pressure on the reproductive and/or mortality characteristics of a population when the density is low. In short, equilibrium or mean of r at zero is maintained by factors which change their influence as the density varies, a conclusion reached long ago by Nicholson (1933), Cole (1948), Ricker (1954a), Lack (1954), and others.

The second question—what influences reduce r to zero at K or M—is one less generally perceived and asked. Having discerned that density-dependent processes exist and must be responsible for the maintenance of equilibrium, some authors conclude that these processes are the only ones of importance in limiting population density (cf. Lack, 1954; Wynne-Edwards, 1962; Brian, 1965; Tanner, 1966). Density-independent influences are rationalized unimportant, often to the dismay of the entomologists to whom it has long been empirically evident that physical factors are important to their poikilothermic subjects (cf. Andrewartha and Birch, 1954).

The problem here would appear to be semantic, and can perhaps be resolved by reference to Fig. 3. In a given environment, any species presumably has a genetically fixed, maximum rate of increase termed by Chapman (1931) the "biotic potential." Some entomologists (e.g., Andrewartha and Birch, 1954) have questioned the value of this concept on the grounds that insects rarely occur in an environment where physical factors are totally favorable. They prefer to think of a maximum increase rate relative to some physical environment, the preferred concept being the intrinsic rate of natural increase (Birch, 1948; Lotka, 1956) or the innate capacity for increase (Andrewartha and Birch, 1954).

While this concept seems to be a valuable one, and the rejection of Chapman's term in insect usage understandable, there would still seem to be merit in the concept of a biotic potential in birds and mammals. A. Leopold (1933) used the terms "potential rate of increase" and "reproduction potential" as synonymous with Chapman's term and concluded that the parameter ". . . is a conventional fixed datum by which diverse actual conditions can be measured and compared . . ." in game populations.

A population growing exponentially according to its biotic potential would have an $r - N$ relationship characterized by the line $r_b - r_{b_n}$. A population growing exponentially according to its intrinsic rate of increase would have an $r - N$ relationship characterized by the line $r_n - r_{n_2}$. An equilibrium population would have an $r - N$ line with a negative slope, as discussed above, and would cross the $r = 0$ line at its equilibrium density. However, its Y-intercept seemingly would be determined by the degree to which density-independent influences reduce r, the reduction by definition being as great at low densities as at high. Hence, the Y-intercept in a population influenced solely by density-dependent factors would be r_b in Fig. 3. The Y-intercept in a population in which density-independent, physical factors were operative might be at r_n.

The $r - N$ regression line in a population with substantial density-independent pressure (lowered Y-intercept) will cross the $r = 0$ line and come to equilibrium at a lower density than will a population with little or no density-independent pressure unless the $r - N$ slope is totally flexible and some function of its Y-intercept. Wagner *et al.* (1965) have suggested that this slope represents the response of a species, in a given environment, to its density and may be substantially an inherent characteristic of the species. In three United States pheasant populations which differed from each other in density by nearly an order of magnitude, and presumably with markedly different Y-intercepts, the slopes were quite similar. Morris (1963a) concluded that insect populations fluctuate geometrically rather than numerically, and the responses are therefore to logarithmic rather than arithmetic differences in density, a conclusion reached by Wagner *et al.* for the pheasant. These lines of evidence would seem to support the view that the $r - N$ slope is not totally flexible, and that the point at which it crosses the $r = 0$ (equilibrium) line is related to the Y-intercept. This effect is shown by the lines $r_b - r_5$ and $r_n - r_2$.

The role of density-independent influences in reducing r and the equilibrium density is now evident. These influences in Fig. 3 are responsible for reducing r at K from r_{b_1} to r_{n_1}. The remaining reduction from r_{n_1} to r_1 is accomplished by density-dependent influences. It is in this way that both types of factors can influence population density, both types being partially responsible and neither being solely influential.

We may now summarize: Equilibrium (mean of $r = 0$) is maintained by factors which intensify their effect when density increases, ease their effect when density decreases, i.e., density-dependent influences. The equilibrium phenomenon has elicited a great deal of interest in the population literature (cf. Nicholson, 1933; Haldane, 1953; Deevey, 1962; Klomp, 1966); its maintenance has explicitly been given the term regulation by recent authors.

Determination of mean density (reduction of r to zero) is effected by any factor which reduces r at its equilibrium density below what r would be in the absence of that factor. Lack of precise statement on this phenomenon in demographic terms has perhaps partly been responsible for its lesser emphasis in the population literature. Yet it is often the one of greatest practical interest in wildlife management. We often need to know what influences depress populations of desirable wildlife species or permit excess numbers of undesirable ones, and how those influences can be manipulated to achieve a desired effect.

In recent years, speculation on whether or not the influence of various factors is related to the density of a population has come to be considered trite or lacking in theoretical promise by many investigators. This has apparently resulted from the impression that so much of the discourse in the literature is either circular or becomes lost in semantic limbo. It seems likely, however, that interest in this general subject will witness a revival with the growing use of predictive models. As discussed above, model development uses such demographic parameters as rate of population change (or its component reproductive and mortality rates) as dependent variables. The form of the mathematical function relating each, influential environmental factor to such rates must also be known and incorporated into the model. To be able to say at least that such a function is positive or negative (positively or negatively density-dependent) or is without mean slope (density-independent) is a first step in elucidating the form of the function.

2. PATTERNS OF FACTOR EFFECTS IN DENSITY DETERMINATION

While density-independent factors cannot maintain a population at equilibrium, they reduce the rate at which a population changes at any given density (Fig. 3). Where density-dependent influences are added, a population will grow at a declining rate until that rate is reduced to zero at what becomes the equilibrium density. Unless the slope of the $r - N$ regression line is totally flexible and inconstant, it will cross the $r = 0$ line at a lower density when it is reduced by density-independent influences, as discussed above. It is this way in which the latter influences limit density along with density-dependent factors.

Seemingly, the importance of any given factor or combination of factors in determining mean density (e.g., density-dependent vs density-independent) is measured by the degree to which it or they reduce r from the biotic potential to 0. Some undoubtedly are of greater importance than others in the reduction of r. This r-reduction criterion of factor importance is somewhat different from the criterion which F. E. Smith (1961) and Watt (1968, p. 60) use as a gage of the importance of density-dependent influences. While again there may well be a problem with semantic precision here, these authors appear to infer the degree of density-dependent restraint from the variability of actual population numbers. In the context of Fig. 3, this constitutes the horizontal array of points on the abscissa.

If the hypotheses here set forth are correct, however, factors operate on rates of change and the degree of density-dependent influence is determined by the amount of reduction in r: The $r_{bn} - r_4$ influence is greater than the $r_{n_1} - r_1$ influence. Thus in three pheasant populations reviewed by Wagner *et al.* (1965) their densities appeared to be correlated inversely with the weight of density-independent influence. The weight of density-dependent influence varied between the three, being positively correlated with equilibrium density. Yet the coefficients by which each population varied were quite similar. Conceivably, the variations in numbers of a population might rather be a function of the variability of the environment including density-independent influences.

Although we have talked from Fig. 3 largely of linear modes of density-independent and density-dependent action, there are nonlinear effects such as Holling (1959) and Tinbergen (1960) have shown for insect predators. Predators may be positively density dependent in the lower range of prey density. But they fall behind the growth of faster-increasing insect populations when the latter reach moderate- and high-density ranges and at that point become negatively density dependent. Among vertebrates, both Lee M. Talbot and Peter Greig Stewart have told me that the African lion (*Panthera leo*) tends to be territorial. As ungulate prey increase, lion densities do not follow suit, and the predators tend to take a declining fraction of the prey. Their effect is thus negatively density dependent.

These inverse and nonlinear effects are among the main reasons why the model in Fig. 3 is oversimplified. Again, it is only ventured here as a framework which hopefully will make some of these concepts more explicit. Conceivably, it may not be applicable in populations, such as those of insects described by Morris (1963a,b) and Watt (1968), in which there is more than one equilibrium density.

Measurement of the relative effects of environmental factors will evidently require simultaneous measurement of r values while the factors are changing by measured amounts. Such measurement would seem to be possible under three circumstances: (1) in the course of natural variation in factors as with climatic variation (cf. Wagner *et al.,* 1965, p. 85); (2) with purposive manipulation of factors, as in predator-control experiments; and (3) by comparison of populations in different environments with different combinations of factors (cf. Wagner *et al.,* 1965, pp. 136–138).

3. Factors Important in Natural Populations

Probably no species, except perhaps the spruce budworm, is so thoroughly understood that we can clearly state the quantitative importance of each factor that has some influence. Yet some of the most important factors have been identified in some species and generalizations are needed from both theoretical and applied considerations.

Among the most stimulating general statements on this subject have been the papers of Hairston *et al.* (1960) and Slobodkin *et al.* (1967). These authors have presented deductive hypotheses that will challenge further reasoning and empirical test for many years to come. The accretion of new knowledge from their stimulus will undoubtedly be great.

Slobodkin *et al.* (1967) firmly pointed out that their hypotheses were presented to explain limitation of the dominant species in a community and not necessarily the limitation patterns of species in general. However, it is easy to gain the latter impression on reading Hairston *et al.* first, both from the title and the text of the paper, as both Murdoch (1966) and Ehrlich and Birch (1967) did.

In the view of Hairston *et al.* and Slobodkin *et al.,* dominant primary consumers are limited at densities below which they damage the vegetation, apparently by predation and parasitism, while secondary consumers in total are limited by intraspecific competition for their own food supply. In the context of the present discussion, predators in the first case and food limitation in the second are responsible for all or most of the reduction of r to zero at K. Lack's general view (1954, 1966) that food is the most important limiting influence on animal populations, particularly birds that are largely insectivorous, would seem to relate to these views.

The conclusion that herbivores are predator limited is primarily deductive although Hairston *et al.* cite as empirical evidence the escape of pests when transported to new continents where their indigenous predatory controls are absent. This is persuasive, but two tentative considerations can be mentioned. The first is the point made above that the response of a population to the removal of a factor does not necessarily imply that the factor is the sole or most important agent operating in the population. Removal of others could conceivably have elicited an equal or greater response.

The second point is that in transporting an exotic, predators are not the only part of the community left behind. Interspecific competition, at least that of the indigenous ecosystem, has been removed. This may be significant, for as Elton (1958) has pointed out, exotics tend to succeed in man-disturbed environments where natural biotas have been altered and, in effect, new niches created. At the same time, the game technician does not see any dearth of predation pressure on exotic game species. Wagner *et al.* (1965) compiled a lengthy list of predators which affect pheasants and pointed out that in many areas, predation is the most important cause of nest destruction. In the Intermountain Region of western United States, I have observed what appears to be substantial pressure by raptors on the chukar partridge (*Alectoris graeca*).

If we cannot rule out, by exclusion, factors other than predation for herbivores and food shortage for carnivores, what other influences appear significant? One is social behavior, as pointed out to me recently in connection with the Hairston *et al.* and Slobodkin *et al.* papers by Robert Hoffman. Its role in population limitation has received increasing interest in recent years, particularly in birds and mammals. Wynne-Edwards' work (1962) represents the most exhaustive analysis of this factor. While one does not need to accept the total, self-limitation view espoused by this author (all reduction in r effected by behavior), some substantial influence from behavior in the reduction of r seems likely.

Perhaps significant to the hypotheses of Hairston *et al.* (1960), Wynne-Edwards postulated the importance of behavior in holding animal populations below the point where they overutilize their food supply and ruin their environment. Again, while one need not accept behavior independent of other factors as accomplishing this effect alone, it is true that the role of behavior as one influence in population limitation has increasingly pressed itself upon population students in recent years.

Perhaps significantly, behavior has received greatest attention in birds and mammals. The widespread view on the general importance of food in limiting fish has persisted strongly among fishery biologists, and has been reiterated recently by LeCren (1965). The indeterminate growth rate and adult body size confer upon fish an advantage where food is short. Growth merely slows and starvation is not an imminent threat. Among homoiothermic birds and mammals, the constant energy need for body-temperature maintenance and the essentially determinate growth rates and body size make food shortage a hazard of considerable danger. Hence, behavior, most often implicated in the population mechanisms of warm bloods, may be an adaptation to deter population growth below levels where starvation is likely to occur.

Among game species, the interrelationships between habitat structure and behavior have long been considered important and were early expressed in the edge principle by A. Leopold (1933). Ducks are well known to require minimal, linear distances of shoreline as part of their territories. Andersen (1953) and Jenkins *et al.* (1963) have implicated movement as a population-limiting influence in Danish roe deer (*Capreolus capreolus*) and Scottish red grouse. And while generally advocating food as a limitation on freshwater fish populations, LeCren (1965) indicated that some stream and limnetic-zone fish may be limited by habitat and behavioral interrelationships.

Another source of limitation, at least on some primary consumers, was mentioned above: qualitative food shortage or nutrition. Although its general significance is yet to be learned, it is clearly important in black-tailed deer and red grouse, as discussed earlier. Work now underway in my own institution by Donald Beale, George M. Scotter, and Arthur D. Smith is implicating the importance of nutrition in population densities of pronghorn antelope (*Antilocapra americana*) in western United States.

Although rationalized unimportant by some workers, including Hairston *et al.* (1960), the role of weather and climate has been empirically implicated too often in population mechanisms to bear light dismissal. Most often implicated for insects, the operation of weather on wildlife populations—e.g., bobwhite, (*Colinus virginianus*) by Kozicky and Hendrickson (1952), wildebeest (Talbot and Talbot, 1963b), white-tailed deer as described in the last section, and pheasant (Wagner *et al.,* 1965)—often in connection with food, has been widely reported.

For these reasons, our future generalizations will probably describe the complex of factors operating on populations, and the relative importance of each factor. We may well compare these complexes between taxonomic groups, between different ecosystems (e.g., tropical vs temperate, early vs late successional, aquatic vs terrestrial), and perhaps between trophic groups as Hairston *et al.* (1960) have done. And we shall stress prediction, particularly of the responses one could expect from manipulating the factors present naturally, or from the imposition of artificial control efforts and exploitation.

REFERENCES

Allee, W. C., E. Emerson, O. Park, T. Park, and K. P. Schmidt. 1949. "Principles of Animal Ecology." Saunders, Philadelphia, Pennsylvania. 837 pp.

Allen, D. L. 1947. Hunting as a limitation to Michigan pheasants. *J. Wildlife Management* **11,** 232–243.

Allen, D. L. 1954. "Our Wildlife Legacy." Funk & Wagnalls Co., New York. 422 pp.

Allen, D. L. 1956. The management outlook. *In* "Pheasants in North America" (D. L. Allen, ed.), pp. 431–466. The Stackpole Co., Harrisburg, Pennsylvania, and the Wildlife Management Inst., Washington, D.C.

Allen, D. L. 1966. Yes! Go ahead and hunt. *Utah Fish Game* **22,** 3–5 and 26. (Reprinted from The National Wildlife Magazine.)

Andersen, J. 1953. Analysis of a Danish roe-deer population (*Capreolus capreolus* (L)) based upon the extermination of the total stock. *Danish Rev. Game Biol.* **2,** 127–155.

Andrewartha, H. G., and L. C. Birch. 1954. "The Distribution and Abundance of Animals." Univ. of Chicago Press, Chicago, Illinois. 782 pp.

Anonymous. 1961. "Management of Yellowstone's Northern Elk Herd." U. S. Natl. Park Serv. No. N16, 22 pp. (mimeo.).

Bannikov, A. G. 1961. L'ecologie de *Saiga tatarica* L. en Eurasie, sa distribution et son exploitation rationnelle. *Terre Vie* **1,** 77–85.

Bannikov, A. G., L. V. Zhirnov, L. S. Lebedeva, and A. A. Fandeev. 1961. "Biology of the Saiga." Izdatel' stvo Sel' skokhozyaistvonnoi Literatury, Zhurnalov i Plakatov, Moskva. (Transl. by M. Fleischmann. Israel Program for Sci. Transl., Jerusalem, 1967. 252 pp.)

Baranov, T. I. 1918. "On the Question of the Biological Basis of Fisheries." Izv. Nauchn. Issled. Ikliol. Inst. No. 1, 71–128. (English transl. by W. E. Ricker with assistance of Natasha Artin. 53 pp. mimeo.)

Bartlett, C. O. 1958. A study of some deer and forest relationships in Rondeau Provincial Park. *Ontario Dept. Lands Forests, Tech. Bull. Wildlife Ser.* No. 7, 172 pp.

Beverton, R. J. H., and S. J. Holt. 1957. "On the Dynamics of Exploited Fish Populations." Vol. XIX. Min. Agr., Fisheries Food, Fishery Invest., Ser. II. H. M. Stationery Office, London. 533 pp.

Birch, L. C. 1948. The intrinsic rate of natural increase of an insect population. *J. Animal Ecol.* **17,** 15–26.

Bohart, G. E., and F. H. Wagner. 1967. (Review of) Brian, M. V., "Social Insect Populations." Academic Press, New York, 1965. *Bull. Entomol. Soc. Am.* **13**, 246–247.

Bourlière, F. 1963. Observations on the ecology of some large African mammals. *In* "African Ecology and Human Evoloution" (F. C. Howell and F. Bourlière, eds.), pp. 43–54. Aldine Publ. Co., Chicago, Illinois.

Brian, M. V. 1965. "Social Insect Populations." Academic Press, New York. 135 pp.

Buechner, H. K., and H. C. Dawkins. 1961. Vegetation change induced by elephants and fire in Murchison Falls National Park, Uganda. *Ecology* **42**, 752–766.

Chapman, R. N. 1931. "Animal Ecology with Special Reference to Insects." McGraw-Hill, New York. 464 pp.

Cole, L. C. 1948. Population phenomena and common knowledge. *Sci. Monthly* **67**, 338–345.

Cook, C. W. 1954. Common use of summer range by sheep and cattle. *J. Range Management* **7**, 10–13.

Craighead, F. C., Jr., and R. F. Dasmann. 1966. "Exotic Big Game on Public Lands." U.S. Dept. Interior, Bur. Land Management. 26 pp. multilith.

Craighead, J. J., and F. C. Craighead, Jr. 1956. "Hawks, Owls and Wildlife." The Stackpole Co., Harrisburg, Pennsylvania, and the Wildlife Management Inst., Washington, D.C. 443 pp.

Dahlberg, B. L., and R. C. Geuttinger. 1956. "The White-Tailed Deer in Wisconsin." Wisconsin Conserv. Dept., Game Management Div., Tech. Wildlife Bull. No. 14, 282 pp.

Darling, F. F. 1960. "An Ecological Reconnaissance of the Mara Plains in Kenya Colony," Wildlife Monograph No. 5. 41 pp.

Darling, F. F. 1964. Conservation and ecological theory. *Brit. Ecol. Soc. Jubilee Symp., London, 1963* pp. 39–45. Blackwell, Oxford.

Dasmann, R. F. 1962. Game ranching in African land-use planning. *Bull. Epizootic Disease Africa* **10**, 13–17.

Dasmann, R. F. 1964a. "Wildlife Biology." Wiley, New York. 231 pp.

Dasmann, R. F. 1964b. "African Game Ranching." Macmillan, New York. 75 pp.

Dasmann, R. F., and A. S. Mossman. 1961. "Commercial Utilization of Game Mammals on a Rhodesian Ranch." Presented before Calif. Sect. Wildlife Soc. 1961 11 pp. mimeo.

Dasmann, R. F., and A. S. Mossman. 1962. Abundance and population structure of wild ungulates in some areas of Southern Rhodesia. *J. Wildlife Management* **26**, 262–268.

Davis, D. E., and F. B. Golley. 1963. "Principles in Mammalogy." Reinhold, New York. 335 pp.

Deevey, E. S. 1962. Animal populations. *In* "Frontiers of Modern Biology" (G. B. Moment, ed.), pp. 18–26. Houghton, Boston, Massachusetts.

Eberhardt, L. 1960. "Estimation of Vital Characteristics of Michigan Deer Herds." Mich., Dept. Conserv., Game Div. Rept. No. 2282, 192 pp.

Edminster, F. C. 1938. Productivity of the ruffed grouse in New York. *North Am. Wildlife Conf., Trans.* **3**, 825–833.

Ehrlich, P. R., and L. C. Birch. 1967. The "balance of nature" and "population control." *Am. Naturalist* **101**, 97–107.

Elton, C. S. 1958. "The Ecology of Invasions by Animals and Plants." Methuen, London. 181 pp.

Elton, C. S. 1966. "The Pattern of Animal Communities." Methuen, London. 432 pp.

Engelmann, M. D. 1966. Energetics, terrestrial field studies, and animal productivity. *Advan. Ecol. Res.* **3**, 73–115.

Errington, P. L. 1945. Some contributions of a fifteen-year local study of the northern bobwhite to a knowledge of population phenomena. *Ecol. Monographs* **15**, 1–34.

Flook, D. R. 1964. Range relationships of some ungulates native to Banff and Jasper National Parks, Alberta. *In* "Grazing in Terrestrial and Marine Environments" (D. J. Crisp, ed.), pp. 119–128. Blackwell, Oxford.

Gause, G. F. 1934. "The Struggle for Existence." Williams & Wilkins, Baltimore, Maryland. 163 pp.

Geis, A. D., and R. D. Taber. 1963. Measuring hunting and other mortality. *In* "Wildlife Investigation Techniques" (H. S. Mosby, ed.), 2nd ed., pp. 284–298. Edwards, Ann Arbor, Michigan.

Glover, J. 1963. The elephant problem at Tsavo. *E. African Wildlife J.* **1**, 1–10.

Gulland, J. A. 1962. The application of mathematical models to fish populations. *In* "The Exploitation of Natural Animal Populations" (E. D. LeCren and M. W. Holdgate, eds.), pp. 204–217. Wiley, New York.

Hairston, N. G., F. E. Smith, and L. B. Slobodkin. 1960. Community structure, population control, and competition. *Am. Naturalist* **94**, 421–425.

Haldane, J. B. S. 1953. Animal populations and their regulation. *New Biol.* **15**, 9–24.

Hickey, J. J. 1952. "Survival Studies of Banded Birds." U.S. Fish Wildlife Serv., Spec. Sci. Rept.: Wildlife. No. 15, 177 pp.

Hickey, J. J. 1955. Is there a scientific basis for flyway management? *North Am. Wildlife Conf., Trans.* **20**, 126–150.

Hochbaum, H. A. 1947. The effect of concentrated hunting pressure on waterfowl breeding stock. *North Am. Wildlife Conf., Trans.* **12**, 53–62.

Holling, C. S. 1959. The components of predation as revealed by a study of small-mammal predation of the European pine sawfly. *Can. Entomologist* **81**, 293–320.

Holling, C. S. 1966. The strategy of building models of complex ecological systems. *In* "Systems Analysis in Ecology" (K. E. F. Watt, ed.), pp. 195–214. Academic Press, New York.

Holsworth, W. N. 1960. Interactions between moose, elk, and buffalo in Elk Island National Park, Alberta. M.S. Thesis, British Columbia. 92 pp.

Hutchinson, G. E. 1957. Concluding remarks. *Cold Spring Harbor Symp. Quant. Biol.* **22**, 415–427.

Hutchinson, G. E. 1965. "The Ecological Theater and the Evolutionary Play." Yale Univ. Press, New Haven, Connecticut. 139 pp.

Huxley, J. 1961. "The Conservation of Wildlife and Natural Habitats in Central and East Africa." UNESCO Mission Rept., Paris. 113 pp.

Ivlev, V. S. 1961. "Experimental Ecology of the Feeding of Fishes." Yale Univ. Press, New Haven, Connecticut. 302 pp.

Jenkins, D., A. Watson, and G. R. Miller. 1963. Population studies of red grouse, *Lagopus lagopus scoticus* (Lath) in north-east Scotland. *J. Animal Ecol.* **32**, 317–376.

Klomp, H. 1966. The dynamics of a field population of the pine looper, *Bupalus piniarius* L. (Lep., Geom.). *Advan. Ecol. Res.* **2**, 207–305.

Kluijver, H. N. 1951. The population ecology of the great tit, *Parus m. major* L. *Ardea* **39**, 1–135.

Knowlton, F. F. 1964. Aspects of coyote predation in south Texas with special reference to white-tailed deer. Ph.D. Thesis, Purdue. 188 pp.

Kozicky, E. L., and G. O. Hendrickson. 1952. Flucuations in bob-white populations, Decatur County, Iowa. *Iowa State Coll. J. Sci.* **26**, 483–489.

Lack, D. 1954. "The Natural Regulation of Animal Numbers." Oxford Univ. Press (Clarendon), London and New York. 344 pp.

Lack, D. 1964. A long-term study of the great tit (*Parus major*). *J. Animal Ecol.* **33**, 159–173.

Lack, D. 1966. "Population Studies of Birds." Oxford Univ. Press (Clarendon), London and New York. 341 pp.

Lambrecht, F. L. 1966. Some principles of tsetse control and land-use with emphasis on wildlife husbandry. *E. African Wildlife J.* **4**, 89–98.

Larkin, P. A. 1963. Interspecific competition and exploitation. *J. Fisheries Res. Board Can.* **20**, 647–678.

Lauckhart, J. B., and J. W. McKean. 1956. Chinese pheasants in the Northwest. *In* "Pheasants in North America" (D. L. Allen, ed.), pp. 43–89. The Stackpole Co., Harrisburg, Pennsylvania, and the Wildlife Management Inst., Washington, D.C.

LeCren, E. D. 1965. Some factors regulating the size of populations of freshwater fish. *Mitt. Intern. Ver. Limnol.* **13**, 88–105.

Leopold, A. 1933. "Game Management." Charles Scribner's Sons, New York. 481 pp.

Leopold, A., L. K. Sowls, and D. L. Spencer. 1947. A survey of overpopulated deer ranges in the United States. *J. Wildlife Management* **11**, 162–177.

Leopold, A. S. 1952. Ecological aspects of deer production of forest lands. *Proc. U. N. Sci. Conf. Conserv. Util. Resources, Lake Success, N.Y. 1949* Vol. 7, pp. 205–207.

Linder, R. L., D. L. Lyon, and C. P. Agee. 1960. An analysis of pheasant nesting in south-central Nebraska. *North Am. Wildlife Conf., Trans.* **25**, 214–230.

Lotka, A. J. 1956. "Elements of Mathematical Biology." Dover, New York. 465 pp.

MacArthur, R. H. 1960. On the relative abundance of species. *Am. Naturalist* **94**, 307–318.

MacArthur, R. H., and J. H. Connell. 1966. "The Biology of Populations." Wiley, New York. 200 pp.

McAtee, W. L. 1932. Effectiveness in nature of the so-called protective adaptations in the animal kingdom, chiefly as illustrated by foods of nearctic birds. *Smithsonian Inst. Misc. Collections* **85**, 201 pp.

McEwan, E. H., and A. J. Wood. 1966. Growth and development of the barren ground caribou. I. Heart, girth, hind foot length, and body weight relationships. *Can. J. Zool.* **44**, 401–411.

Macfadyen, A. 1964. Energy flow in ecosystems and its exploitation by grazing. *In* "Grazing in Terrestrial and Marine Environments" (D. J. Crisp, ed.), pp. 3–20. Blackwell, Oxford.

Martin, P. S. 1964. "Animal Introduction in the New World/A Prospectus." Geochronology Lab., University of Arizona. 11 pp. mimeo.

Moisan, G., R. I. Smith, and R. K. Martinson. 1967. "The Green-Winged Teal: Its Distribution, Migration, and Population Dynamics." U.S. Fish Wildlife Serv., Spec. Sci. Rept.: Wildlife No. 100, 248 pp.

Morris, R. F. 1963a. 40. Resume. *In* "The Dynamics of Epidemic Spruce Budworm Populations" (R. F. Morris, ed.), pp. 311–320. Mem. Entomol. Soc., Canada.

Morris, R. F. 1963b. 18. The development of predictive equations for the spruce budworm based on key-factor analysis. *In* "The Dynamics of Epidemic Spruce Budworm Populations" (R. F. Morris, ed.), pp. 116–129. Mem. Entomol. Soc., Canada.

Murdoch, W. W. 1966. Community structure, population control, and competition—a critique. *Am. Naturalist* **100**, 219–226.

Murie, O. J. 1951. "The Elk of North America." The Stackpole Co., Harrisburg, Pennsylvania, and the Wildlife Management Inst., Washington, D.C. 376 pp.

Murphy, G. I. 1966. Population biology of the Pacific sardine (*Sardinops caerulea*). *Proc. Calif. Acad. Sci., Fourth Ser.* **34**, 1–84.

Murphy, G. I. 1967. Vital statistics of the Pacific sardine (*Sardinops caerulea*) and the population consequences. *Ecology* **48**, 731–736.

Nelson, K. L. 1965. Status and habits of the American buffalo (*Bison bison*) in the Henry Mountain area of Utah. *Utah State Dept. Fish Game Publ.* **65-2**, 142 pp.

Nicholson, A. J. 1933. The balance of animal populations. *J. Animal Ecol.* **2,** 132–178.

Odum, E. P. 1959. "Fundamentals of Ecology," 2nd ed. Saunders, Philadelphia, Pennsylvania. 546 pp.

Ovington, J. D., D. Heitkamp, and D. B. Lawrence. 1963. Plant biomass and productivity of prairie, savanna, oakwood and maize field ecosystems in central Minnesota. *Ecology* **44,** 52–63.

Paine, R. T. 1966. Food web complexity and species diversity. *Am. Naturalist* **100,** 65–75.

Pearl, R., and L. J. Reed. 1920. On the rate of growth of the population of the United States since 1790 and its mathematical representation. *Proc. Natl. Acad. Sci. U.S.* **6,** 275–288.

Pearsall, W. H. 1964. The development of ecology in Britain. *Brit. Ecol. Soc. Jubilee Symp., London, 1963* pp. 1–12. Blackwell, Oxford; Suppl. to *J. Ecol.* **52,** and *J. Animal Ecol.* **23,** 1–244.

Peterle, K. J., and W. R. Fouch. 1959. "Exploitation of a Fox Squirrel Population on a Public Shooting Area." Mich., Dept. Conserv., Game Div. Rept. No. 2251, 4 pp. (mimeo.).

Petrides, G. A. 1956. Big game densities and range carrying capacity in East Africa. *North Am. Wildlife Conf., Trans.* **21,** 525–537.

Petrides, G. A., and W. G. Swank. 1958. Management of the big game resource in Uganda, East Africa. *North Am. Wildlife Conf., Trans.* **23,** 461–477.

Petrides, G. A., and W. G. Swank. 1966. Estimating the productivity and energy relations of an African elephant population. *Proc. 9th Intern. Grasslands Congr., São Paulo, 1965* pp. 832–842.

Raymont, J. E. G. 1966. The production of marine plankton. *Adv. Ecol. Res.* **3,** 177–205.

Ricker, W. E. 1946. Production and utilization of fish populations. *Ecol. Monographs* **16,** 373–391.

Ricker, W. E. 1954a. Effects of compensatory mortality upon population abundance. *J. Wildlife Management* **18,** 45–51.

Ricker, W. E. 1954b. Stock and recruitment. *J. Fisheries Res. Board Can.* **11,** 559–622.

Ricker, W. E. 1963. Big effects from small causes: Two examples from fish population dynamics. *J. Fisheries Res. Board Can.* **20,** 257–294.

Robinette, W. L. 1956. Productivity—the annual crop of mule deer. *In* "The Deer of North America/The White-Tailed, Mule and Black-Tailed Deer, genus *Odocoileus*/Their History and Management" (W. P. Taylor, ed.), pp. 415–429. The Stackpole Co., Harrisburg, Pennsylvania, and the Wildlife Management Inst., Washington, D.C.

Schaller, G. B. 1967. "The Deer and the Tiger—A Study of Wildlife in India." Univ. of Chicago Press, Chicago, Illinois. 370 pp.

Scott, R. F. 1954. Population growth and game management. *North Am. Wildlife Conf., Trans.* **19,** 480–503.

Slobodkin, L. B. 1954. Population dynamics in *Daphnia obtusa* Kurz. *Ecol. Monographs* **24,** 69–88.

Slobodkin, L. B. 1961. "Growth and Regulation of Animal Populations." Holt, New York. 184 pp.

Slobodkin, L. B. 1964. Experimental populations of hydrida. *Brit. Ecol. Soc. Jubilee Symp., London, 1963* pp. 131–148. Blackwell, Oxford; Suppl. to *J. Ecol.* **52,** and *J. Animal Ecol.* **33,** 1–244.

Slobodkin, L. B., F. E. Smith, and N. G. Hairston. 1967. Regulation in terrestrial ecosystems, and the implied balance of nature. *Am. Naturalist* **101,** 109–124.

Smith, A. D. 1949. Effects of mule deer and livestock upon a foothill range in northern Utah. *J. Wildlife Management* **13,** 421–423.

Smith, A. D. 1965. Determining common use grazing capacities by application of the key species concept. *J. Range Management* **18**, 196–201.

Smith, F. E. 1952. Experimental methods in population dynamics: A critique. *Ecology* **33**, 441–450.

Smith, F. E. 1961. Density dependence in the Australian Thrips. *Ecology* **42**, 403–407.

Smith, F. E. 1963. Population dynamics in *Daphnia magna* and a new model for population growth. *Ecology* **44**, 651–663.

Smith, S. H. 1966. Species succession and fishery exploitation in the Great Lakes. *Symp. Overexploited Animal Population, 1966* Am. Assoc. Advance. Sci., Washington, D.C., 28 pp. mimeo.

Smith, S. H. 1968. Species succession and fishery exploitation in the Great Lakes. *J. Fisheries Res. Board Can.* **25**, 667–693.

Solomon, M. E. 1957. Dynamics of insect population. *Ann. Rev. Entomol.* **2**, 121–142.

Solomon, M. E. 1964. Analysis of processes involved in the natural control of insects. *Advan. Ecol. Res.* **2**, 1–58.

Southern, H. N. 1965. The place of ecology in science and affairs. *New Zealand Ecol. Soc., Proc.* **12**, 1–10.

Steyn, T. J. 1966. Game farming and hunting areas. *Flora Fauna* **17**, 1–3.

Taber, R. D., and R. F. Dasmann. 1958. The black-tailed deer of the chapparral/its life history and management in the north coast range of California. *Calif. Dept. Fish Game Bull. No.* 8, 163 pp.

Talbot, L. M. 1967. Personal communication.

Talbot, L. M. 1962. Food preferences of some East African wild ungulates. *E. African Agr. For. J.* **27**, 131–138.

Talbot, L. M. 1963. Comparison of the efficiency of wild animals and domestic livestock in the utilization of east African rangelands. *I.U.C.N.* [N.S.] No. 1, 328–335.

Talbot, L. M., and M. H. Talbot. 1963a. The high biomass of wild ungulates on East African savanna. *North Am. Wildlife Conf., Trans.* **28**, 465–476.

Talbot, L. M., and M. H. Talbot. 1963b. "The Wildebeest in Western Masailand, East Africa," Wildl. Monograph No. 12. 88 pp.

Talbot, L. M., M. H. Talbot, W. J. A. Payne, H. P. Ledger, and M. Verdcourt. 1965. The meat production potential of wild animals in Africa/A review of biological knowledge. *Commonwealth Agr. Bur. Tech. Commun.* **16**, 42 pp.

Tanner, J. T. 1966. Effects of population density on growth rates of animal populations. *Ecology* **47**, 733–745.

Teer, J. G., J. W. Thomas, and E. Walker. 1965. "Ecology and Management of White-Tailed Deer in the Llano Basin of Texas," Wildlife Monograph No. 15. 62 pp.

Tinbergen, L. 1960. The natural control of insects in pine woods. 1. Factors influencing the intensity of predation by songbirds. *Arch. Neerl. Zool.* **13**, 265–343.

Uhlig, H. G. 1956. "The Gray Squirrel in West Virginia." West Virginia Conserv. Comm., Charleston, West Virginia. 83 pp.

Verhulst, P. F. 1838. Notice sur la loi que la population suit dans son accroisement. *Corresp. Math. Phys.* **10**, 113–121.

Volterra, V. 1926. Variations and fluctuations of the number of individuals in animal species living together. *Atti Accad. Nazl. Lincei, Mem. Classe Sci. Fis., Mat. Nat.* [6] **2**, 31–113. (English transl., R. N. Chapman, 1931, "Animal Ecology with Especial Reference to Insects," pp. 409–448. McGraw-Hill, New York.)

Wagner, F. H., and A. W. Stokes. 1968. Indices to overwinter survival and productivity with implications for population regulation in pheasants. *J. Wildlife Management* **32**, 32–36.

Wagner, F. H., C. D. Besadny, and C. Kabat. 1965. "Population Ecology and Management of Wisconsin Pheasants." Wisconsin Conserv. Dept., Tech. Bull. No. 34, 168 pp.

Watt, K. E. F., ed. 1966. "Systems Analysis in Ecology." Academic Press, New York. 276 pp.

Watt, K. E. F. 1968. "Ecology and Resource Management—A Quantitative Approach." McGraw-Hill, New York. 450 pp.

Wynne-Edwards, V. C. 1962. "Animal Dispersion in Relation to Social Behavior." Oliver & Boyd, Edinburgh and London. 653 pp.

Chapter IX Ecosystem Models in Watershed Management

CHARLES F. COOPER

I. INTRODUCTION

A watershed (or catchment, to use the more descriptive and accurate British term) is a specific segment of the earth's surface, set off from adjacent segments by a more or less clearly defined boundary, and occupied at any given time by a particular grouping of plants and animals. This is almost a rewording of the classic definition of an ecosystem. As such, a catchment of convenient size is useful for studying interactions among plants and animals and their nonliving environment. The report

309

of Bormann and Likens in Chapter IV is an excellent example of the way in which detailed evaluation of the relations among the physical and biological components of a catchment can lead to new understanding of the processes of nature.

But our concern here is primarily with natural resource management. In this context a catchment is a producer of goods and services desired by mankind. A nonurban, nonagricultural watershed is an integrated system or machine for transforming solar radiation, precipitation, other environmental factors, labor, and capital into wood products, livestock products, wildlife, recreational and esthetic satisfactions, and water. Such a managed watershed is more than just a discrete ecosystem; it is also a component of a larger social and economic system.

Systems, social and biological, exist at many levels of integration. The social and economic system within which resource management decisions are made represents a higher level of integration than that of the ecosystem, narrowly defined; the latter, in turn, is a level of integration that encompasses the plants, animals, soils, microorganisms, and physical features of a catchment. The usefulness of the level of integration concept is that the goal of any given level of integration is manifested at the level above, while its mechanism is derived from the next lower level of integration (Feibelman, 1955). Thus the goal of a resource management system is generation of maximum sustained outputs of human satisfactions over time, a goal that has meaning only when expressed at the level of integration represented by the human social system.

The mechanism of a managed watershed lies at the next lower level of integration, comprised of the forest management subsystem, the grazing subsystem, the recreation use and development subsystem, and the water management subsystem. These subsystems interact to produce the vegetation, animal, and soil conditions that in part govern the yield and quality of water as well as of the other products and services of the watershed. For analysis of any particular system, three levels of integration must be considered: its own, the one above, and the one below (Feibelman, 1955).

The ecological theory that attempts to describe the responses of plant and animal communities to management practices deals at present with individuals or populations, not with systems as wholes. This has been reasonably satisfactory as long as the goals of management have been relatively simple—production of usable crops of wood or livestock without severely impairing water quality, for instance. Management, however, is manipulation of an entire ecosystem. The interconnections among the plants, animals, and physical features of a watershed are so complex that modification of any one component automatically affects all the others to a greater or lesser degree. Nevertheless, the ecological theory upon which management practices are based remains principally a collection of facts about isolated populations, not a set of predictive statements about integrated systems.

As increasing human populations require more efficient use of natural resources, and as the public becomes ever more aware of environmental quality, the implications for other aspects of the system of manipulations of specific components become increasingly important. Thus, the only level of ecological theory that will ultimately provide the necessary guidance to management is a theory of ecosystems.

II. THE SYSTEMS APPROACH TO PROBLEM SOLVING

"Ecosystem" and "systems approach" have been much in the news of late, but neither is new. All of us regularly practice the systems approach in our personal and professional lives. We treat our family as a system, relating the aspirations and needs of each member to the whole, and deploying our financial and human resources to benefit the entire family as much as any of its members. As resource managers, we treat resources as a system when we enforce stringent and costly erosion control methods in logging operations, or when we attempt to balance wildlife use against grazing by domestic livestock on ranges critical for the wintering of both. We usually get no farther than this, because the issues become too complex and the many factors that have to be taken into account overwhelm us with too much information to handle as individuals (Michaelis, 1968).

A. New Tools for Optimal Resource Management

Powerful new tools—in logic, in statistical analysis, and especially in the development of high-speed computers—now make it possible to handle large masses of information in a very short time. A computer can be programed to demonstrate on paper the consequences of alternative courses of action, and to evaluate how any one of the thousands of system components would be affected if a major change were made in any of the others. The objective is to determine the optimum course of action, given the economic, institutional, and ecological constraints that bound the system.

The design of optimal resource management systems involves elements of the problems of the engineer and of the academic biologist. The engineer's task is to design an antiaircraft gun that will meet or exceed performance specifications at no more than a given cost. He can incorporate commercially available hardware or design his own to do the job. In either case, the composition and performance of the individual components is known. The biologist's assignment, by contrast, is to find out, with a minimum of destructive testing, how the other man designed the gun (Machin, 1964). By observing the operation of the system under a variety of imposed stresses, he should eventually be able to understand how it is put together. The resource manager must combine both of these approaches to design an efficient system for producing the maximum flow of desired goods and services, using components whose functioning he only partly understands and whose performance he can predict only within wide limits. Concurrently with his design problem, he must try to find out more about the components from which he must build his management system.

B. The Systems Approach

The systems approach involves several successive steps:

(1) Establish the goals or objectives of the projected system in the context of the larger system in which it is embedded. This is a continually changing process requiring mature judgment. Thus we hear much of an impending critical national water shortage, and of the urgent need to extract more water from the major watersheds of the United States. Yet a report sponsored by the National Academy of Sciences (White, 1966) begins with the statement that "there is no nationwide shortage of water and no imminent danger of one." There is a shortage of water of desired quality at the places where people want it, at a price they are willing to pay. The concept that additional water is in itself a worthy aim of watershed management needs constant reexamination. In fact, it is precisely this stage of goal definition that is at once the most critical and the most difficult aspect of the systems approach to watershed management.

(2) Determine the relationships among the variables of the system, and between these variables and the objectives previously identified. This is termed constructing a model of the system. The development of system models proceeds in several stages. They are initially built up from simple mathematical statements and statistical distributions which represent functions and interrelationships derived from measurements on actual systems, or from plausible estimates where real data are lacking. These statements are translated into computer language, permitting simulation of system behavior on the machine. The results of this simulation are compared with the outcome of concurrent experimentation and observation in the real world; points of agreement and divergence are examined to refine and improve the model, and to identify aspects of the system about which more empirical information is needed. System modeling is thus a continuous, circular feedback process linking field measurement

and observation with computer simulation or analytical mathematical solutions. The resulting model is an abstraction and simplification, mirroring only those aspects of the real system which are deemed of particular interest.

(3) Quantify the various outputs or results that the system can achieve. These outputs must be expressed in common numerical terms, and in our economic world these common terms are usually dollars and cents. This imposes another difficulty, for many of the outputs from complex watersheds are almost incommensurable. Recreation values can seldom be accurately expressed in the same units as timber harvest; the value of water itself at the source is a matter of extensive controversy among economists.

(4) Quantify in a similar fashion the functional relationships among the elements of the model and all the inputs or resources that are needed to build each of the alternative systems that might be specified. These inputs and resources are the costs.

(5) Combine the two preceding steps to determine the input-output, or better the cost-benefit relationships of the particular model being investigated. This cost-benefit analysis must take into account the dynamic nature of the real world; calculation in terms of a flow of costs and benefits over time rather than as a static equilibrium solution is necessary to take account of possible changes in technology and in the desires of society in the future.

(6) Finally, determine from the cost-benefit relationships that choice among all possibilities that provides the most favorable ratio of benefits to social costs, including opportunity costs and the value of options foregone. At this stage, linear programing, dynamic programing, and related mathematical optimization techniques come into play.

The systems approach, as here outlined, is not a formula for replacing decision-makers by machines or for predicting coming events. Rather, it provides a method for indicating the future consequences of present policy decisions, for anticipating future problems, and for designing alternative solutions so that society has more options and freedom of choice than is likely when problems descend unnoticed and demand an immediate response. It aims to suggest a range of "alternative futures" among which choice is possible (Bell, 1967).

III. GOALS OF WATER MANAGEMENT

As already indicated, goals for management of forest and range catchments are continually evolving. Formerly plausible objectives become irrelevant in the light of new possibilities revealed by comprehensive analysis. It is in the realm of goals, perhaps, that natural resources management differs most strikingly from military and industrial operations in amenability to the procedures of systems analysis. In the latter instance, it is usually possible to specify a fairly clear objective that changes relatively little over time—maximum profit or greatest cost-effectiveness, for instance. In natural resource management, goals are constantly changing, both because people's wants and desires change and because increasing ecological knowledge indicates new opportunities or restrictions.

The goals of water management are usually several, which may or may not be compatible with each other: maximization of total water yield; maintenance of maximum water quality with respect to suspended and dissolved matter and to temperature; reduction of flood peaks; and augmentation of low flows. When desires for maximum yields of high quality forest products, wildlife, livestock, and esthetic and recreational opportunities are added to this list, a complex of objectives appears that clearly cannot be fully satisfied. By considering the needs and desires of a region as a whole, it may be possible to establish more rational aims for management of individual catchments than when each is considered by itself.

IV. MODELS FOR DESIGN

Managed forests and ranges of the future will have to be designed to meet specific objectives in water yield, water quality, and water control (Anderson, 1966). Such design will require synthesis and interpretation of data on hydrological processes and on the past performance of many watersheds in diverse climatic and geological situations. We cannot afford to test all the possible combinations of management policies that might be tried—not even by taking full advantage of efficient factorial experimental designs to get the maximum of information from the minimum number of tests. This sort of experimentation would take so long that the cost in money, land, and time would be prohibitive. Except for a limited number of small catchments, with an even more limited variety of plant covers and land usages, our present knowledge of the quantitative effects of various forms of land management on the water balance is still relatively meager (Reynolds and Leyton, 1968).

A. Empirical Models

Existing data, or data that in principle can be obtained in the field, can be used in an organized way to make predictions about the hydrological effects of alternative methods of management. Although a profound understanding of all the relevant processes and interactions would almost certainly help in this type of prediction, it is often possible to develop a predictive system adequate for most purposes without detailed understanding (Slobodkin, 1968). The procedure consists of measuring those observable variables in the hydrological cycle which appear pertinent, and then attempting to establish algebraic relations between them. It is hoped that these relationships will hold within the range of conditions normally encountered in practice (Amorocho and Hart, 1964).

Anderson (1966) has carried this process to a high level of accomplishment. Using factor analysis and related procedures, he has predicted sediment discharge and snow accumulation, in separate studies, from measured properties of a large number of catchments. Thirty physical, biological, and land use characteristics of each catchment were used in the sediment study, and 45 in the snow cover analysis. The mathematical screening procedure eliminated most of these variables as either irrelevant or redundant, i.e., closely correlated with other, apparently more important characteristics. In the streams tested, sediment yield could be closely predicted from six variables: rain area, fires, unimproved roads, poor logging, steep grasslands, and mean annual water flow.

This approach does not require that there be any previously known logical or empirical relationship whatever between the various measurements, but it does require that the precision of the measurements be fairly high. The resulting predictions may be reasonably good, but they do not take the place of actual models or theories when the unusual or unexpected occurs (Slobodkin, 1968). Anderson did use known or assumed relationships in the selection of the variables to use in his screening procedure, but the results of his analysis do not add greatly to our understanding of hydrological processes despite their high predictive value. It is questionable how helpful Anderson's factor analysis model would be for predicting the hydrological effects of a new perturbation such as deliberate precipitation increase through cloud seeding. It does not take into account the dynamic nature of vegetation as it might be affected by a long, continued program of weather modification, nor how the resulting vegetation changes would affect hydrological behavior.

B. Predictive Models

Predictive models based on ecosystem processes, on the other hand, appear to hold the greatest promise for the future. These models are built around the flow of matter (particularly water) and energy through the ecosystem. No model of this sort has yet reached the level of prediction achieved by Anderson's factor analysis scheme, and some of the methods he pioneered will doubtless eventually be incorporated into complete watershed ecosystem models.

Once the system to be studied has been defined, an abstraction or simplification is constructed in the form of a system model which identifies as accurately as possible the system components and their functional interactions. Initially the model might be a block diagram with blocks representing the major components and lines representing the flows of matter or energy from one component to another (Fig. 1). The construction of such a block diagram is a useful exercise in problem formulation in its own right; the mere act of describing a system in terms of block diagrams and feedback mechanisms requires precise identification of previously vague concepts and may lead to a degree of understanding attainable in no other way. To those who have not tried it, the construction of a block diagram encompassing each of the separate elements of a system is highly recommended as a means of forcing rigorous conceptualization of ideas and formulation of the relations between subsystems.

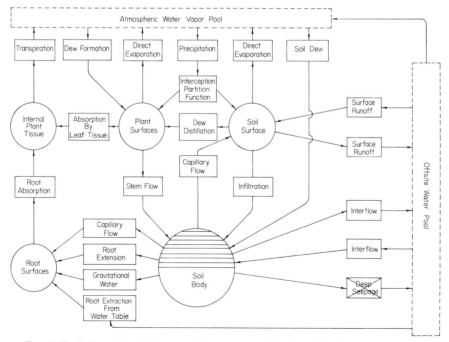

FIG. 1. Preliminary block diagram of the processes affecting soil moisture in semidesert soils. Circles represent storage reservoirs, lines are flow pathways. Boxes represent transformations that regulate flow quantities. The next step would be to diagram the controls and feedback mechanisms that govern each transformation.

Ultimately, though, the model will evolve into a mathematical representation. Ecosystem models will of necessity have to be designed for computer usage, since digital computer simulation is the only technique known today which can adequately represent the complex interactions of a dynamic and constantly varying system such as an ecosystem.

C. Some Steps in Modeling

The first step in model construction is to identify the important components of the system. A component is a single living species or a nonliving substance, such as soil or atmosphere. It may not be practical to take account of every component individually; consequently, functionally similar ones may be placed together and treated as a single component. Thus at the preliminary stage of modeling a range watershed ecosystem, it might be appropriate to lump annual grasses, perennial grasses, and shrubs into only three separate components.

Matter and energy flow in identifiable pathways from some specific components to others. The second step is to determine the quantitative functional relationships of the components, i.e., how the flows depend upon the quantities and the attributes of each of the several components. The information necessary to specify the form of the functional relationship between components, and to obtain preliminary estimates of their coefficients, will come from prior knowledge available in the literature and in the experience of the individuals concerned (G. G. Marten, personal communication). The relationships need not be exact to be useful for a model of the ecosystem as a whole. They only have to be close enough to give better results than could be obtained by common sense without the mathematics (Simon, 1960). This is because the behavior of the whole is dominated by the pattern of connections between the component parts and processes, rather than by the exact nature of any one component.

The model is then run or experimented with to arrive at deductions regarding the system. The initial model gives the rates of flow only at one brief instant when all the components are essentially fixed. As time passes, however, the stock of matter or energy in the various components changes as a consequences of flows in and out. The soil dries, and vegetation is grazed. The flow rates themselves, which depend upon the component quantities, change accordingly. It is necessary to use mathematical equations or a simulation program to indicate what these changes will be, since unaided intuition cannot cope with simultaneous changes in a large number of interacting components.

The deductions from the earliest iterations of the model are compared with the real world system. Undoubtedly major discrepancies will be discovered—points at which the response of the model fails to conform to the responses of the system. Efforts are then made to improve the model to make it agree more closely with reality.

At this stage the analyst is prepared to undertake a sensitivity analysis of the system model. This is intended to determine the relative sensitivity of the system to variations in specific inputs. Sensitivity analysis may demonstrate that an appreciable change in one input variable or management practice has comparatively little effect on the system's output, whereas another may produce a major change in system response.

The particular type of sensitivity analysis to be employed depends on the mathematical structure of the system model. This will usually be some form of computer simulation program. Each of the system's parameters is varied in succession, while the others are held constant. The output of the simulated system, which might be in the form of water yield or sediment production, is followed for different values of each successive input variable. It may develop, for instance, that the output changes only slightly as some input is progressively changed within a particular range. As a threshold is exceeded, however, further changes in the given input variable may produce important changes in output.

The value of sensitivity analysis is twofold. In the first place, it can help to determine the limits within which certain management practices can be varied without causing appreciable deterioration in watershed performance. In the second place, it can help to spotlight those aspects about which more needs to be known before system performance can adequately be predicted. Much orthodox scientific thought has held that all aspects of a system must be equally understood before real progress can be made; sensitivity analysis leads to concentration on the truly pertinent aspects of a system's behavior. Preliminary computer experiments have shown that hypothetical simple systems are far more sensitive to changes in the organizational structure of relations between components than to changes in the values of the components themselves. This suggests that field research should be directed first at definition of system components and of links between them, second at the processes that influence transfer rates between components, and only third toward measurement of the actual quantities of the various components (F. E. Smith, personal communication).

Forms of sensitivity analysis other than simulation are useful where the mathematical formulation of the model permits. In particular, if the system or an important subsystem can be formulated as a linear programming problem, powerful and efficient tools of sensitivity analysis based on mathematical optimization procedures are available (Dantzig, 1963; Navon and McConnen, 1967).

D. The State of the Art

No complete ecosystem models of whole catchments yet exist, but several partial models have been developed. The Stanford Watershed Model (Crawford and Linsley, 1966) is a digital simulation model which uses precipitation and potential evapotranspiration as the basic inputs. Actual evapotranspiration, streamflow, and soil moisture levels are obtained as the outputs. Streamflow is calculated at several locations along the stream channel, resulting in a series of hydrographs representing runoff contributions from a specific portions of the basin. Infiltration and related influences of vegetation and land use are read in as data; there is no attempt to simulate the dynamic ecological process that might alter the response of the catchment over time. Close agreement can be obtained between simulated and observed hydrographs from specific storms in particular basins. Several similar models are under development elsewhere by hydrologists and engineers. In very few cases are quantitative data about vegetation effects on the water balance included in such analyses (Reynolds and Leyton, 1968), and most have concerned themselves entirely with water yield and flow rates to the exclusion of water quality. One of the objectives of the analysis of ecosystems project of the International Biological Program is the construction and testing of general ecosystem models of a sort not now available.

V. COSTS, BENEFITS, AND OPTIMIZATION

Ecosystem models are not justified if they merely mimic a population system already in existence. Rather, as resource managers we wish to determine, more cheaply and quickly than by actual testing in the field, what will be the probable consequences of particular management strategies. More specifically we wish to determine if certain strategies or combinations of strategies are consistently superior or inferior to others (Watt, 1968).

Superiority or inferiority implies a standard of value. This standard has usually been a measure of economic return, but as Gross (1966) has pointed out, overreliance upon economic data because they seem precise and are readily available often leads to a narrow and unbalanced view of the state of a system. In addition to its economic aspects, every situation has political, social, and cultural aspects. Moreover, qualitative information may be fully as important as quantitative information. Gross (1966) has suggested that advances in the social sciences during the past decade have made it possible to think of powerful models that bring economic, political, sociological, and cultural variables together into a general systems framework to yield workable social indicators as measures of accomplishment. This is not the place to consider these developments in "social accounting" except to suggest that resource managers would do well to be aware of them.

A. A General Procedure for Optimization

A general procedure for optimizing the cost-benefit relations of any management problem has been rather simply described by Simon (1960):

(1) Construct a mathematical model of the system, as discussed in the preceding section.

(2) Define an objective function, the measure that is to be used for comparing the relative merits of possible courses of action. Where a watershed is expected to produce a variety of goods and services, an appropriately weighted objective function must be found that will adequately represent the value placed on each.

(3) Obtain empirical estimates of the numerical parameters that specify the particular concrete situation being considered, and insert these estimates into the previously determined model.

(4) Carry through the mathematical procedures for finding the course of action which, for the specified parameter values, maximizes the objective function.

B. Constraints on the Procedure

Certain conditions must be satisfied to apply this recipe:

(1) A quantitative objective function must be found. If the problem is so hopelessly qualitative that it cannot be described even approximately in terms of quantitative variables, the approach fails. As already indicated (Gross, 1966), further research is needed to develop methods of dealing with multiple social objectives. Existing procedures can optimize only one objective at a time, although the single-valued objective function to be maximized may itself be a weighted composite of several others.

(2) There must be ways of estimating the numerical values of appropriate parameters of system structure with sufficient accuracy for the task at hand. Proper estimation of these values depends largely on the knowledge and experience of the man in the field.

(3) The specifications of the model must fit the available mathematical tools. In some instances, linear programing, dynamic programing, and related efficient mathematical programing methods can be used. More often, model complications will require use of simulation methods. Simulation problems are characterized by being mathematically intractable

and having resisted solution by analytical methods. They involve many variables, functions which are not mathematically well behaved, and important random components.

(4) The problem must be small enough that the calculations can be done in reasonable time and at reasonable cost. Continual improvement in computers is steadily raising the limit of feasibility, but not eliminating it.

To Simon's conditions may be added a fifth: That the results have sufficient generality to be applied to other systems than the specific one for which the model was developed. Much manpower, time, and money can be expended in collecting data from a relatively small area which is bound to be more or less unique, and from which extrapolations are difficult if not impossible. Models are usually truly useful only if they are susceptible of quantitative generalization or application to other areas without needing much more information (Reynolds and Leyton, 1968).

C. Multiple Use Optimization

We need optimization models, not just of single catchments but of the resource management systems of whole regions. Such a regional model might lead in some instances to abandonment of the multiple use concept in favor of devoting some basins wholly to water production, perhaps through application of the water-harvesting procedures being developed by Myers (1963). This is the process of collecting, conveying, and storing water from an area that has been treated to increase the runoff of rain and snowmelt by artificially preventing infiltration, transpiration, and erosion.

A principal advantage of the systems approach to watershed management is that by directing attention to all aspects of a problem it reduces the likelihood of suboptimization. Suboptimization of management planning for a single catchment might lead, for instance, to concentration on maximum livestock production at the expense of water quality, or vice versa. On the other hand, optimization of joint production within a catchment might itself be suboptimal in a regional systems setting. Regional systems models might therefore lead to more efficient management of the aggregate of all catchments.

VI. CONCLUSIONS

There may seem to be an excessive reliance here on computers and on abstract system models, to the exclusion of biological and physical processes actually taking place on catchments in the real world. This is

intentional, for the higher the level of integration, the less tightly organized and coherent is the system and the greater is the complexity of its
components (Feibelman, 1955). The information needed to keep track
of the variables in such a system is so great as to overwhelm the unaided
human mind.

A complex system such as a managed catchment integrates vegetation,
animals, soils, topography, and climate. Each of the subsystems involves
rather specific relationships among its components, which can be studied
by detailed field observation. As we begin to consider more fully the relations between subsystems, the situation becomes less specific and more
abstract. Any rich understanding of a complex system's behavior requires
one to run repeatedly up and down the ladder from abstraction to specificity. This is the only way to avoid the twin dangers of decreased precision as one goes up the ladder and of increasing concentration upon
specific events of decreasing relevance for the whole as one moves down-
ward (Gross, 1966).

Watershed managers and watershed researchers have long made use
of ecosystem concepts. It would be pointless to review the many competent studies that have assessed the impact of animals upon vegetation,
and of the latter upon water quality and water yield. But even the best
of these attempts have dealt with only a relatively few variables and
interactions.

It will be necessary to continue to do so for some time to come. We do
not yet have sufficient information or sufficient understanding of many
processes to create fully adequate ecosystem models. For instance, we
still do not know how to translate observations on water use from small
plots to catchments of a few acres, nor how to quantitatively translate
hydrological observations from small catchments to stream basis of 10
to 400 square miles. Until more is known about the interactions of geology, climate, and channel hydraulics, it is not possible to extend the results of small catchment studies with real accuracy to basins of a size
that are of common interest (Cooper, 1963).

Nevertheless, it is the combination of high-speed computers, optimization methods, and new ecosystem theory that promises the most exciting
developments for watershed managers in the future. These developments
will be fully exploited by a new generation with the kind of education
described in Chapter X.

VII. SUMMARY

An uncultivated watershed is an integrated system that transforms precipitation, solar radiation, other environmental variables, labor, and capital into wood products, livestock products, wildlife, recreational and

esthetic satisfactions, and water. The forest management subsystem, the grazing subsystem, the recreation use and development subsystem, and the water management subsystem interact to produce the vegetation, animal, and soil conditions that govern the yield and quality of its products and services. The only level of ecological theory that can effectively guide management of such complex systems is a theory of ecosystems.

An ecosystem model is built up from mathematical statements and statistical distributions derived from field measurements. These statements are translated into computer language and the operation of the system is simulated on the machine. System modeling is a circular feedback process linking observations in the real world with computer simulation, leading to progressive improvement of the model.

System models enable resource managers to define, more quickly and cheaply than by actual testing in the field, certain strategies that are likely to be consistently superior to others. The most promising strategies can then be selected for fuller field evaluation. Proper formulation of the entire system in model form reduces the likelihood of suboptimization—concentration on livestock production at the expense of water quality, or vice versa. Perhaps most important, it permits preliminary testing of the system's sensitivity to changes in inputs. This may help to determine the limits within which management practices can be varied without appreciable deterioration in watershed performance. It can also help to spotlight those aspects of system behavior about which more must be known before its performance can adequately be predicted. The addition of high-speed computers, optimization methods, and new ecosystem theory to mature judgment and biological understanding promises a new level of accomplishment for natural resource management in the future.

REFERENCES

Amorocho, J., and W. E. Hart. 1964. A critique of current methods in hydrologic systems investigation. *Am. Geophys. Union, Trans.* **45**, 307–321.

Anderson, H. W. 1966. Watershed modeling approach to evaluation of the hydrological potential of unit areas. *In* "International Symposium on Forest Hydrology" (W. E. Sopper and H. W. Lull, eds.), pp. 737–748. Pergamon Press, Oxford.

Bell, D. 1967. The year 2000—the trajectory of an idea. *Dædalus* **96**, 639–651.

Cooper, C. F. 1963. Investigational methods in forest hydrology. *Australian Forestry* **27**, 93–105.

Crawford, N. H., and R. K. Linsley. 1966. Digital simulation in hydrology: Stanford Watershed Model IV. *Stanford Univ., Dept. Civil Eng., Tech. Rept.* **39**, 210 pp.

Dantzig, G. S. 1963. "Linear Programming and Extensions." Princeton Univ. Press, Princeton, New Jersey. 632 pp.

Feibelman, J. 1955. Laws of integrative levels of nature. *Brit. J. Phil. Sci.* **5**, 59–66.

Gross, B. 1966. The state of the nation: Social systems accounting. *In* "Social Indicators" (R. A. Bauer, ed.), pp. 154–271. M.I.T. Press, Cambridge, Massachusetts. 357 pp.

Machin, K. E. 1964. Feedback theory and its application to biological systems. *Symp. Soc. Expts. Biol.* **18**, 421–445.

Marten, G. G. Personal communication.

Michaelis, M. 1968. Can we build the world we want? *Bull. At. Scientists* **24**, 43–49.

Myers, L. E. 1963. Water harvesting by catchments. *Ariz. Watershed Symp., Proc.* **7**, 19–22.

Navon, D. I., and R. J. McConnen. 1967. Evaluating forest management policies by parametric linear programming. *U.S. Dept. Agr., Forest Serv. Res. Paper* **PSW-42**, 1–13.

Reynolds, E. C. R., and L. Leyton. 1967. Research data for forest policy: The purpose, methods, and progress of forest hydrology. *9th Brit. Commonwealth Forestry Conf., 1968.* Preprint, pp. 1–16.

Simon, H. A. 1960. "The New Science of Management Decision." Harper, New York. 50 pp.

Slobodkin, L. B. 1968. Aspects of the future of ecology. *BioScience* **18**, 16–23.

Smith, F. E. Personal communication.

Watt, K. E. F. 1968. "Ecology and Resource Management—A Quantitative Approach." McGraw-Hill, New York. 450 pp.

White, G. F. 1966. Alternatives in water management. *Natl. Acad. Sci.—Natl. Res. Council, Publ.* **1408**, 1–52.

SECTION IV **INSTILLING THE ECOSYSTEM CONCEPT IN TRAINING**

The first nine chapters of this book show numerous ways that ecosystem concepts have been or can be useful in both research and management in natural resource sciences. The last chapter by Van Dyne primarily concerns ways of implementing the ecosystem concept in training in natural resource sciences. The complexity of ecosystems is clearly shown in the first nine chapters and in several places systems approaches, systems techniques, and systems tools, especially computers, are alluded to as prerequisites for ecosystem research or management. Van Dyne suggests three major ways of incorporating this kind of training in natural resource sciences. Van Dyne received his undergraduate degree in agriculture at Colorado A&M College in 1954, his master's in animal husbandry at South Dakota State University in 1956, and his doctorate in nutrition, with minors in biochemistry and biometry, at the University of California in 1963. After his master's degree he was successively an instructor in forestry and range management at Colorado State University, an assistant professor in the Animal and Range Science Department at Montana State University, and an assistant research nutritionist at the University of California at Davis. In these positions his work involved research in range nutrition, range measurements, and soil–plant relationships and teaching in various areas of range management. After his doctorate he was employed by Union Carbide Corporation as a health physicist in the Oak Ridge National Laboratory, where his work was primarily concerned with modeling and analysis of biological systems and in herbage dynamics in old-field ecosystems and radioactively contaminated areas. Simultaneously he was a Ford Foundation professor in botany at the University of Tennessee, and taught graduate work in systems ecology. Since 1966 he has been employed in the College of Forestry and Natural Resources at Colorado State University holding a joint appointment in range science and fishery and wildlife biology. His teaching has been concerned primarily with modeling and analysis of ecological systems, as described in Chapter X. His research has concerned modeling of forest productivity, analytical study of numerical methods for systems analysis of ecological problems, and direction of the grassland biome studies in the United States International Biological Program.

Chapter X Implementing the Ecosystem Concept in Training in the Natural Resource Sciences

GEORGE M. VAN DYNE

327

I. INTRODUCTION

The preceding chapters clearly show there are many applications of ecosystem concepts in the natural resource fields. They also show that there are major differences in definitions of ecosystems and ecosystem concepts. The title of this chapter includes the terms "ecosystem," "training," and "natural resource sciences." In order to prevent ambiguity, the uses of some of these terms are defined below. Authors of several of the preceding chapters use various kinds of models to illustrate properties of ecosystems or entire ecosystems. In this chapter, I shall show how models can be used more in training in natural resource sciences. The diversity of disciplines, skills, and ideas that characterize good ecology are clearly illustrated in preceding chapters. The need for this diversity itself represents a dilemma—that of interactions which are necessary in our training and working approaches. These items are considered briefly in this section.

A. Definitions

Let me define the space scale, i.e., the fields that are discussed. Reference is made herein to several renewable natural resource fields—such as forest, watershed, wildlife, fisheries, and range management. For jobs in these fields students receive their training in various agricultural, natural resource, and liberal arts college departments.

Let me define the time scale. Reference will be made primarily to training natural resource scientists, ecologists, and managers who probably will not be effective in shaping major resource management decisions for ten years, but who will be sorely needed sooner. The titles of natural resource scientists, ecologists, and managers will be used somewhat interchangeably. It is the author's view that the top-level resource manager in, say, fifteen years may be trained very much like the scientist.

We have about worn out the word ecosystem in this book. You have read several definitions—both of the ecosystem as a concept and of the ecosystem as a spatial unit. For purposes of this chapter, I shall use Tansley's definition (1935) of the term ecosystem, "a system resulting from the integration of all living and nonliving factors of the environment." The term ecosystem also is used to describe the concept or approach of studying biotic–abiotic complexes. In the sense that the term ecosystem implies a concept and not a unit of landscape or seascape, the emphasis is that the biologist must look beyond his particular biological entity and must consider the interrelationships among these components and their environment.

B. Simplified Models

Consider the etymology of the word ecosystem. "Eco" implies environment. The term "system" implies an interacting, interdependent complex. The complex can be viewed in a simplified way (Fig. 1). An ecosystem is an integrated complex of living and nonliving components. Each component is influenced by the others, with the possible exception of macroclimate, and now man is on the verge of exerting meaningful influence over macroclimate.

A system is an organization that functions in a particular way. The functions of an ecosystem include transformation, circulation, and accumulation of matter and the flow of energy through and within living organisms by means of their activities and natural physical processes. Some specific functional processes include photosynthesis and decomposition, and more general functional processes include herbivory, carnivory, parasitism, and symbiosis.

A system functions only if there is a driving force. Consider, for example, a very simplified view of an ecosystem (Fig. 2) with a biological

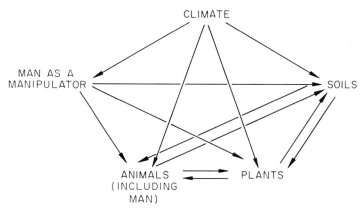

FIG. 1. Understanding ecosystems means understanding interactions among components. Each component is influenced by the others with the possible exception of macroclimate. Man is now on the verge of exerting meaningful influence over macroclimate.

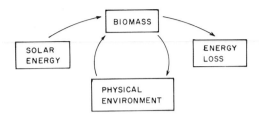

FIG. 2. Essentially ecosystems consist of interacting biological and physical components with inputs and outputs of energy.

component and a physical component with arrows denoting their inter-action. Also, there is a driving force, solar energy, and, to comply with the laws of thermodynamics, an energy loss via respiration.

At another level of complexity (Fig. 3), the biomass is segregated into three categories—producers, consumers, and decomposers—and man becomes a special fourth category of harvesters and manipulators. Still the physical environment is lumped under the category of mineral reser-voirs and inputs. Still there is an energy input, a flow of energy through the system, and a leakage of energy or respiratory loss. Now nutrient flow

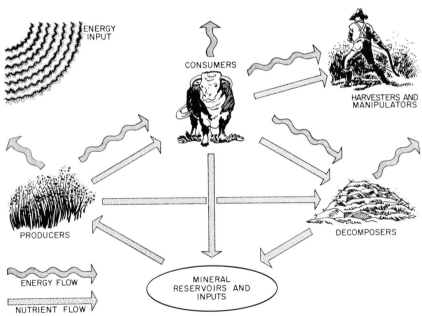

FIG. 3. Major ecosystem biocomponents are producers, consumers, and decomposers. Mineral reservoirs and inputs are important in terrestrial ecosystems. Man is both a spec-tator and a participant in the functioning of ecosystems.

or cycling of nutrients through components in the system has been added. Furthermore, the basic characteristic of natural resource management is included, i.e., man is both a spectator of and a participant in the functioning of ecosystems. Man has manipulated ecosystems to maximize the flow of nutrients and energy to him from the producers and primary consumers. He has attempted to minimize the respiratory losses of energy from producers, consumers, and decomposers.

Even this is a simplified diagram because ecosystems possess several other important characteristics. Ecosystems are characterized not only by a high magnitude of complexity, but also by a specific kind of complexity. Historical events are important so that events at any moment depend on previous circumstances as well as on existing conditions. Individuals, populations, and communities are not static entities; they change in directions dictated, at least partly, by their history. Spatial effects are similarly important. Ecosystems are open and interact with each other.

Physical and biological transport systems, together with inevitable physical and biological diversity (even over small areas), produce consequences that are essential properties of ecological systems. In addition to historical and spatial effects, the prevalence of nonlinear relations, thresholds, limits, and discontinuities are structural features that further characterize the complexity of describing ecological systems.

The magnitude and distinctive character of this complexity cannot be realistically treated in a fragmented manner. Our new ecology, in Odum's words (1964), must be a "systems ecology," even though the complexity presents massive problems of analysis and synthesis. Training scientists and managers to understand such complexity is the central theme of this chapter.

C. Need for the Ecosystem Concept in Training

Ecology has long been recognized as a multidisciplinary and integrative science. To understand the interaction between an individual organism and its environment requires an integration of knowledge from a variety of disciplines, e.g., genetics, physiology, biochemistry, morphology, and behavior. But ecology is more than the study of isolated individuals and their physical environments. There are higher levels of organization where individuals form populations, and populations interact to form communities. The emergent properties that appear at these higher levels of organization have generated additional disciplines: population ecology and genetics, trophic-dynamics, bioenergetics, and community organization. It is necessary to study the ecosystem as a whole in order to understand energy transformations, the hydrological cycle, or cycles of carbon, nitrogen, phosphorus, or other elements.

The ecosystem is a fundamental unit of study in basic ecology. As has been amply demonstrated, the ecosystem concept is useful, directly or indirectly, in many areas of applied ecology. For example, it is useful in the management of renewable resources such as forests, ranges, watersheds, fisheries, wildlife, and agricultural crops and stock (e.g., see Cole, 1958; Dyksterhuis, 1958; Leopold *et al.*, 1963; Lutz, 1963; Ovington, 1960; Pechanec, 1964). Additionally, understanding the ecosystem concept not only is useful but also is required for rational disposal of radioactive wastes and for analysis of environmental pollution.

Schultz (1967) indicates that this concept is useful in two major ways, and both of these have implications for training scientists and managers. First, the ecosystem concept includes and integrates several important subconcepts which provide useful models for research in resource fields. Second, this concept provides the basis or framework for critical evaluation of the impact of various practices and policies which seem good on the surface and continue to go unchallenged. These reasons, combined with many available well-documented examples, accentuate the need not only for sound overall ecological training but especially for implementation of the ecosystem concept in training in the natural resource sciences. With this background, some of the trends and unique ways of implementing the ecosystem concept in training are discussed.

Basically there are two major problems in implementing the ecosystem concept in training. Perhaps the first problem is securing an adequate base or fund of knowledge. The second problem is equally important, i.e., the integration of this knowledge. The training problems are complex. There has been a lack of integration of biological and physical disciplines in our training programs. The persistent lack of integration stems from the character of ecology as well as from its historical development, as will be considered further in the next section.

D. The Second Dilemma

Malthus was concerned by the fact that populations grow faster than do their means of subsistence. Malthus' dilemma, i.e., the first dilemma, is, from one point of view, an imbalance between the energy available to man and the energy he requires. Consider a second dilemma that was overlooked by Malthus when he foresaw the threat of uncontrolled population growth. The second dilemma concerns the increase in complexity, in the proliferation of the semantic environment, which accompanies the growth of population (Weinberg, 1967). For example, when the population of an area increases, the number of contacts between members of that population increases in proportion to the square of the size of the population. The greatly increased number of contacts and interactions soon stresses our communication, transportation, and psychological systems. We may find that we may be able to delay greatly the first dilemma, i.e., the energy imbalance, through such means as atomic energy. We have not found the solution to the second dilemma, the one caused by interactions.

Similar interactions make our task difficult in ecology. Yet the study of interactions of organisms with organisms and with their environment is the heart of ecology and is a concept which must be instilled in our teaching. Consider the problem of studying an entire ecosystem containing many interacting components. In a cohesive study many scientists must interact in studying these components, and the number of contacts among scientists, as well as among organisms in the ecosystem, grows approximately as the square of the number of scientists.

Our ability to condense and synthesize the body of information available to us may well limit the growth of ecology. We need more theorists and synthesizers to compact the literature and to make it more available to the scientific community. There is a tendency for scientists to work isolated from other disciplines, i.e., to eliminate contacts. Perhaps this is because it is easier, and perhaps it is a reflection of their training. Specialization predominates in our academic curricula; balanced generalization is rare. There is a tendency in training to stress generation of more information (although much of it is duplicating existing information) rather than to condense the information now available. Reversing, or at least balancing, this tendency is one way of improving many training programs.

II. THE ECOLOGICAL REVOLUTION

We have only recently begun a revolution in ecology that soon will have implications in the training of resource managers. When ecology emerged from its initial descriptive phase two routes were taken. One evolved toward evolutionary theory and the other toward functional ecology.

A. Historical Approaches in Ecology

The functional ecologist concentrated on the analysis of mechanisms underlying the action of ecological systems. In the 1920's and early 1930's the ecologists began to develop analytical models of their systems using the mathematical languages and techniques of classic physics (Holling, 1968). It soon became apparent, however, that if the models were based on realistically complex assumptions the models became completely intractable mathematically. Therefore, the model builder quickly learned to develop models based on a small number of simple assumptions and to live with the fact that his models did not correspond very closely to the real world. While the models were thereby holistic and integrative, they suffered the major drawback that they were unrealistic.

Functional ecologists who lacked a mathematical bent consequently felt uneasy with these simple models. They concentrated instead on experimental and field studies of isolated fragments of ecological systems or upon laboratory populations of organisms that, superficially, seemed simpler and more tractable analogs of natural populations (Holling, 1968).

Although both approaches have been rewarding, they are clearly limited. Fragmental studies, no matter how challenging, are still not holistic. The original and persistent problems in ecology arose from the frustrating attempts to study these enormously complex systems with tools and languages designed for much simpler ones. Significant concepts nevertheless have emerged, and studies of community structure, energy flow, biogeochemical cycles, and homeostatic mechanisms all produced significant constructs. In the last few years, however, new experimental approaches have been developed to permit integrated analysis of whole systems, not just fragments. For example, sophisticated field-sampling procedures and devices to monitor climatic changes and animal activity have evolved to provide data of the quantity and quality required to study natural communities (Watt, 1966). Multivariate statistical techniques also have been refined into powerful tools. Finally, and perhaps most important, the development of systems analysis and of the allied computer techniques and languages, for the first time, presents the ecologist with a language of synthesis specifically suited to handle both the magnitude and kind of complexity found in ecological systems.

Monographs and books displaying and discussing uses of these techniques in natural resource problems are only beginning to appear. At the present time, these potentially integrative concepts, which have emerged from separate branches of ecology, and the new techniques of analysis and synthesis of complex systems which have become available are not often being put together. The techniques and concepts are sufficiently new that few resource scientists have the necessary background and experience to exploit them. The scientists who do have these skills are widely separated, and many of these men have few regular contacts with students and even fewer contacts with resource management agency personnel. The small supply of these quantitatively trained scientists is now the rate-limiting factor required for expansion of this integrated ecology or for full implementation of the ecosystem concept.

B. A Feedback from Training

New quantitative methods and new, complex instruments are slowly being introduced into use in the natural resource fields. Some of these have been mentioned above and others will be referred to later in this chapter. How do these methods and instruments ensure, or relate to, better resource use? The thesis of this chapter is that better-trained natural resource scientists and managers, who have command of these concepts and techniques, will lead to new applications and better resource use (Fig. 4). In turn, these better-trained natural resource scientists and managers will develop new methods and new instruments. These new methods and instruments will lead to new applications and hence better resource use. The broken arrows in Fig. 4 represent feedbacks, and as we use our resources better (and we will have to if we are to survive) the entire cycle shown will be repeated, i.e., the development of even more better-trained scientists and managers, more new methods, and new instruments.

C. A Problem-Solving Approach in Ecology

Let me discuss training by using the direct approach of problem solution. This is an engineering approach, for one must resolve the problem before one can apply his tools and techniques to it. Let us assume we have a resource problem and are faced with the problem of solving it. I think we have three broad categories of tools with which we can attack resource problems (Fig. 5): conceptual or methodological tools, mechanical tools, and mathematical or analytical tools. As some of these tools are described, the connection to structuring training programs will begin to become clear.

Basically, our conceptual or methodological tools for solving natural

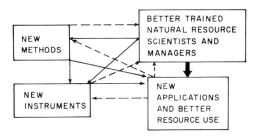

FIG. 4. Better resource use will be brought about by use of new methods and new instruments implemented by better-trained natural resource scientists and managers. There are a great number of feedbacks among the components of the natural resource problem.

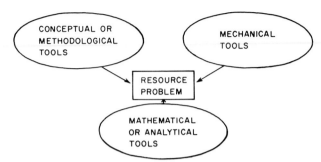

FIG. 5. A problem-solving approach in resource management requires first isolating and defining the problem and then applying to it conceptual, mechanical, and analytical tools for its solution.

resource problems include clear definition of the problem, definition of a model, and application of scientific method. The fundamentals of the scientific method will not be repeated here since most readers are well aware of this approach.

A basic conceptual requirement in solving natural resource ecosystem problems is clear definition of the problems. It is axiomatic that ambiguous use of terminology and an ambiguous statement of the problem lead to ambiguities of thought as well. Consider, for example, the importance of clear definition of the time scale with which we are concerned. In Fig. 6 the magnitude of some hypothetical ecosystem property is plotted

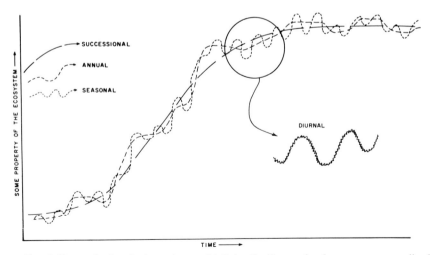

FIG. 6. To emphasize the importance of defining the time scale of concern, a generalized property of an ecosystem is plotted against time. Depending upon the time span of interest, the functional form between the property and time varies widely.

against time. Depending upon the time span, several functional forms can be discerned, varying from a rhythmic sinusoidal fluctuation on a short time span to a general sigmoid response over a long period of time. Clearly, ambiguous definiton of time span here could lead to highly varying conclusions about the time response of the property.

More and better use of logic and scientific and statistical methods is needed. We also need a clear definition of a model if we are going to design research to test a hypothesis. There is perhaps a tendency for resource scientists to pass over this phase of analysis, i.e., condensing what is known into a model. The failure of scientists to comprehend what is known already is understandable, in part, because of the volume of material to be covered. Inadequate examination of facts and data and inadequate formulation of hypotheses lead to uncritical selection of experiments, which test poorly formulated hypotheses. Resource scientists often are at fault here. The experimental design is, essentially, the plan or strategy of the experiment to test hypotheses.

Our usual approach in studying ecosystems or in conducting experiments is shown on the left in Fig. 7 and represents one of two main ways

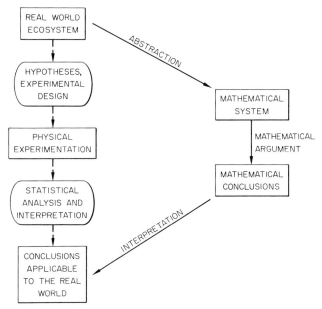

FIG. 7. The conventional approach in studying ecosystems or ecosystem components is shown on the left and includes processes of formulating hypotheses, designing and conducting experiments, and analyzing and interpreting results. A second and new approach to studying natural resource problems involves the abstraction of the system into a model, application of mathematical argument, and interpretation of mathematical conclusions.

of experimenting with ecosystems. The conventional process involves formulating hypotheses, designing and conducting experiments, and analyzing and interpreting results. The second method, on the right, involves abstraction of the system into a model, application of mathematical argument, and interpretation of mathematical conclusions. More often we will be required to seek the route shown on the right, for in many instances we will not be able to experiment. We may not be able to test a situation that does not exist but which may become real, e.g., widespread fire, thermonuclear war, and wide-scale environmental pollution. Where experimentation is too costly, mathematical modeling or mathematical experimentation may be especially useful.

This procedure of mathematical modeling is somewhat new to many resource scientists and managers and, in part, is just as much art as science. This procedure of modeling, interpretation, and verification is used in many engineering and scientific disciplines. The success of the procedure, however, depends upon the existence of an adequate fund of basic knowledge about the system. This knowledge, of course, is developed largely by the biologists and not by the analysts. This knowledge permits predictive calculations.

My second construct is that we need to master a variety of mechanical tools to solve natural resource problems (see Fig. 5). We shall need tools such as digital and analog computers, electrical, mechanical, and hydraulic simulation devices, and artificial populations. The act of expressing or testing biological problems with such analogs often reveals some unsuspected relationships and leads to new approaches and investigations. In conducting tomorrow's experiments in natural resource sciences, more refined chemical and analytical equipment, such as gas chromatographs, infrared gas analyzers, and recording spectrophotometers will be needed. This is not to imply, however, that complex instrumentation is a substitute for critical thought.

The positions of these typical tools, the types of resource biologists, and the level of organization they are studying are shown in Fig. 8. In the near future forest, range, wildlife, and fisheries scientists, as well as the bioclimatologists and systems ecologists, will make increasing use of such tools as computers and telemetry in studying levels of organization from the organisms on up to the ecosystem. This topic will be dealt with more fully in Section III.

In continuing discussion of mathematical tools, let us examine the foundations of the working methods. These foundations are sets, functions, and relations. Strange as it may seem, these foundations of mathematics are now being introduced at two extremes in the educational process. Children are studying them in primary grades, and some students are studying them at the Ph.D. level!

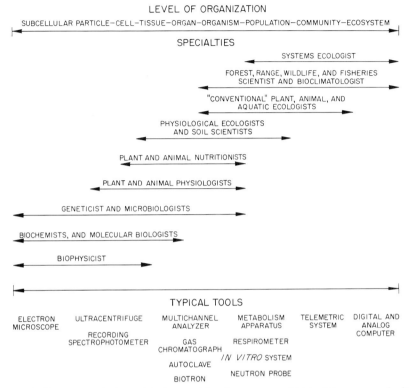

FIG. 8. Systems ecologists might be considered as a group of biologists concerned with levels of organization primarily from populations to ecosystems. The double-ended arrows indicate general positions of the specialities. Tools, especially, and to some extent the media overlap widely for different specialties.

These foundation concepts are used in the common mathematical working methods that are taught in undergraduate science programs. These common working methods increasingly are permeating curricula in natural resource sciences. Calculus, probability theory, and statistics have been basic for some time. Now differential equations are becoming increasingly important, and matrix algebra extremely so.

These mathematical techniques are used in the new tools we are hearing about. For some time we have been using least-squares models in statistical analyses in resource management fields. Mathematical programing techniques now are being applied increasingly in resource analysis and management; included here are linear, nonlinear, and dynamic programing. Several other tools from systems analysis and operations research are important—especially various kinds of compartmental model approaches which, essentially, are means of representing dynamic systems by systems of differential or difference equations. Solution of these equations then describes the state and changes of state of the system over time.

III. THE COMPUTATIONAL REVOLUTION

Science now is in the early stages of a major revolution caused by the increase in computational and mathematical power afforded by computers. It is evident that the magnitude of this increase is so large that it is rather difficult to comprehend and to appreciate. There has been an approximately 100,000-fold increase in speed of numerical computation in the last 50 years (Smith, 1966). Costs of computation also have decreased greatly. Such sciences as mathematics, statistics, astronomy, physics, and biomedicine are making good use of techniques and approaches offered by computers. Although there is a general awareness of the importance of computers in biology, and especially in ecology, there is a shortage of natural resource ecologists adequately trained in the use of computers.

A. Needs for Computer Training

Although review of major ecological and natural resource journals shows an increasing incorporation and use of the mathematical, statistical, systems-engineering, and computer techniques by natural resource ecologists in their research and studies, only a small proportion of the papers contain results of such usage. It was estimated early in this decade that we must train several thousand additional students at the graduate level in biophysics, biomedicine, bioengineering, biomathematics, and related interdisciplinary, computer-supported fields by 1970 (Schmitt and Caceres, 1964). In the 1960–1970 decade the needs for teaching and research facilities to accomplish such training in these interdisciplinary areas will double every three years. There now is an immediate need for training natural resource ecologists in the use of computers as a research tool and the situation could become critical within a decade. Computers will play increasingly important roles in quantitative ecology in the near future, and methods of analyses from statistics, operations research, and systems analysis will be used increasingly. Most resource biologists now in top management positions have no familiarity with computers, or they are acquainted only with first- or second-generation hardware and software. Computers in tomorrow's technology will have much larger and faster memories, remote consoles, and time-sharing systems.

It is essential that natural resource ecologists receive instruction early in their training in the use and limitations of a variety of computational techniques, and students should gain familiarity with both the software and hardware. Although training in natural resource ecology has improved, the needs are great, and more students with quantitative interests need to be attracted to the field. Of the total graduate enrollment in all biological sciences less than 1% are in ecology (Tolliver, 1965). Only a few of these students now receive sound statistical, mathematical, and computer science training (Watt, 1965). Most of the important natural resource problems reside at the ecosystem level of organization, a level whose study is much aided by computer-implemented methods from systems analysis and operations research. For significant advances in the future, natural resource ecologists will require even greater proficiency and sophisticated levels of training in chemistry, mathematics, and systems analysis (Miller, 1965).

It is also evident that there is increased need for interdisciplinary programs for training applied ecologists in coupling systems techniques and approaches with ecological methods. Many ecologists in natural resource management and agriculture have drawn upon ecological principles in the planning, execution, and interpretation of their research. There are increasing needs for ecologists to interact with systems engineers; there are great needs for engineers to utilize ecological techniques and principles in interdisciplinary studies. Training programs need to be designed to challenge students from both classic and applied ecological areas. Interdisciplinary training programs to integrate the efforts of bioenvironmental scientists and engineers are needed to provide the multidisciplinarians to solve tomorrow's complex resource problems.

B. Some Approaches to Computer Training for Natural Resource Sciences

Computers are being used increasingly in basic and applied ecological courses in various universities (e.g., Alaska, British Columbia, California-Davis, Colorado State, Cornell, Georgia, Tennessee, Washington, and Washington State). The experience is, however, that in most universities it is difficult to find more than one staff member in a given department with adequate formal training in depth in basic ecology, resource management, basic mathematics, applied mathematics, systems engineering, and computer sciences. To provide adequate instruction to students from varying disciplines (biology to bioengineering) and for reciprocal intellectual stimulation, it appears necessary to have in a teaching team several men with training covering the above fields. Another difficulty with many of the above programs is that computational training and use

have generally been introduced at the graduate rather than the undergraduate level. Thus students seldom have the opportunity to compare
the approaches, philosophies, and methods of various instructors over a
period of time. The students commonly are forced to rely primarily on the
insight of one individual.

IV. AN UNDERGRADUATE CURRICULUM

Much can be done to reshape some of our training programs in the
natural resource sciences and some ideas are given in the following sections. The suggested curricula in the following sections do not apply to
any specific school nor to any specific resource management field, although specific examples are included in the discussions. To change a
curriculum, especially undergraduate, in a major way in our resource
management schools is a major undertaking requiring much deliberation,
evaluation, and compromise. Thus, the example curricula which follow
are not considered final, the best, or the only approach—they represent
only a point of departure. They do not represent college or departmental
opinion. The author has had the good fortune of working and teaching in
colleges of liberal arts, agriculture, and natural resources at several
universities. Hopefully, the following ideas include some of the better
approaches from each field.

In this section a general outline is presented of the type of curricular
program needed to cover background and skills that today's undergraduate students will need in tomorrow's natural resource management, but
first we shall briefly mention some problems.

A. Some Common Problems

A National Academy of Sciences committee reports some of the common problems in our current curricula for undergraduate education in
renewable natural resources (Panel on Natural Resource Science, 1967).
Based on efforts of outstanding educators, the report pinpoints five curricular problems: (1) excessive emphasis on narrow vocational training;
(2) barriers that make it necessary to offer specialized courses in professional areas; (3) too much emphasis on practices, too little on principles; (4) insufficient progress in revising academic programs to reflect
newly emerging areas for professional employment; and (5) excessive
proliferation of curricula and courses.

We have much opportunity in natural resource management for curricular modemization. Consider the curricular blocks outlined in Fig. 9
and the numbers of quarter credits assigned to these. Obviously, such

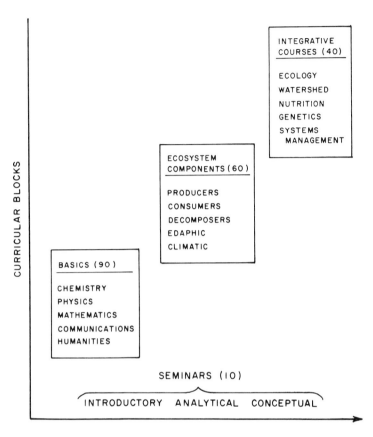

FIG. 9. A suggested composition of an undergraduate, 4-year program for training natural resource ecologists. Curricular blocks are shown in approximate time sequence in a 4-year program and approximate quarter credits assigned to each block are given in parentheses.

requirements may not fit the patterns developed in many institutions. It is important to consider both the total number of hours required for the undergraduate program and, more important, the relative distribution of subject matter areas within this total. This program totals to 200 quarter credits, a figure which is within the range required by many institutions.

B. Basics on Which to Build

My fundamental tenet is that our present-day curricula in natural resource management areas lack adequate basics. Some 45% of the credits in the undergraduate program should be used in a curricular block designed to give a thorough foundation in the basics—chemistry, physics, mathematics, communications, and the humanities (Fig. 9).

Training should include a solid series in general or inorganic chemistry, at least a survey course if not a sequence in organic chemistry, and a course in quantitative or qualitative analysis to develop laboratory skills and techniques. Students interested in research careers or graduate work should include physical chemistry, biochemistry, and radiochemistry. Generally, about 20 credits would be allocated to chemistry.

Considering the importance of electronics, atomic physics, thermodynamics, etc., it is essential that students in natural resource sciences take a basic course sequence in physics. Such material usually can be acquired in 10 to 15 credits.

There is an increasing awareness and use of mathematics in both basic and applied areas of biological sciences. Fortunately, students entering from high school often are trained in mathematics equivalent to the first year of college of several years ago. A basic working knowledge of mathematics includes algebra, trigonometry, and analytical geometry and calculus. However, the classic mathematics sequences are designed primarily for engineers and physical science majors. Curriculum modernization can produce substantial gains in efficiency. For example, at Colorado State University a biological science student who has had algebra may take a biomathematics sequence of 12 credits and cover working techniques through differential equations. Admittedly, most students completing this sequence will not develop or prove many new theorems, but they will have a substantial working knowledge of mathematics. Of course, undergraduate students in natural resource fields should also become acquainted with probability and statistical methods.

Let me emphasize here the importance and needs for mathematical training. Consider the four combinations generated by deterministic and stochastic phenomena, each with few or many variables (Duren, 1964). The tools required in study of organized simplicity, i.e., deterministic with few variables, include the classic analytical geometry—calculus sequence and difference and differential equations. Disorganized simplicity, i.e., stochastic with few variables, requires probability and statistics for analysis. Organized complexity requires linear algebra and many-variable advanced calculus. Study and use of complex stochastic models are needed for analysis of phenomena characterized as disorganized complexity.

I have not included advanced calculus and stochastic models in the essential minimum credits for mathematics, but students interested in quantitative graduate programs should include them. Computers are especially important in the last two areas, and computing practice and numerical analysis are equally important. Natural resource scientists should have knowledge and understanding of computers, at least in their graduate training and hopefully soon in their undergraduate training, especially in natural resource fields. We shall return to this point when discussing curricula around the nation. About 20–25 credits should be devoted to mathematics.

Communications are of obvious importance and would include English, technical writing, and public speaking. Humanities are important, but most students select only a "course here and a course there" in humanities and never develop a good understanding of any of these fields. A course in logic and scientific method would be especially desirable for natural resource science students. Areas from which students should select a block of courses could include sociology, psychology, philosophy, political science, and economics. About 35 to 40 credits would be devoted to communications and humanities.

These basic subjects should and could occur in the conventional four-year undergraduate program. However, to accomplish this might mean dropping some of the how-to-do courses characteristic of many of our natural resource schools. Also, it would mean condensing and revising considerably many of the management courses.

C. Training at Ecosystem Component Level

The center block in Fig. 9 refers to courses on specific ecosystem components. About 30% of the undergraduate effort should be expended in this area so that students could obtain training concerning basic biology and the structure and function of each major component of the ecosystem. These are listed in five categories: producers, consumers, decomposers, edaphic components, and climatic components. These courses should occur primarily at an intermediate point in the training of students. Training in these ecosystem component areas will not be detailed except to emphasize the importance of decomposers. Nutrients do not "sink" into one compartment of the ecosystem, but they are cycled repeatedly. This cycling is brought about by the decomposition of organic plant and animal materials and an understanding of decomposition processes is essential. Natural resource scientists should have training in general microbiology, soil microbiology, and soil zoology.

D. Integrative Courses

Natural resource majors should take courses in their junior and senior years which integrate basic sciences and components of the ecosystem (Fig. 9). About 20% of the undergraduate training should come in this area. There are several resource management courses which cut across disciplines and which include both biotic and abiotic components of the ecosystem, such as watershed management, and which can be excellent integrative courses. Basic nutrition can represent another integrative course. Courses in general plant and animal ecology are included under different components of the ecosystem, but there are many other ecology courses available in land-grant institutions which come in the integrative course block. Courses on natural resource management should be included in the integrative courses, but their content should be restricted to principles. A new approach is needed in natural resource management courses. First, less emphasis should be given to separate management courses for each resource field. The principles from the separate fields should be drawn together into cohesive, comparative courses. Second, such courses could include natural resource ecosystem structure, function, and management. In my opinion, many systems analysis techniques, such as simulation and gaming, could profitably be incorporated into these integrated management courses.

E. Seminars and Colloquia

The inclusion of three levels of seminars or colloquia (Fig. 9) differs from conventional approaches where seminars are restricted to the senior year. Seminars and colloquia should represent at least 5% of the undergraduate program. The university objective of providing the students with a broad ecological sensitivity for the complex problems of mankind should be introduced and fulfilled in introductory seminars in the freshman year and partly in conceptual seminars in the senior year. The seminars should be led by senior professors and should examine ecologically fundamental topical problems and, at the introductory level, they should provide awareness of the multiple dimensions of these problems. At the terminal level these same problems would be treated in depth and perspective. The intermediate level represents seminars that are analytical in nature. Some of the colleges at the new University of Wisconsin at Green Bay are utilizing this approach.

F. Some Problems of Implementing Change

Unfortunately, this general kind of program is not being implemented for several reasons. The idea of requiring blocks or areas of courses rather than specific courses requires much more counseling time than is normally given to our undergraduate students in many land-grant schools. Yet such an approach to education is being implemented at the University of Wisconsin at Green Bay where the colleges are being organized around themes (Sargent, 1968). For example, there is a College of Environmental Science including the basic sciences, earth sciences, ecology, engineering sciences, and agricultural sciences. Essentially, the college will be developing people well-trained to tackle natural resource problems. Their curricula are being designed to provide the student with a broad ecological sensitivity for the complex problems of mankind. Second, the students must develop a proficiency in environmental sciences.

Admittedly, there are many weaknesses in the curricular approach just outlined; and in our usual land-grant schools we often do not have enough facilities, adequately trained manpower, or finances to educate students as well as we would like to. However, it is all too easy to fall into the trap of using the same old tattered lecture notes year after year; in some natural resource fields there are very few textbooks upon which to rely. Look at range science, for example. How many fresh and original texts are there in the field? How many recent, detailed monographic treatments of important range problems are there? Implementing the above concepts in undergraduate programs would require major revision and innovation (and perhaps revolution) in courses in resource management fields.

V. A DOCTORAL PROGRAM

Maybe it is overoptimistic to think that such depth and breadth as outlined in Fig. 9 can be accomplished in an undergraduate degree, and perhaps it is too early. Perhaps such programs could first be introduced in graduate work. There are two types of Ph.D. programs suitable for systems ecology (Fig. 10), again suggesting basic building blocks rather than individual specific courses. The major controversial characteristic of these systems ecology programs is the amount of time required in formal classwork to get training in depth in several areas. These programs (Fig. 10) also assume a sound undergraduate background and that the student can plan his entire graduate program at one time. If these assumptions are not met, the programs outlined could still be useful guidelines.

A. Experimental versus Analytical Emphasis

Two types of emphasis in systems ecology have been outlined. The first is for the graduate student who is training for a career in experimental ecology. This man needs depth in understanding different components of the ecosystem, their interaction, and the theory, as well as the necessary skills and knowledge to communicate with other specialists. Tomorrow's

PhD PROGRAM IN SYSTEMS ECOLOGY

	EXPERIMENTAL	THEORETICAL
ECOLOGY AND PHYSIOLOGY		
PRODUCERS	12	4
CONSUMERS	12	4
DECOMPOSERS	12	4
ABIOTIC COMPONENTS		
CLIMATIC	6	3
EDAPHIC	9	6
HYDROLOGICAL	3	3
ADVANCED ECOLOGY		
PRINCIPLES AND THEORY	12	12
SYSTEMS	10	15
MATHEMATICS, STATISTICS AND LOGIC	12	39
INSTRUMENTATION AND METHODS	6	3
PHYSICAL SCIENCES	3	3
SEMINARS	3	4
COURSE TOTAL	100	100

FIG. 10. A comparison of two suggested approaches for doctoral coursework in systems ecology. Approximate quarter credits are given for blocks of courses. Training multidisciplinarians for work in interdisciplinary teams may require more than usual emphasis on formal coursework in graduate programs.

natural resource scientist must be a specialist in generalization; he must be an interdisciplinarian and a multidisciplinarian. The experimental systems ecologist will take about twice as much coursework concerning abiotic and biotic components of ecosystems as the ecologist emphasizing theoretical problems (Fig. 10).

The theoretical systems ecologist will work primarily in the modeling and analysis of natural resource problems, also as a member of an interdisciplinary team. He too will have to be a multidisciplinarian, but his depth and strength will lie in the modeling and analysis fields. Both the experimental and theoretical systems ecologists need to be well based in the ecology and physiology of the biocomponents as well as understanding the abiotic components of ecosystems and their interactions with the biocomponents. Two types of advanced-level, formal ecological training are needed: a sequence of courses on principles and theory, and a sequence related to systems, modeling, and analysis; they will be dealt with in a later section.

Advanced mathematical techniques, skills in instrumentation and methods, and further training in physical sciences as well as seminars also should be included in the training of systems ecologists (Fig. 10).

B. Ecological Principles

There is a type of course that is sorely missing in natural resource management, or ecology, in our land-grant universities, where many of our resource scientists are trained, i.e., a course on ecological principles and theory. This should be a graduate-level course requiring as background plant ecology, animal ecology, or both, and natural resource management. Perhaps this is not unique. However, the course should be team taught, i.e., by scientists from a wide variety of disciplines. Each scientist would be assigned specific principles closely related to his own research and interests. The course should cover the classic and current literature and it should not be organism-specific. The course should require synthesis of topics by the students.

C. Systems-Oriented Courses

The following comments are not based on an exhaustive survey of departments or catalogs, but are simply based on personal experience and a brief review of systems courses. Consider courses related to systems per se; call it systems ecology if you will, or natural resources ecosystems, or what have you. Two approaches may be taken in courses of this nature: a conceptual approach or an analytical approach. Each approach may be implemented at the undergraduate or graduate level, generating four combinations.

I know of no strong analytical, systems-oriented courses of this nature at the undergraduate level in environmental biology or natural resource sciences in universities in this nation. Only one course emphasizes the conceptual, systems-oriented approach at the undergraduate level; this is a course on ecosystems and resource management that could be used as an example for future course developments. It is given in the Resource Planning and Conservation Department of the School of Natural Resources at the University of Michigan. The course requires senior standing in the biological sciences or natural resources, and its purpose is to introduce students to some of the newer techniques being developed for the analysis, design, and control of complex goal-seeking systems. In particular, the course seeks to acquaint students with the methods of thinking that have been stimulated by recognition of the similarities among a variety of large-scale systems. The aims in this course are that all students will learn the language that they will increasingly hear from systems analysts in future years and that a few students will be motivated to become actively interested in the applications of systems analysis to natural resource problems.

There are at least two conceptual, systems-oriented graduate courses in United States' universities. The first of these is a course concerning natural resource ecosystems given in the School of Forestry of the University of California at Berkeley. The purpose in this course is to discuss relationships between disciplines, philosophical and ecological background, topics from organization theory, systems analysis, and synthesis and characteristics of ecosystems. Another course of this type is given at the University of Georgia in the Zoology Department. The purpose of this course is to emphasize bioenergetics and ecosystems. Most of the students in this course come from liberal arts areas, but some students from wildlife and forestry are now having the benefit of training in this course which is taught by a team of three instructors.

There are at least five courses at the graduate level that can be classified as analytical, systems-oriented courses. These courses are taught at Colorado and Washington State Universities and at the Universities of Tennessee, Georgia, and California at Davis.

In the Information Science Department at Washington State University there is a course concerning the modeling and simulation of biological systems. Unfortunately, few biological science students and even fewer natural resource science students have the benefit of this training. The Tennessee course is a year-long sequence in systems ecology which originally was taught by three instructors in the Botany Department (Patten, 1966). The author had the good fortune of working with this course in its initial two years. Another of the instructors has since initiated similar coursework at the University of Georgia. At the University of California at Davis in Zoology there is a solid biomathematics course, requiring calculus and statistics as prerequisites. This course is patterned after the last two-thirds of the recent and interesting text on quantitative ecology in resource management (Watt, 1968). The approach in that course can be understood by examining the last section of that text.

D. A Systems Ecology Sequence

The three courses that have concerned us have been given college-wide numbers in the College of Forestry and Natural Resources at Colorado State University. These courses are systems ecology, ecological simulation, and natural resource models; each is five quarter-credits. This year-long sequence is team-taught by five men from departments of Forest and Wood Sciences, Fishery and Wildlife Biology, Range Science, Mathematics and Statistics, and Electrical Engineering.

The objectives of this course sequence are to provide students from a variety of disciplines an integrated, computer-assisted program designed

(1) to develop and investigate ecological principles in a quantitative manner, (2) to develop skills in the use of modern high-speed digital and analog computers, (3) to provide an introduction to probabilistic and deterministic methods from statistics, operations research, and systems analysis, and (4) to integrate these skills, concepts, and methods in the study of classic and applied ecological problems. We were fortunate to obtain National Science Foundation support to help purchase an analog computer for class use and to help purchase extra digital computer time; this external financial assistance was necessary to initiate the series. The students now average about one turn-around per day on the digital computer on a batch-processing basis, and they also use a remote console. A course of this nature, especially in its initial years, also requires more instructor time than do many other areas of teaching. Some of our funds supported this extra teaching time. In three years we have increased from 8 to 24 students in the first course of the sequence. Our students represent ten different majors from throughout the university. In each of these courses there is parallel discussion of biological and mathematical topics as outlined briefly here. Naturally, since the sequence is relatively new there is still much addition and deletion of topics and shifting of topics from course to course.

In systems ecology the major biological topics are succession and accumulation, ecosystem energetics, and site quality and relations. These topics are used to introduce mathematical techniques which are implemented by use of computers. Programing of both digital and analog computers is introduced in this course. Some of the mathematical topics taught include study of states and changes of state of systems by means of Markov processes, matrix algebra, multiple regression, and interative nonlinear regression.

Ecological simulation, based upon the course in systems ecology, concerns problems of mathematical models, especially those implemented by analog computers. Topics such as competition and succession are discussed by simulation. Optimization techniques are introduced for study of ecological problems, and topics of invasion, dispersal, and diversity are considered. These biological topics are used as examples to evaluate systems with both deterministic and stochastic approaches. Compartmental models are studied and explored using both analog and digital computers. Linear and nonlinear programing techniques are studied and used. Students do textbook-type problems in mathematical programing as well as solving problems generated from their own fields.

The third course in this sequence, natural resource models, reviews general modeling concepts and covers some of the biological topics previously discussed, but emphasizes interrelations of structure, function, and stability. Considerable emphasis is given to systems engineering techniques and approaches, especially transform systems, in studies of ecosystem dynamics. Some of the mathematical topics that are included are transformation and feedback processes, network theory, and response surface examination.

How much background do students need for these courses? We require a year each of calculus, statistics, and ecology. Many natural resource students had calculus as undergraduates, but they are not conversant with it. Therefore, we provide out-of-class work sessions to polish their skills in these areas. But this is not the solution to the problem. The solution lies in increasing use of mathematical and statistical methods in our intermediate and upper undergraduate courses in the natural resource fields. We do not require differential equations, matrix algebra, and computer programming as prerequisites for the course sequence described, although many of the students have some background in one of these three areas. We do cover many topics from these areas during the year-long sequence.

Experience with this sequence has influenced my belief that there will soon be increasing use of models in understanding natural resource phenomena. The students have shown great interest and ingenuity in applying analytical techniques and models to their own fields. In time these models will change in size, in complexity, in structure, in randomness, and in scope. This is illustrated in Fig. 11, which is based in part on some ideas presented by Clymer (1966).

The models will become larger and more realistic (Fig. 11). Size will be reflected in more variables and more effects. The models will start with static linear systems as a first approximation, but will lead to dynamic nonlinear systems which will better represent the complexity of life. I foresee more logic operations and more loops and subroutines in natural resource models and thus more digital computation, especially via remote terminals and consoles. There will be increased incorporation of randomness in these models, both in the system itself and in the system inputs, allowing a more realistic simulation of real-life processes which are probabilistic, not deterministic, in nature. The scope of the models will increase in detail, and in defining and developing these models more and more disciplines (and interactions among disciplines) must be involved. It is hoped that these models will simulate or analyze at least simplified resource ecosystems and that they will be used increasingly in undergraduate programs and by some practicing land managers.

All of these factors point to the need of more thorough, more detailed, and more exhaustive training of natural resource scientists. Truly, the natural resource scientist of tomorrow must be a multidisciplinarian for

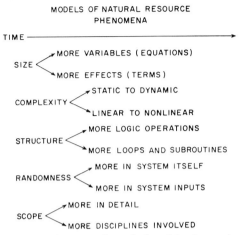

Fig. 11. With time and experience, mathematical models of natural resource phenomena will increase in size, complexity, and random components. The models will involve more detailed structure and have greater scope.

interdisciplinary teams. Can these men be trained and these teams be developed in time to solve many of our pressing natural resource problems?

E. Training Multidisciplinarians for Interdisciplinary Teams

The doctoral programs outlined in Fig. 10 call for about 100 quarter-credits of formal work beyond the bachelor's degree. Obviously, this is much more coursework than is required in graduate programs in many foreign countries, and it is more than for many programs in the United States. If we are to train competent multidisciplinarians perhaps we must increase the formal coursework component and decrease the research component in the doctoral program. There are two main reasons for this. First, in order for most students to receive more than a superficial understanding of the many fields mentioned above, they need the assistance an instructor can provide. That is, most students cannot gain adequate skill in, and understanding of, these subjects by self-teaching alone in a reasonable period of time. Consider, for example, the necessity of an instructor's assistance in studying many areas of advanced applied mathematics. Consider also the necessity, at least in most of our land-grant schools, of formal courses to support the acquisition, maintenance, and instruction in use of complex laboratory equipment. Second, postdoctoral opportunities are greatly increasing for well-trained natural resource science students. Some aspects of research training normally provided in the doctoral research could be accomplished on a postdoctoral basis.

The systems ecologist, either experimentally or analytically oriented, must be prepared to produce in the environment of team research and team management. Imagination and inventiveness are required for success, and these traits are difficult to develop by training. A successful systems ecologist will be one who has the imagination to perceive an important problem before other members of the research team or management group do. He must have the inventiveness to devise and weigh alternatives for its solution. Although some of the required traits perhaps are inborn, these problems emphasize the necessity of multidisciplinary training. In order to contribute effectively in the interdisciplinary team the systems ecologist must have sufficient depth in more than one specialty in order to make significant contributions to the solution of the problem. Thus, systems ecologists will need both breadth and depth of training and interests.

In many respects, research teams working on major problems have individuals contributing at least at three scientific levels: the laboratory or field scientists working largely alone on a specific problem, a leader coordinating efforts of a group, and a third level of workers coordinating the efforts of groups leaders. The level of abstraction increases with each level of effort. In some branches of science the theoretical scientist works at the highest level of abstraction. He may spend much of his time examining the pieces of experimental data amassed by the others and attempting to fit these data into a broader framework. The theoretical physicist does this job for physics; to a lesser extent, the theoretical chemist does this job for chemistry; and to an even lesser extent the theoretical biologist does this job for biology (Weinberg, 1967). The theoretical ecologist or natural resource scientist is rarer yet, partly because there are relatively few resource scientists, partly because their fields are relatively young, and partly because the task of compacting the literature (in the broad sense) is so difficult. The efforts of the theorists and the experimentalists complement one another. There are few well-developed teams working on natural resource problems, and the rate of development of these teams awaits well-trained systems ecologists.

VI. CONTINUING EDUCATION—COOPERATIVE TRAINING PROGRAMS

Even if a few natural resource departments being to train doctoral-level people with the skills and concepts outlined above, probably few of these early graduates will find their way into the resource management agencies. This is because the agencies have not thus far put a priority on hiring advanced degree persons for their top management positions. The agencies have relied almost entirely on men brought up through the ranks, and few of the men in the ranks have this kind of background. Furthermore, people with the type of training outlined above already are in great demand in universities and in the research agencies in the United States and in other countries. The small supply is emphasized by Slobodkin (1965), who states that "the number of good quantitative ecologists is in the thirties or forties for the entire world. . . ."

Thus, we cannot afford to wait for a few undergraduate programs in natural resource sciences to be modified and begin to trickle a few bachelor's level ecosystem-oriented and analytically oriented people into the management positions. We cannot wait for even fewer, much-needed, doctoral-level scientists and managers to be accepted quickly into top management positions. Two supplementary types of short-cut programs are suggested here which could involve at the outset a large number of present top managers.

A. Combined Applied Research—Management Training Programs

Resource management agencies are not without capable young men with strong biological backgrounds; these men probably soon will fill top management positions. Universities and research agencies are not without a few capable, quantitatively oriented individuals who can converse, interact, and work with the above management people. Unfortunately, in the past these groups of individuals have been well isolated! The researchers often are interested in many phases of applied research, and many of the young, key management personnel are interested in and could profit by advanced training. The management agencies are faced with many problems to which some answers could be obtained from carefully designed, short-term, applied research programs. Certainly natural resource management is a complex problem, especially when overall, quantitative, and comparative decisions must be made considering both short-term and long-term effects.

A systems analysis approach can be used when bringing together these management and research specialists to focus upon important problems. One useful procedure would be to develop models to condense information about various resource management problems and to use these simulation models in management games. Such models should be based on ecosystem concepts and a systems approach would require at least the following major steps (Brooks, 1967). The variables in the natural resource ecosystems representing man's inputs must be specified, i.e., the management parameters. Obviously, this would require direct interaction between the managers and modelers. The various outputs from the ecosystem of interest to man must be specified, and a crude initial model coupling the inputs and outputs must be constructed as a starting point. Although crude, even the initial model should be a tremendous asset to thinking, planning, and to research design. Certainly many parameters and functions will only be estimated in the initial model, and this will direct attention to specific researches needed to obtain improved estimates. With an initial model on hand for exploration, it would be useful to examine more carefully management goals and objectives and see how they relate to system output. Alternative management strategies could be explored and a framework developed for comparing these sets of strategies and systems outputs in a quantitative manner.

Knowledgeable resource management is a complex process involving simultaneous consideration of many variables. All of the information available to evaluate a single-use management scheme properly is difficult enough to assemble and study. When multiple use is practiced the interactions of the several uses produces a truly staggering array of problems to be solved. There are increasing demands for resource managers to assess quantitatively the influence of alternative management practices, and now resource managers often are required to show optimum return in many resource uses for a given input of management funds. These predictions and calculations are fraught with changing and uncertain resource use values.

Although the difficulties of resource use and management evaluation are real enough, ordinary procedures of solution fall far short of the goal. Each new set of environmental conditions, or a change in valuation of the products or in management practices, often requires a whole new set of base information and a whole new set of evaluations. A device or procedure which would simulate action of a multiple-use system would offer a fast, economical means of evaluating the implications of alternate management decisions with a minimum of field effort. Such a procedure is valuable for decision makers managing arid rangelands, for example, because of the large and diverse areas being managed by a single individual and because of the fragility and potential future importance of these resources.

Decisions of whether or not to control juniper trees on semiarid rangelands are examples of management problems which lend themselves to a simulation process. Major benefits of such improvement practices are increased forage production, improvement of wildlife habitat, and soil protection. A simulator including these outputs would have great value because once models are developed they can often be adapted to a wide range of problems. Thus, a simulation scheme developed primarily for use in the piñon-juniper type would have many components suitable for use in the sagebrush type. Such simulators also can be expanded continually to incorporate new and different information as it becomes available, and they should be very useful as training devices for resource managers. In addition, a simulator can be a very powerful tool for pointing out research and other information needs. Any simulation scheme developed, however, would be incomplete without a means of testing the simulator against actual field results, redesigning the simulator based on these tests, and retesting. Furthermore, the value of such a simulator would be greatly enhanced if it were used to train field personnel in its use as a management tool and if agency personnel participated in helping provide parameters and their variances and functional relationships in the model.

Several methods of optimizing complex management decisions are just beginning to be put into use today. Some examples of optimization techniques, simulation models, and systems procedures beginning to be used in natural resource management are shown in the following references:

Amidon (1966)	Map display system
Arcus (1963)	Grazing management simulation
Broido et al. (1965	Operations research applications
Ferrari (1965)	Models in agriculture research
Goodall (1967)	Range management simulation
Gould and O'Regan (1965)	Simulation in forest planning
Gulland (1962)	Models of fish populations
Hool (1966)	Dynamic programing in forest production control
Larkin and Hourston (1964)	Salmon biology simulation model
Nautiyal and Pearse (1967)	Optimum forest conversion analysis
Newnham (1966)	Simulation of forest harvest
Van Dyne (1966a)	Linear programing in range resource analyses
Watt (1964)	Computers in insect pest control

Although simulators have been developed for some management situations by researchers, and linear programing techniques have been merged with computer mapping techniques, there are apparently no simulation models being used directly by resource management personnel. Thus, the necessary feedback from resource manager to simulator, which is necessary to add realism, is lacking. To solve their unique management problems, decision makers in the natural resource management agencies should apply modern systems analysis techniques to the multiple-use problem. Techniques developed in research studies should be modified and adapted to management problems of real-life complexity. Coupled with conventional and time-tested resource management tools, these new techniques and approaches would provide the decision maker a powerful array of methods for solving complex management problems. Cooperative projects can be developed to accelerate the incorporation of scientific knowledge into resource management in such a way that careful analysis of alternatives and optimization for multiple long-term use can be considered in the decision-making process.

B. A Case Example

An example follows of a resource management problem and a plan for a combined training-research program developed by a colleague, D. A. Jameson, and the author. This example is included with only a brief explanation to show how existing management parameters and knowledge can be coupled with specially designed applied researches to lead to the development of simulation models of resource systems. Such models can be used in training decision makers and in improving resource management decisions. The example is outlined here in the general format of a research proposal.

Consider the application of these approaches and techniques to studying management of the piñon-juniper type. The objectives of such a combined research-training–management game approach to resource management would be (1) to develop a computer simulation model to consider simultaneously influences on forage production, wildlife responses, sediment production, and other land uses as a result of piñon-juniper type manipulation and conversion, (2) to utilize this simulator to point out needs for specific biological and physical information, (3) to utilize the simulator as a training tool for land managers in resource decision-making, (4) to conduct necessary extensive and intensive research, in cooperation with land managers, to measure parameters needed for objectives (2) and (3), and (5) to redesign the simulator in light of the knowledge gained in objective (4). The general flow and feedback of field data and information, modeling results, and gaming are shown in Fig. 12.

A useful approach in piñon-juniper range management situations would be to develop a computer simulator which behaves much like the natural system. A digital computer program can be prepared which will represent a multiple-use decision problem involving forage production, wildlife, and soil erosion. In this phase of the work detailed analyses can be made of the literature on the resource type in question (Fig. 12). These summaries and interpretations can be closely cross-checked with management agency field personnel in the areas involved to ensure the accuracy of both quantitative and qualitative interpretations. Initially, a plan would be to evaluate the results of a township-size stand of piñon-juniper, with management prescription developed for each of the 576 forty-acre blocks within the township. Simultaneously with this development careful and extensive inquiries of field personnel would be made to ascertain types

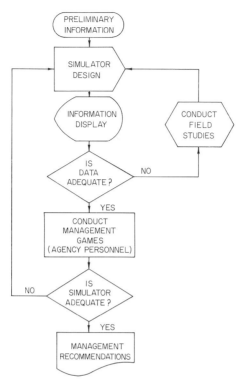

FIG. 12. A simplified flow chart outlining the development and use of a simulation model for natural resource management.

of decision-making situations encountered by field personnel. Two classes of information are needed. First, broad-scale records are needed on example grazing allotments (or portions thereof), such as for seasonal stocking rates and yield by domestic and wild animals, estimated water yield and quality, etc. This information may be obtained best through cooperative efforts with regional technicians of the management agencies. Second, specific parameters and their variances and the functional relationships between these parameters will need to be determined from short-term experiments involving detailed studies. Many of these studies could be suitable for graduate student research projects, especially if close cooperation can be maintained in their design and conduct with the management agency personnel.

Segments for such a simulation model have been developed for making decisions on piñon-juniper control to increase forage production (Jameson, 1969). Basic information is also available on the expected response of deer and elk to piñon-juniper control (Reynolds, 1964). Some estimates of soil losses can be obtained from watershed studies. Other functional relationships must be derived from field inspection and research. Some examples of the functional relationships for which some information is already available and which are required for construction of the simulator include rate of tree growth; relationships between forage production and tree cover; the rate of forage decline due to tree growth; the rate of forage increases following control; the relationship of treatment cost to tree cover; the rate of change of treatment costs with tree growth; the border effect and spatial relationships of tree control with wildlife movements; the effect of slope and aspect on wildlife populations; and the effect of juniper control on soil losses.

Simulation runs can be made on a computer to test the model. Sensitivity analyses can be made to evaluate the influence on the production criterion of variations in the controlling and dependent variables of the model. As soon as the initial development and testing of the simulator is completed, a 1-week or 2-week workshop, utilizing selected field management personnel as decision makers, can be conducted to test the simulator and to train field personnel for its use. Personnel for the workshop should be selected from among those who show promise of developing into high-level decision makers. These men, in a sense, would use the simulator to play management games against one another (Fig. 12). Probabilistic elements, such as variable climatic conditions, could enter into these games. It is expected these managers will pinpoint weaknesses in the simulator. From the results of this workshop, additional experiments can be designed in order to refine functional relationships or provide new ones which appear to be needed.

The processes of modifying the model to make it more realistic and more inclusive, of integrating new information and ideas provided by the resource managers in the field and in the management-game workshops, and of integrating new information derived in the specific field research studies must be repeated through several cycles in developing successively improved models (Fig. 12). At this stage the model can be expanded to couple the management decisions for one specific ecosystem type (e.g., piñon-juniper) to those for contiguous ecosystem types (e.g., sagebrush, forested range, seeded pastures, haylands, etc.).

In such cooperative studies the management agency would be asked both to sponsor and to participate in a bold new approach to improve and quantify resource management decisions. Management, research, and university personnel together can develop procedures and methods which have great potential in resource management, but they will be pioneering these efforts. To ensure success of such a project, in addition to the active participation of key personnel from one or more management agencies, the work would involve significant amounts of time of the university investigators on a continuing basis. Graduate research assistants could be used on specific field studies, and there are good possibilities that some agency personnel could be involved in such efforts as part of an on-the-job phase of an advanced-degree program in resource management.

Resource management agencies should also consider maintaining their own private, selected, training programs, but on an accelerated basis. As their personnel gain experience and skills with quantitative management methods and ecosystem concepts, they should consider using new kinds of consultants in their programs. For example, a few full-time or part-time consulting environmental engineers are becoming established. Although thus far they have been used mainly in industries utilizing natural resources, they could provide effective services to our state and federal resource management agencies until more quantitatively oriented resource managers become available.

C. Interdisciplinary Dialogues

The above example of a kind of continuing education, which involves developing a resource simulation model and using it in management games, shows the necessity of incorporating our conventional resource management wisdom with new methods and techniques to optimize resource use. Implicit in such a plan are the new ideas that we must utilize professionals of many disciplines in the program, and that the dialogue must be continuing. The free communication between the workers must be permanently maintained, not just in conferences and symposia, but on a day-to-day working basis (Slobodkin, 1968).

A second kind of continuing education would use many of the techniques mentioned above, but essentially it would be a continuing interdisciplinary dialogue on important natural resource problems. Equally important is that we utilize an ecosystem approach in such dialogues and not focus on narrow problems. This is essentially a new type of continuing education that has not been recognized in the planning of staff time in many of our state or federal resource management agencies or in many of our land-grant institutions housing our natural resource departments. Because of incipient environmental problems (Cassidy, 1967), personnel from many different federal and state management and legislative bodies should be involved in the dialogue. The legislators are of obvious importance when we consider the direct impact they can have on use of our resources. This range of experts must be brought together at a convenient location at frequent and continuing intervals to discuss natural resource management problems. The groups should be small so that informality can be maintained, and the location should be sufficiently isolated so that each member can concentrate on the problem at hand, i.e., he must escape interruptions common at his home office. Thus, the facilities for such dialogues should contain adequate living, eating, drinking, and meeting facilities at a single location. The problems to be discussed should be of practical significance and should have political, social, engineering, and scientific ramifications (Slobodkin, 1968). Thus, a diverse group of participants would be needed at each dialogue session, and some individuals should participate in several sessions to provide continuity.

VII. SOME CONCLUSIONS REGARDING ECOLOGICAL TRAINING

A. Some Problems in Training

The needs increase for sound ecological training, with emphasis on new quantitative techniques, but there is failure in meeting these needs, primarily because of the shortage of experienced teachers and programs in this area. In both academic and governmental institutions there are but few scientists having ecological knowledge and first-hand experience in utilizing new tools and techniques for solution of real-life problems. The paradox is that computational and analytical methods are available and that many resource scientists are aware of these tools, and recognize intuitively their potential, but that there are relatively few instances of these tools being applied to important natural resource management problems.

Why have the resource scientists and the analysts not joined forces? Perhaps our conventional, departmentally oriented teaching and research programs have suppressed cross-fertilization of ideas. Perhaps it is because the task is great, for although classroom generalizing is easy, it is quite another task for a single individual to find the time and resources to apply analytical techniques and ecological principles in analysis of realistically complex resource problems. Unique training programs based on concrete, real-life problems and interaction of faculty and students from different disciplines are needed immediately. But combining research and education across disciplines and across training levels, e.g., faculty and students, meets with many complications that are not easily overcome in the usual sources of financing training programs.

B. Ecology: The Sobering Science

Man is a vital part of most major ecosystems, and awareness is increasing of his part in them and his influence on them. Humans are both parts of and manipulators of ecosystems, and induced instability of ecosystems is an important cause of economic, political, and social disturbances throughout the world. In altering his environment in order to overcome its limitations to him, man often is faced with undesirable consequences. Man's activities affect his environment in a way he cannot predict. Sometimes, it seems he does not care about possible consequences, and he has changed his environment both intentionally and accidentally. Society is continually pressing for changes, and some of our scientists, engineers, and administrators seem to be "in a contest as to who can promise the greatest immediate results in modifying the weather, warming the Arctic, moving the rivers, and mining the oceans" (Ackerman, 1967). Man also encounters difficulties when he attempts to return ecosystems to their native state, to preserve vegetation by development of national parks, or to preserve herbivore populations by control of predators. We still need to know the long-term effects and profits of ecosystem manipulation, such as even further shortening of food chains, as human populations continue to increase exponentially and to impose greater stresses on our world ecosystem. Knowledge about the entire ecosystem has become so important that ecologists can no longer be satisfied to be concerned with specific individual species or populations. In addition to plant ecologists, animal ecologists, microbial ecologists, etc., we must now train more and more young ecologists to confront the entire complexity of the ecosystem. They must be systems ecologists.

Some examples of undergraduate and graduate programs have been outlined, but it is embarrassing that in many ways we are not prepared to train enough young people in sufficient time to help solve tomorrow's problems. Alternately, there is increasing interest in a variety of disciplines in the application and understanding of ecosystem principles. We need to draw together scientists and managers from various disciplines for cooperative use of techniques and methodologies from various fields in analyzing economically and sociologically important problems in natural resource manipulation and management.

C. The Need for Predictability

There is an increasingly closer relationship in the approaches of many biological scientists and physical scientists. In part, this has resulted in the expression "environmental science" gaining vogue (Cain, 1967b). Concern with the environment is the vogue of the day, and ecology must be the wave of tomorrow (Ackerman, 1967). Systems approaches, borrowed from industry, military, and government, are having increased application and potential in ecology. To apply these new methodologies and concepts we must depend upon a new breed of resource manager who must come from today's and tomorrow's classrooms.

We need to be able to do more than to characterize the environment to the extent needed to exploit it wisely. We need to know its inner workings to manage it wisely and to live with the consequences of our manipulation. We must, in fact, anticipate the consequences of manipulation. Thus, we need a more fully developed "predictive science" of ecology. Today, we are not yet able to make such predictions, but we are beginning to ask some of the relevant questions (Cooper, 1968). Because of the inherent variability of ecological processes, it is unlikely that strictly deterministic predictions of environmental manipulation can ever be made.

D. The Need for Generalization

We have long been in a period of increasing specialization in many fields of human endeavor. But now science is in the early stages of a major revolution. Many efforts are emerging toward integration, coordination, and generalization (Cain, 1967a). This is evident at all levels. For example, the systems approach was used by Secretary of Defense McNamara in programing, planning, and budgeting in the Department of Defense. Now PPBS approaches are being initiated in many agencies.

In effect, the rise of ecological thinking in science represents a more system-oriented approach than that of the specialists, for the ecologist is concerned with interrelations, not merely with things. As Cain (1967a) aptly stated, "There is a movement from the narrow to the broad, from the specific to the general, from the individual to the social, from independence to interdependence." This movement is gratifying and necessary if we are to avoid the chaos that threatens our lives and country from growth and diversity.

The increased implementation of systems techniques has been allowed by the increase in computational and mathematical power afforded by computers. Many of these systems approaches require assembling vast amounts of data, analysis and reduction of these data, and interpretation of the complexity of ecological systems with powerful mathematical techniques. Both digital and analog computers are necessary and essential tools in these processes.

E. A Balance of Undergraduate, Graduate, and Continuing Education

Some suggestions have been made on how we might accelerate and implement training for solution of complex resource problems and undergraduate and graduate programs have been outlined. Such programs must be initiated soon, but even so they will not supply the demand rapidly enough. Although some educators recognize these needs, there are many obstacles to developing such new curricula with the resources currently available. In general, this means crossing departmental lines, even crossing institutional lines, generating support beyond our normal institutions, and devoting ourselves to learning from other educators and students, as well as sharing our knowledge by teaching other educators and students.

If one is to consider problems of real-life magnitude and complexity, such learning and teaching activities must be combined with research. This level and type of effort are not easily defended nor easily supported in our conventional training programs in either the natural resource disciplines, from which the problems may arise, nor in the basic or classic ecological and computational disciplines, from which the principles and methods for solution may arise. One solution to these problems is the development of integrated workshop-research programs, such as the example outlined.

ACKNOWLEDGMENTS

The course sequence described herein at Colorado State University was initiated through National Science Foundation Grant GZ-991. Many concepts described herein were developed in a team research project on grassland ecosystems sponsored by National Science Foundation Grant GB-7824 and in an ecological modeling workshop program supported by the Ford Foundation. Several ideas presented in an essay of limited distribution have been incorporated in this chapter, including Figs. 1, 3, 7, and 8 (Van Dyne, 1966b). Constructive criticisms of the manuscript by D. A. Jameson and P. T. Haug are much appreciated.

REFERENCES

Ackerman, W. C. 1967. Ecology—the sobering science. *Trans. Geophys. Union* **48,** 809–810.

Amidon, E. L. 1966. MIADS2: An alphanumeric map information assembly and display system for a large computer. *U.S. Dept. Agr., Forest Serv., Res. Paper,* **PSW-38,** 12 pp.

Arcus, P. L. 1963. An introduction to the use of simulation in the study of grazing management problems. *New Zealand Soc. Animal Prod., Proc.* **23,** 159–168.

Broido, A., R. J. McConnen, and W. G. O'Regan. 1965. Some operations research applications in the conservation of wildland resources. *Management Sci.* **11,** 802–814.

Brooks, D. L. 1967. Environmental quality control. *BioScience* **17,** 873–877.

Cain, S. A. 1967a. *Ann. Honors Convocation School Nat. Resources, Univ. Michigan, Ann Arbor, Mich. 1967* 8 pp. USDI News Release.

Cain, S. A. 1967b. *Alma Coll. Am. Assembly Population Dilemma, Alma, Mich., 1967* 11 pp. USDI News Release.

Cassidy, H. G. 1967. On incipient environmental collapse. *BioScience* **17,** 878–882.

Clymer, A. B. 1966. Trends in computer applications. *Symp. Trends Computer Technol., Columbus, Ohio, 1966* (unpublished).

Cole, L. C. 1958. The ecosphere. *Am. Scientist.* **198,** 83–92.

Cooper, C. F. 1968. Needs for research on ecological aspects of human use of the atmosphere. *In* "Task Group. Human Dimensions of the Atmosphere," pp. 43–51. Natl. Sci. Found.

Duren, W. L., Jr., Chairman. 1964. "Tentative Recommendations for the Undergraduate Mathematics Program of Students in the Biological, Management, and Social Sciences." Committee on Undergraduate Program in Mathematics, Mathematical Association of America. 32 pp.

Dyksterhuis, E. J. 1958. Ecological principles in range evaluation. *Botan. Rev.* **24,** 253–272.

Ferrari, T. J. 1965. Models and their testing: Considerations on the methodology of agricultural research. *Neth. J. Agr. Sci.* **13,** 366–377.

Goodall, D. W. 1967. Computer simulation of changes in vegetation subject to grazing. *J. Indian Botan. Soc.* **46,** 356–362.

Gould, E. M., Jr., and W. G. O'Regan. 1965. Simulation: A step toward better forest planning. *Harvard Forest Papers* **13,** 1–86.

Gulland, J. A. 1962. The application of mathematical models to fish populations. *In* "The Exploitation of Natural Animal Populations" (E. D. LeCren and M. W. Holdgate, eds.), pp. 204–217. Wiley, New York.

Holling, C. S. 1968. The tactics of a predator. *In* "Insect Abundance" (F. R. E. Southwood, ed.), Vol. 4, pp. 47–58. Roy. Entomol. Soc., London.

Hool, M. N. 1966. A dynamic programming-Markov chain approach to forest production control. *Forest Sci. Monographs* **12,** 1–26.

Jameson, D. A. 1969. Optimum timing for juniper control on southwestern woodland ranges. *J. Farm Econ.* (submitted for publication).

Larkin, P. A., and A. S. Hourston. 1964. A model for simulation of the population biology of Pacific salmon. *J. Fisheries Res. Board Can.* **21,** 1245–1265.

Leopold, A. S., S. A. Cain, C. H. Cottam, I. N. Gabrielson, and T. L. Kimball. 1963. Wildlife management in the national parks. *Am. Forests* **69,** 32–35 and 61–63.

Lutz, H. J. 1963. Forest ecosystems: Their maintenance, amelioration, and deterioration. *J. Forestry* **61,** 563–569.

Miller, R. S. 1965. Summary report of the ecology study committee. *Bull. Ecol. Soc. Am.* pp. 61–62.

Nautiyal, J. C., and P. H. Pearse. 1967. Optimizing the conversion to sustained yield—a programming solution. *Forest Sci.* **13,** 131–139.

Newnham, R. M. 1966. A simulation model for studying the effect of stand structure on harvesting pattern. *Forestry Chron.* **42,** 39–44.

Odum, E. P. 1964. The new ecology. *BioScience* **14,** 14–16.

Ovington, J. D. 1960. The ecosystem concept as an aid to forest classification. *Silva Fennica* **105,** 73–76.

Panel on Natural Resource Science. 1967. Undergraduate education in renewable natural resources—an assessment. Commission on Education in Agriculture and Natural Resources, Agricultural Board, Division of Biology and Agriculture. *Natl. Acad. Sci.— Natl. Res. Council Publ.* **1537,** 1–28.

Patten, B. C. 1966. Systems ecology: A course sequence in mathematical ecology. *BioScience* **16,** 593–598.

Pechanec, J. F. 1964. Progress in research on native vegetation for resource management. *North Am. Wildlife Natural Res. Conf., Proc.* **29,** 80–89.

Reynolds, H. G. 1964. Elk and deer habitat use of a pinyon-juniper woodland in southern New Mexico. *North Am. Wildlife Natural Res. Conf., Trans.* **29,** 438–444.

Sargent, F. 1968. Personal communication.

Schmitt, O. H., and C. A. Caceres, eds. 1964. "Electronic and Computer-Assisted Studies of Biomedical Problems." Thomas, Springfield, Illinois. 314 pp.

Schultz, A. M. 1967. The ecosystem as a conceptual tool in the management of natural resources. *In* "Natural Resources: Quality and Quantity" (S. V. Ciriancy-Wantrup and J. J. Parsons, eds.), pp. 139–161. Univ. of California Press, Berkeley, California.

Slobodkin, L. B. 1965. On the present incompleteness of mathematical ecology. *Am. Scientist* **53**, 347–357.

Slobodkin, L. B. 1968. Aspects of the future of ecology. *BioScience* **18**, 16–23.

Smith, F. G. 1966. "Geological Data Processing: Using FORTRAN IV." Harper, New York. 284 pp.

Tansley, A. G. 1935. The use and abuse of vegetational concepts and terms. *Ecology* **16**, 284–307.

Tolliver, W. E. 1965. "Trends in Graduate Enrollment and Ph.D. Output in Scientific Fields," Appendix 4, Resources for Med. Res. Rept. No. 6. U.S. Dept. Health, Education and Welfare, Washington, D.C.

Van Dyne, G. M. 1966a. Application and integration of multiple linear regression and linear programming in renewable resource analyses. *J. Range Management* **19**, 356–362.

Van Dyne, G. M. 1966b. Ecosystems, systems ecology, and systems ecologists. *U.S. At. Energy Comm., Oak Ridge Natl. Lab. Rept.* **ORNL 3957**, 1–31.

Watt, K. E. F. 1964. The use of mathematics and computers to determine optimal strategy and tactics for a given insect pest control problem. *Can. Entomologist* **96**, 202–220.

Watt, K. E. F. 1965. An experimental graduate training program in biomathematics. *BioScience* **15**, 777–780.

Watt, K. E. F., ed. 1966. "Systems Analysis in Ecology." Academic Press, New York. 276 pp.

Watt, K. E. F. 1968. "Ecology and Resource Management—A Quantitative Approach." McGraw-Hill, New York. 450 pp.

Weinberg, A. M. 1967. "Reflections on Big Science." M.I.T. Press, Cambridge, Massachusetts. 182 pp.

Author Index

Numbers in italics refer to the pages on which the complete references are listed.

369

Subject Index

A

Abiotic components, 31
Abiotic studies, 42
American Society of Range Management, 102
Arctic ecosystem, 24
Autecology, 5
Available organisms, manipulation of, 166

B

Bioecological equivalence, principle of, 199, 207
Biogeocoenose, 15, 190
Biotic potential, 295
Block diagram construction, 316

C

Catchment, definition, 310
Change
  directional, 121
  exploitation induced, 276
  fluctuation, 125
  nondirectional, 120
  spatial, 117
  temporal, 119
Clear-cutting, forest experiment, 69
Climate, manipulation of, 163
Climax, 123, 128
Competition, 118, 210
  exploitation and niche concept, 271
  intraspecific, 299
  interactions, 275
Computers, 311, 322, 340–341
  digital simulation, 312
Consumers, 31, 35
Courses, systems oriented, 349–352

Curricula
  continuing education, 354–362
  doctoral, 347–354
  problems
    of change, 346–347
    in natural resource sciences, 342
  undergraduate, 342–347
Cycles
  calcium in forest ecosystem, 66
  lemming, 86–88
  mineral, 17
  nutrient, 87–88

D

Decomposers, 31
Density dependence, 290
  influences, 295
Density independence, 290
  influences, 296
Diets, cattle and sheep, 140
Dilemma
  first, 332
  second, 332–333
Dimensional analysis, 252
Diversity
  ecological, 228
  microsite, 228
  taxonomic, 227
Drought, 132

E

Ecology, 331
  growth of, 333
  historical approaches, 333–334
Ecological revolution, 333–340
Ecosystem concepts, 2, 12–14, 50, 247, 250
  in fishery versus wildlife biology, 260
  in forestry, 95